Basic Fabrication and Welding Engineering

F.J.M. Smith A.Weld.I., M.I.S.M.E.

Senior Lecturer in Fabrication and Welding Engineering, Department of Engineering, Tottenham College of Technology.

Cartoons by Mervyn Hughes, M.I.S.M.E.
Lecturer in the Department of Engineering at Henley College of Further Education, Coventry.

Longman

Acknowledgements

LONGMAN GROUP LIMITED
London
Associated companies, branches and representatives throughout the world

© Longman Group Limited 1975

All rights reserved. No part of this publication may be reproduced, stored in a retrieval system, or transmitted in any form or by any means, electronic, mechanical, photocopying, recording, or otherwise, without the prior permission of the Copyright owner.

First published 1975
Second impression 1978

ISBN 0 582 42431 3
Library of Congress Catalog Card Number 73 — 86514
Set in IBM Century typeface

Printed in Hong Kong by
Wing Tai Cheung Printing Co. Ltd.

We are grateful to the following for permission to reproduce copyright photographs and drawings:

AGA (UK) Ltd
Douglas Barnes (Machinery) Ltd
The British Oxygen Company Ltd
British Standards Institution
Carver & Co. (Engineering) Ltd
Duplex Electric Tools Ltd
F. J. Edwards Ltd
Elliott Machine Equipment Ltd
GKN Lincoln Electric Ltd
Hatfield Machine Tool Co. Ltd
L. J. Hydleman & Co. Ltd
I.P.C. Business Press Ltd
The Kingsland Engineering Co. Ltd
A. J. Morgan & Son (Lye) Ltd
The National Federation of Building Tra Employers (Constructional Safety)
Neldco Ltd
James North & Sons Ltd
Philpott & Cowlin Ltd
William Press & Son Ltd
Rockweld Ltd
Rushworth & Co. (Sowerby Bridge) Ltd
The Welding Institute
Wolf Electric Tools Ltd

Contents

Chapter	Contents
1	Safety
2	Communications
3	Workshop calculations
4	Basic engineering science
5	Measurement and marking out
6	Materials
7	Material removal
8	Movement, restraint and location
9	Workshop operations
10	Fabrication processes
11	Welding

List of tables

1.1 Arc-welding hazards
1.2 Secondary welding cables
1.3 Filters for manual as arc-welding
1.4 Filters for gas welding and cutting
1.5 Welding gas mixtures

2.1 Reading drawings in first angle projection
2.2 Colour coding— compressed gas cylinders
2.3 Colour bands—hazards, gas cylinder contents

3.1 Mass/weight conversion
3.2 Densites of common engineering materials
3.3 Mass/unit area for sheet metal
3.4 Mass/metre run for mild steels bars

4.1 Common applications of fluxes for gas-welding
4.2 Effects of varying flame conditions when welding low-carbon steel (oxy-acetylene process)
4.3 Coefficients of linear expansion

5.1	Data for marking out pipe flanges	9.8	Sequence of operations for forming external angle ring
5.2	Constants for bolt hole location (flanges)	9.9	Sequence of operations for forming the internal angle ring
5.3	Tools used by template makers		
5.4	Materials for templates	10.1	Comparison of 'Vee' die ratios
		10.2	Bending forces required for metals other than mild steel
6.1	Effect of carbon content	10.3	Neutral line data for bending sheet metal
6.2	Rate of cooling	10.4(a)	Calculation—centre line bend allowance
6.3	Tempering temperature	10.4(b)	Calculation—centre line bend allowance
6.4	Common physical properties of some pure metals	10.5	Calculation—neutral line bend allowance
6.5	Effect of welding heat on the temperature of the plate	10.6	Calculation—neutral line bend allowance
		10.7	Calculation—neutral line bend allowance
7.1	Blade clearances for optimum cutting	10.8(a)	Calculation—precision bend allowance
7.2	Cutting speeds and feeds for H.S.S. milling cutters	10.8(b)	Calculation—precision bend allowance
7.3	Typical cutting times (abrasive cutting off machine)	10.9	Calculation—precision bend allowance
7.4	Approximate pressures for hand cutting steel plate	10.10	Calculation—precision bend allowance
7.5	Operating data for portable machines	10.11	Hot-rolled sections for structural work—standard sizes
7.6	Cutting data for profile cutting machines		
		11.1	Sizes of welding hose in general use
9.1	Market forms of supply	11.2	Downhand butt welds in steel
9.2	Classification of cold rolled sheets	11.3	Welding speeds and data for leftward welding
9.3	Classification of hot rolled materials	11.4	Welding speeds and data for rightward welding
9.4	Tools and equipment required to make a rectangular pan	11.5	Welding currents
9.5	Tools and equipment required to make a funnel	11.6	Electrode coating materials
9.6	Abrasive cutting data for some common sizes of angle section	11.7	Coding of coverings and data for six classes of electrodes
9.7	Sequence of operations for forming the frame	11.8	Definitions of welding positions
		11.9	Coding for welding current conditions

Preface

This book has been written to provide a primary source of information for students following the Engineering Craft Studies Course, Part 1 (City and Guilds Course 200). It should be used in conjunction with *Basic Engineering* to provide full coverage of the student's needs.

Basic Fabrication and Welding Engineering closely follows the syllabus published by the Council of Technical Examining Bodies and in order that as much factual information as possible can be included in a book of reasonable length, detailed explanation has been left to the class teacher where it properly belongs. Worked examples are included, but exercise questions have been omitted as these, together with investigation and project sheets, are available as a complementary package.

To make this book as interesting as possible, the written text has been kept to a minimum and extensive use is made of illustrations, diagrams and tables. The cartoon character 'Fred', who introduces each chapter with a topical misadventure, is a warning to those who would make light of their technical education and training.

The Author wishes to thank Mr R. L. Timings of the Henley College of Further Education, Coventry, for his contributions on the mechanical engineering topics to be found in this book.

Finally, the Author wishes to thank his friends and colleagues who have assisted him in writing this book and checking the proofs; to the many Companies and Professional Institutions who have provided up to date technical data, and the publishers for their help and advice in the preparation of the manuscript.

1975   F. J. M. SMITH

FRED—Some biographical notes

It is not that he is daft or even particularly slow. He is just unfortunate to be about forty years of age and to have received only an elementary wartime education. Not for Fred the benefits of the 1944 Education Act and the Industrial Training Act.

His second misfortune was to be born in an industrial town of the less progressive type. He is not a bad craftsman: it is just that in non-progressive industrial towns immediately post war you learned your trade in small family businesses 'sitting next to Nelly'. In the process you learned a lot of bad practices that never get into a modern training centre.

He could have been sorted out if he had gone to night school at the local tech. The money wasn't good so he worked overtime instead. He's not bitter, but he can't stand the superior young apprentices with the airs and graces picked up in the training centre and on day release at tech. He's never more pleased than when he can show them they are wrong. Trouble is they pick up his faults more often. They seem to get all the new machines, too. It is not really fair because he's quite sure that he could understand all about N.C. machining if only some one would tell him. They are always going to tell him — never quite get round to it. Trouble is he can't make head nor tail of the instruction books for these new machines.

He stands as an awful warning to any young apprentice who fails to take advantage of educational and training opportunities available today. Be warned young man! Some day you could be in Fred's position!

1975   R. L. TIMINGS

1 Safety

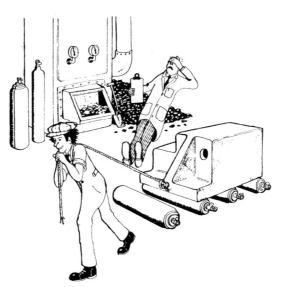

How many hazards can you see?

PART A SAFETY ON SITE

Recommendations for dress and behaviour to be adopted in the factory situation have already been discussed in *Basic Engineering*. However, the welding and fabrication engineer is often expected to work *on site*. This entails working in partly-completed buildings or on equipment being erected out of doors which introduces additional hazards. When working under such conditions additional precautions must be taken. The erector must not only be mindful of the safety codes of his own trade, but also of the *codes of safety for building sites* as well. Sections **1.1** to **1.6** indicate some precautions that should be observed when working *on site*.

1.1 Protective clothing

In recent years, tremendous advances have been made in the design and range of protective clothing. Such very necessary clothing is manufactured to the highest standards with regard to materials, construction, and resistance to wear or to damage.

Examples of protective clothing in everyday use are Wellington boots, overalls, aprons, leggings, headgear, and gloves. Because of adverse weather conditions when working *on site*, protective clothing generally includes jackets and over-trousers made from oilskin, plastic, rubber, or other waterproof materials. Figure 1.1 shows a few of the numerous types of protective garments available.

- Elasticated hood
- **Anorak**
- Zip front
- Elasticated cuffs
- **Overtrousers** have elasticated waist and ankles

(a) **'Foul-weather clothing'**
Made from high quality protective material which has a great resistance to tearing, abrasion, oils, and most chemicals. One such material is a very light but strong 3-ply fabric consisting of two sheets of VINYL laminated together through a high-tenacity NYLON reinforcement, all the seams being electronically welded

- Deep collar which may be turned up for added protection
- Reinforced shoulders and back
- Reinforced cuffs

(b) **'Donkey Jacket'**
Generally made from hard wearing melton cloth reinforced with P.V.C. on shoulders back and cuffs

Fig. 1.1 Protective clothing

Because a workman's hands are in constant use they are always at risk. He has to handle dirty, oily, greasy, rough, sharp, brittle, hot, and maybe corrosive materials. Gloves and 'palms' of a variety of styles and types of materials are available to protect the hands, whatever the nature of the work. Some hand-protection methods are illustrated in Fig. 1.2.

1.2 Head protection (BS2826)

When working *on site*, all persons should wear safety helmets because of the ever present danger from falling objects. *Even small objects such as nuts or bolts can cause very serious head injuries when dropped from heights.* When entering on to a site THE FIRST PRIORITY IS ADEQUATE HEAD PROTECTION. Figure 1.3 shows a typical safety helmet.

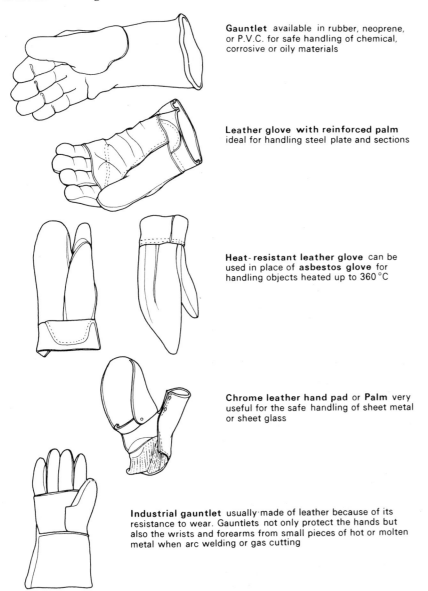

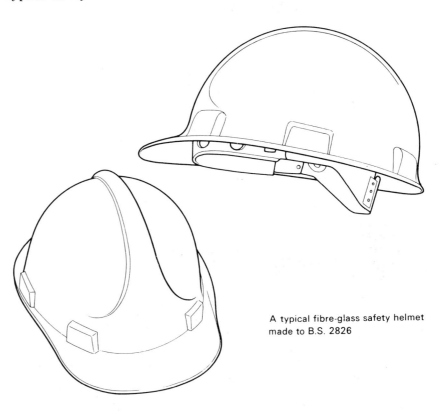

Fig. 1.3 Safety helmets

Fig. 1.2 Hand protection

Safety helmets are made to BS 2826 from moulded plastic or fibre glass reinforced polyester. Colour coded for personnel identification, they are light and comfortable to wear, yet despite their lightness have a high resistance to impact and penetration.

To eliminate the possibility of electric shock, safety helmets have *no metal parts*. The materials used to manufacture the outer shell have to be non-flammable and their electrical insulation must withstand 35 000 volts.

Figure 1.4 shows the harness inside the safety helmet. This provides ventilation and a *fixed safety clearance* between the helmet and the wearer's skull.

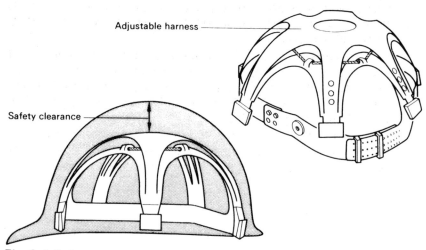

Fig. 1.4 Safety helmet harness

Well-designed SAFETY HELMETS are fitted with a special suspension harness which ensures SAFETY and COMFORT to the wearer. Such harness may be readily adjusted for size, fit, and angle by the individual. Whatever the adjustment made, a FIXED SAFETY CLEARANCE of 32 mm between the wearer's skull and the crown of the helmet must always be maintained. The entire harness can be easily removed for cleaning or sterilising. The special plastics used in the manufacture of safety helmets have to withstand high- and low-temperature requirements for impact penetration. Such plastics are usually made to a special formula which gives extra life to the helmet by reducing the effect of ULTRA-VIOLET LIGHT.

1.3 Foot protection

The practice of wearing unsuitable footwear should always be discouraged. It is not only false economy, but extremely dangerous to wear boots or shoes for work when they have become either shabby or useless for ordinary purposes. NEVER WEAR SOFT FOOTWEAR SUCH AS PLIMSOLLS OR SANDALS — this type of footwear offers no protection against 'crushing', or 'penetration' from underneath.

In safety footwear, protection is provided by a steel toe-cap (inside the boot or shoe) which conforms to a strength specification in accordance with BS 1870. This Standard requires that steel toe-caps must be capable of withstanding a blow of either 134 joules or 200 joules. Footwear in the former category has to be marked GRADE 2, and in the latter GRADE 1. Safety footwear is now available in a very wide range of styles and is of attractive appearance, as illustrated in Fig. 1.5.

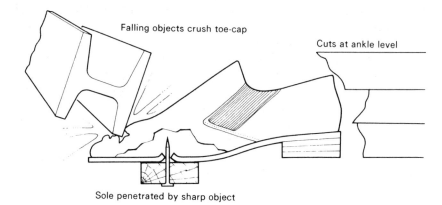

*Light-weight shoes offer **No** protection*

INDUSTRIAL SAFETY SHOE

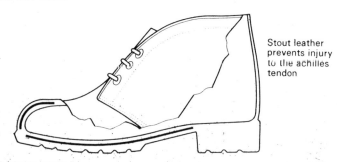

INDUSTRIAL SAFETY BOOT

Fig. 1.5 Industrial safety shoes

The hazard of slipping can be avoided by simply wearing industrial boots or shoes which have reliable non-slip soles. Waterproof rubber ANKLE BOOTS or KNEE BOOTS are also available with special cleated heels and soles.

1.4 Precautions when working aloft

Whenever persons are required to ascend to a height in order to reach the workplace, the following are a few of the many rules or regulations which should be observed for safety:

1. *Always have some breakfast, or at least a hot drink such as a cup of tea or coffee before you go to work* — AN EMPTY STOMACH CAN OFTEN RESULT IN A SUDDEN ATTACK OF FAINTNESS in even the healthiest person.
2. SAFETY BELTS OR HARNESS should be worn whenever possible, and these must be securely anchored. Figure 1.6 illustrates a safety device which can be fitted permanently to any structure where personnel are moving in a vertical plane.

This type of 'SAFETY BLOCK' has many applications in construction work, maintenance, and other engineering operations where workers are moving up or down shafts, chimneys, masts, ladders, cranes, or on the outside of structures. THE BLOCK GIVES COMPLETE SECURITY AGAINST FALLS IN THESE KINDS OF SITUATIONS.

It operates in both directions and the worker is protected when travelling both up and down. Should the worker slip, his fall is stopped within 300 mm with a gradual deceleration that causes no discomfort. The block unlocks automatically when the pull on the rope is released.

3. It is often desirable for a worker wearing safety harness to have a limited safe radius of operation, this can be achieved by use of a special safety block which is illustrated in Fig. 1.7.

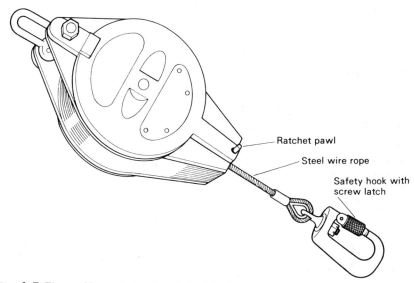

Fig. 1.7 The self-contained safety block

This block is completely self-contained. It holds approximately 5 metres of STEEL WIRE ROPE mounted on a spring-loaded drum. With normal movement the rope pulls out and winds up automatically, keeping the rope taut and giving the user freedom of movement with absolute safety.

The desired limited safe radius of operation can be obtained with the safety block by pulling out the required length of rope and locking it with a ratchet pawl as shown. A sudden pull engages a locking device and the rope is stopped. Because of a friction brake on the rope drum in the block there

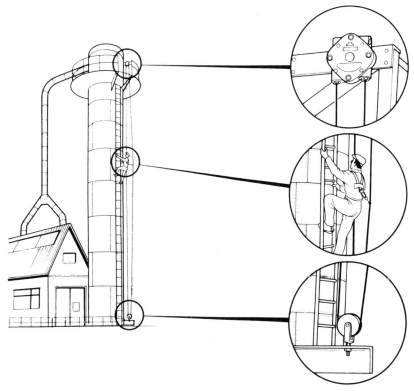

Fig. 1.6 Use of a static safety block

is no jolt on the worker in case of a fall, yet the fall is stopped after about 300 mm. Figure 1.8 shows how the self-contained block is used in practice.

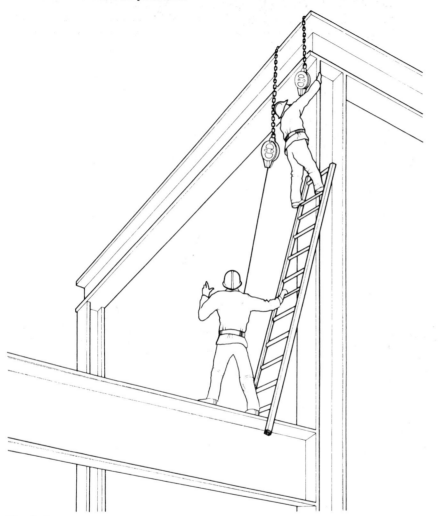

Fig. 1.8 The use of self-contained safety blocks

4. All stairways and walkways should be provided with hand-rails. On stairways which would otherwise be open-sided, a lower rail should also be provided. All overhead platforms or walkways should be fitted, wherever practicable, with permanent toe-boards. These safety practices are shown in Fig. 1.9. It shows how guard rails and toe-boards must be fitted on every side of a working platform from which personnel are likely to fall more than 2 metres. The diagram also shows a safe means of access.

Ladders should always be placed so that there is adequate space behind each rung for a proper foothold. Particular attention should be paid to this point at the landing platform.

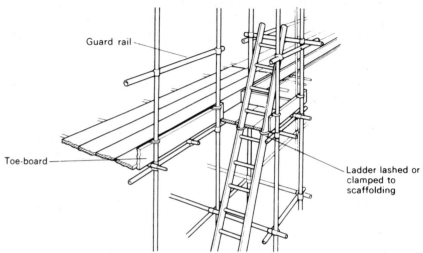

Fig. 1.9 Overhead platforms

5. Always place tools and materials in positions of stability, i.e., away from edges from which they might fall. *A tool box or tool kit properly secured or hooked to a ladder or platform is to be recommended.*
6. Workers should not throw tools or materials down from aloft. *Such irresponsible behaviour can often result in fatal injuries to the persons who happen to be below.* It is considered good practice to make provisions for the lowering of tools or materials, or, if it is safe to do so, take them down.

1.5 Use of ladders

Building and Construction Regulations contain a wealth of information on the safe use of ladders. The following 'Safety Hints' will serve to indicate some of the many elementary precautions which should be regarded as standard procedure:

1. Ensure that ladders are well constructed and of sound material, and are well looked after.
2. DO NOT PAINT LADDERS. They should not be painted because this tends to *hide any defects and conceal danger*. CLEAR VARNISH IS NORMALLY USED TO PROTECT AND PRESERVE TIMBER LADDERS.

3. NEVER USE AN UNSOUND LADDER. No ladder should be used, for example, if it has a missing or defective rung, or when the uprights (or stringers) show signs of splitting. ANY DEFECTIVE LADDER SHOULD BE IMMEDIATELY DESTROYED.
4. *The correct pitch of a ladder must always be observed.* THE VERTICAL HEIGHT FROM THE GROUND OR BASE TO THE POINT OF REST SHOULD BE FOUR TIMES THE DISTANCE BETWEEN THE BASE OF THE VERTICAL HEIGHT AND THE BASE OF THE LADDER, as illustrated in Fig. 1.10.
5. Make certain that the ladder reaches at least 1 metre above the landing platform. *This is to provide a handhold while stepping from the ladder,* as shown in Fig. 1.10.
6. Use the correct length of ladder for the job. NEVER LASH TWO SHORT LADDERS TO MAKE A LONGER ONE.
7. MAKE SURE THAT THE LADDER IS SET ON A FIRM LEVEL BASE. Care must be taken on sloping surfaces — *the packing of ladder feet is a dangerous habit which must be discouraged.* Figure 1.11 shows a few of the safe practices observed when setting ladders.

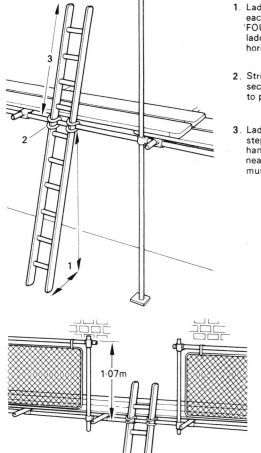

1. Ladders should be set 0·3m out for each 1·2m of height, i.e., use the 'FOUR TO ONE RULE' so that the ladder is at an angle of about 75° to the horizontal.

2. Stringers should be lashed or clamped securely to some convenient anchorage to prevent slipping sideways.

3. Ladders must extend 1·07m above the stepping-off point to ensure adequate handhold. If this is not possible then a nearby handhold of equivalent height must be provided, as shown below.

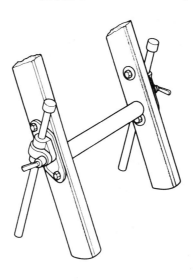

Adjustable safety device to suit any uneven contour of surface. Can easily be fitted to most ladders

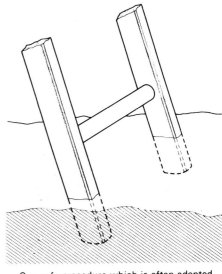

One safe procedure which is often adopted on site is to bury the foot of the ladder in the ground up to the first rung

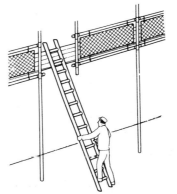

If in doubt have a person 'footing the ladder.'

Fig. 1.11 Safety precautions - ladders.

Fig. 1.10 Setting a ladder

8. Lash the ladder as soon as possible (see Fig. 1.10). Until this is done, ensure that a workmate is standing on the bottom or 'footing it'.

9. Always face the ladder when climbing or descending, and *beware of wet, greasy, or icy rungs.*
10. INSPECT LADDERS BEFORE USE AND REGULARLY WHEN STORED. Store ladders where they will not suffer damage either by the weather or by mechanical means. Always rest ladders, DO NOT HANG THEM FROM THE STILES OR RUNGS.

(Note: The *uprights* (or *stringers*) are also known as *stiles* in some parts of the country.)

The author is indebted to the Royal Society for the Prevention of Accidents for much of the information used in this section of the chapter.

PART B ARC WELDING
1.6 Fusion welding

When metals are joined by fusion welding (*Basic Engineering*, Section 10.27) their edges are heated until they become molten and run together. Additional metal, in the form of a *filler rod*, is added to fill any gaps and make up oxidation losses.

Obviously, heat sources that can operate above the melting points of alloy steels, and that have the energy output to melt thick plate, must be potentially hazardous. Fusion welding equipment will now be considered in greater detail so that the hazards can be identified and suitable precautions taken.

1.7 Arc welding equipment (mains operated)

This equipment is designed to change the high voltage alternating-current mains supply into a safe, low-voltage, heavy-current supply suitable for arc welding. Figure 1.12 shows examples of some typical arc welding sets.

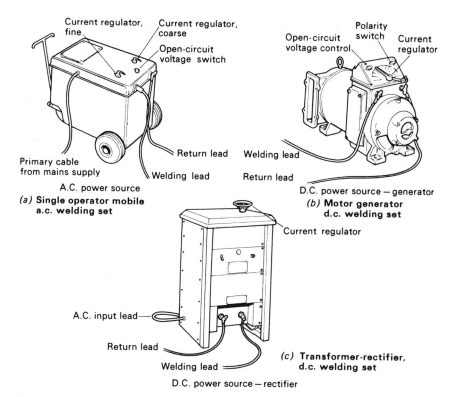

Fig. 1.12 Mains operated arc-welding equipment

It will be seen that the output can have an alternating current waveform or a direct current waveform. For safety, the output voltage is limited to between 50 and 100 V; however, the output current may be as high as 500 A. Figure 1.13, a typical arc welding circuit, shows that a welding set is basically a transformer to break down the high mains voltage, and a tapped choke to control the current flow to suit the gauge of electrode used.

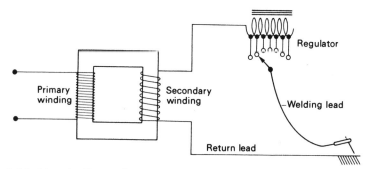

Fig. 1.13 Circuit diagram of an alternating current arc-welding set

1.8 Hazards associated with mains-operated arc welding equipment

The hazards that may arise from mains-operated welding equipment are set out in Table 1.1. To understand these more fully, constant reference should be made to Fig. 1.14 which shows a schematic diagram of a typical manual, metallic-arc welding circuit. It will be seen that the circuit conveniently divides into two parts:
1. The primary (high voltage) circuit, which should be installed and maintained by a skilled electrician.
2. The secondary (low voltage) external welding circuit, which is normally set up by the welder to suit the job in hand.

Table 1.1 Arc-welding hazards

CIRCUIT — HIGH VOLTAGE — Primary	
Fault:	*Hazard:*
1. Damaged insulation	**Fire** — loss of life and damage to property
	Shock — severe burns and loss of life
2. Oversize fuses	**Overheating** — damage to equipment and fire
3. Lack of adequate earthing	**Shock** — if fault develops — severe burns and loss of life
CIRCUIT — LOW VOLTAGE — Secondary (very heavy current)	
Fault:	*Hazard:*
1. Lack of welding earth	**Shock** — if a fault develops — severe burns and loss of life
2. Welding cable — damaged insulation	Local arcing between cable and any adjacent metalwork at earth potential causing **Fire**
3. Welding cable — inadequate capacity	Overheating leading to damaged insulation and **Fire**
4. Inadequate connections	Overheating — severe burns — **Fire**
5. Inadequate return path	Current leakage through surrounding metalwork — overheating — **Fire**

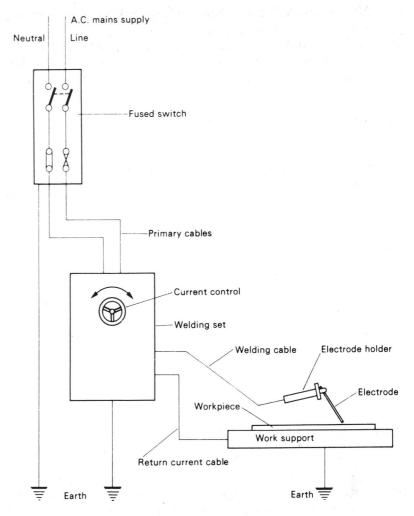

Fig.1.14 Manual metal-arc welding circuit diagram

To eliminate these hazards as far as possible, the following precautions should be taken. These are only the *basic* precautions and a check should always be made as to whether the equipment and working conditions require special, additional precautions:

1. Make sure that the equipment is fed from a switch-fuse so that it can be isolated from the mains supply. Easy access to this switch must be provided at all times.
2. Make sure that the trailing primary cable is armoured against mechanical damage as well as being heavily insulated against the high supply voltage (415 V).

3. Make sure that all cable insulation is undamaged and all terminations are secure and undamaged. *If in doubt, do not operate the equipment until it has been checked by a skilled electrician.*
4. Make sure that all the equipment is adequately earthed with conductors capable of carrying the heavy currents used in welding (*Basic Engineering*, section 11.23).
5. Make sure the current regulator has an 'OFF' position so that in the event of an accident the welding current can be stopped without having to follow the primary cable back to the isolating switch.
6. Make sure that the 'external welding circuit' is adequate for the heavy currents it has to carry.

1.9 The external arc welding circuit

This is normally set up by the welder himself to suit the job in hand. There are three important connections for every welding circuit:

1. *The welding lead* This is used for carrying the welding current from the power source to the ELECTRODE HOLDER.

2. *The welding return cable* This is the connection for carrying the 'return' current between the 'work' and the power source of supply, and is usually attached to the 'work' by a special spring or screw-clamp which is properly secured to one end of the cable, as shown in Fig. 1.15.

3. *The welding earth This is necessary on all welding circuits* to maintain the workpiece and any other conductors or metal structures that may come in contact with it at EARTH POTENTIAL.

The proper selection and care of welding cables is an essential safety requirement, for it is surprising how many faults occur in practice through neglecting to take elementary precautions. The most common faults are:

1. Bad connections;
2. The use of longer cables than necessary (causing excessive 'volt-drop');
3. Defective insulation;
4. The use of secondary cables of incorrect current carrying capacity.

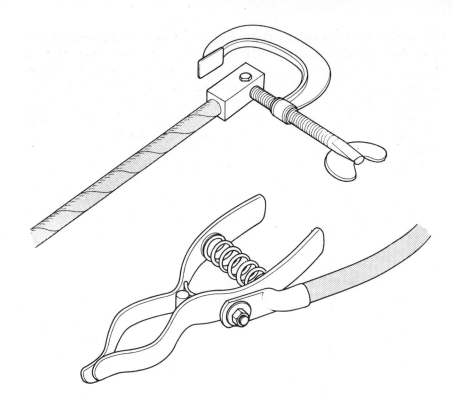

Fig. 1.15 Return current clamps

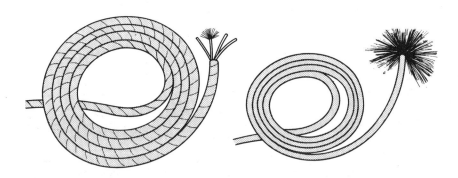

(a) Trailing cables are used for primary connections between the welding equipment and the current supply

(b) Welding leads are multi-st___ ued for extra flexibility

Fig. 1.16 Welding cables

Figure 1.16 shows the fundamental difference between the primary trailing cable and the welding lead. The former has three or four cores of relatively thick stranded wire, whereas the latter derives its flexibility and heavy current-carrying capacity from hundreds of strands of very fine wire. This fine wire is contained within a paper wrapping, to allow them to slip readily when the cable is bent. Wear resistance is usually provided by a tough outer sheath of insulating material, such as tough vulcanised rubber (TVR).

The main hazards associated with the external welding circuit are given in Table 1.1, and are overcome by the use of cables of adequate capacity and terminating them correctly. Table 1.2 gives details of 'secondary welding cables'. Figure 1.17 shows typical connectors for welding and return cables.

Most welding equipment has a duty cycle of 40 per cent at maximum welding current. If welding is carried out with a lower current it is possible to load the equipment longer.

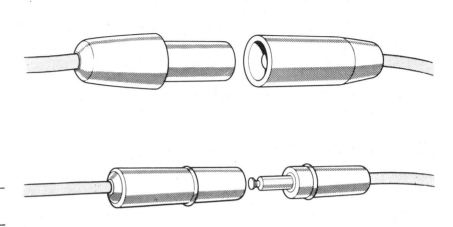

Fig. 1.17 Fully insulated quick-acting cable connectors

Table 1.2 Secondary welding cables

Constructional details and current ratings of rubber insulated/rubber sheathed and rubber insulated/P.C.P. sheathed cable with copper conductors.

CONDUCTOR DETAILS		CURRENT RATINGS (Amperes) at a maximum duty cycle of				
Nominal area (mm^2)	Nominal number and nominal diameter of wires (mm)	100%	85%	60%	30%	20%
16	513 / 0·20	105	115	135	190	235
25	783 / 0·20	135	145	175	245	300
35	1107 / 0·20	170	185	220	310	380
50	1566 / 0·20	220	240	285	400	490
70	2214 / 0·20	270	295	350	495	600
95	2997 / 0·20	330	360	425	600	740
120	608 / 0·50	380	410	490	690	850
185	925 / 0·50	500	540	650	910	1120

The cables connecting the welding set to the electrode holder and the EARTH RETURN connecting the workpiece to the welding set *must be of the appropriate size for the maximum* WELDING CURRENT *of the set.*

Duty cycle — A POWER SOURCE for manual arc welding is never loaded continuously because of the periods spent on such operations as electrode changing, slag removal, etc.
The ratio of the actual welding or arcing time and the total working time expressed as a percentage is called the 'duty cycle'

$$\text{DUTY CYCLE} = \frac{\text{ARCING TIME}}{\text{TOTAL WORKING TIME}} \times 100\%$$

(where the total working time = arcing time + handling time).

1.10 Electrode holders

These should be soundly connected to the welding lead. They should be of adequate rating for the maximum welding current to prevent them heating up and becoming too hot to handle. Many types of electrode holders are available, some are partially-insulated and others fully-insulated, as illustrated in Fig. 1.18.

Electrode holders of the partially-insulated type have, in addition to a handle made of a heat-resisting non-ignitable insulating material, a protective guard in the form of a disc between the handle and the exposed metal parts. This guard has a dual purpose — it affords protection to the operator's hand by preventing it slipping down on to the exposed metal parts, which are both electrically alive and physically hot; it also enables the holder to be placed on the workpiece or supporting structure without the exposed parts 'shorting'.

The fully-insulated electrode holder, as the name implies, has all metal parts protected by an efficient insulating material, with the exception of the small portion where the electrode is inserted.

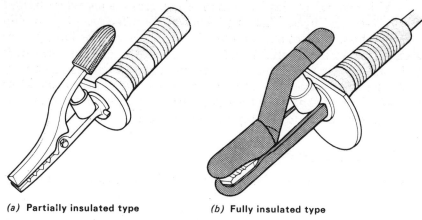

Fig. 1.18 Electrode holders

1.11 Mobile welding plant

Figure 1.19 shows a typical engine-driven mobile welding set. Although there is no high voltage supply associated with this equipment it has hazards of its own:
1. Storage of flammable fuel oil for the engine.
2. Toxic exhaust gases. (Adequate ventilation must be provided to disperse these if the equipment is used indoors.)

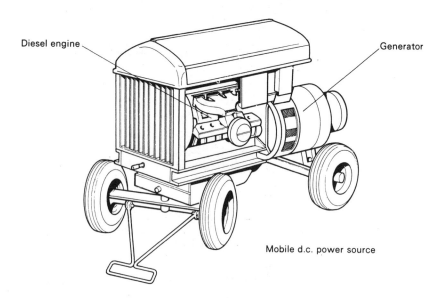

Fig.1.19 A mobile engine-driven direct current welding set

1.12 Protective clothing for the eyes and head

For all arc welding operations it is essential to protect the welder's head from radiation, spatter, and hot slag, and for this purpose either a helmet or a hand shield must be worn. Examples are shown in Fig. 1.20.

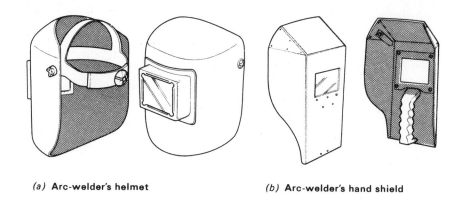

(a) Arc-welder's helmet   (b) Arc-welder's hand shield

Fig. 1.20 Eye and head protection

An arc welder's HAND SHIELD protects one hand as well as the face. It is fitted with a handle which is made of material which insulates against heat and electricity. The handle may either be fixed inside the shield to protect the hand from the heat and rays of the arc, or fixed outside and provided with an effective guard for the same purpose.

Welder's HEAD SHIELDS are usually fitted with an adjustable band to fit the wearer's head. *This band, and the means of adjustment, should be thoroughly insulated from the wearer's head, the insulation being non-absorbent as a precaution against dampness such as perspiration which tends to make it conductive.* Head shields are designed to pivot so as to provide two definitely located positions:
1. Lowered in front of the face — the welding position — for protection; and,
2. Raised in a horizontal position to enable the welder to see when not striking the arc.

Some welders prefer to use a handshield, rather than a headshield because it is the least tiring protection to use. *However, headshields provide better protection and allow the welder the free use of both hands.*

The injurious effect of the radiations emitted by an electric arc are similar whether a.c. or d.c. is used for welding. Exposing the eyes and face to INFRA-RED rays would lead to the face becoming uncomfortably hot and might induce serious eye troubles. If too much ULTRA-VIOLET radiation is received by the welder or anyone in the vicinity, it can cause an effect similar to sunburn on the skin and a condition known as ARC EYE. In addition, too much visible light will dazzle the operator, and too little can cause eyestrain and headaches.

The obvious precaution is to prevent the harmful radiations from the welding arc and the molten weld pool from damaging the eyes. *This is achieved by the use of special glass filters of suitable colour and density, which also reduce the intensity of the visible light rays.* All helmets and handshields are provided with a filter glass and a less expensive protective cover glass on the outside. Table 1.3 gives some examples of the filter glasses recommended in BS 679.

Table 1.3 Filters for manual arc-welding

GRADE OF FILTER REQUIRED	APPROXIMATE RANGE OF WELDING CURRENT (AMPS)
8 / EW	Up to 100
9 / EW	100 to 200
10 / EW	Up to 200
11 / EW	Up to 300
12 / EW 13 / EW 14 / EW	For use with currents over 300

Each filter purporting to comply with BS 679 should be permanently marked as follows:
1. BS 679
2. The certification mark of the British Standards Institute
3. The manufacturer's name, symbol or licence number
4. The figures and letters denoting its shade and type of welding process (gas or electric)

Filter glasses are expensive, therefore they should be used with a clear plain cover glass on the outside in order to protect them from damage by spatter and fumes.
These cover glasses are relatively cheap and easily replaceable.

The removal of slag by 'chipping' can create an accident hazard to the face and eyes. Although welders are often tempted to chip away without protection, the use of a shield or goggles having clear glass windows is essential in all such operations. Where a screen is to be used solely for protection when de-slagging it is advisable to use a shatterproof glass or plastic cover. *When using a handshield or helmet of the dual-purpose type, the change-over device for raising or lowering the dark filter glass should fail to 'safe'.*

1.13 Body protection

Figure 1.21 shows a welder fully equipped with protective clothing.

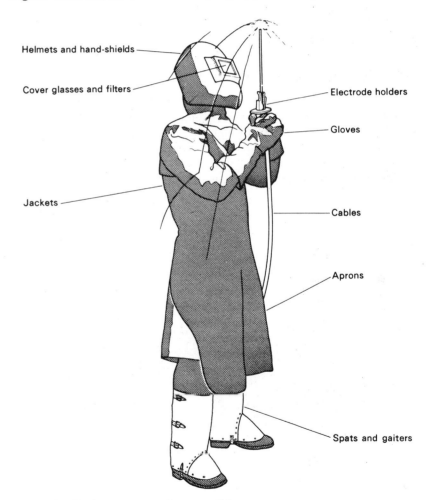

Fig. 1.21 Fully protected arc welder

The welder's body and clothing must be protected from radiation and burns caused by flying globules of molten metal. It may be necessary for a welder to wear an apron, usually of asbestos or thick leather, to protect his trunk and thighs whilst seated at a bench welding. An apron should also be worn if the welder's clothing is made of flammable material.

When deep gouging or cutting is carried out using metal-arc processes, the amount of 'spatter' is considerably greater than that experienced with normal arc welding, and therefore *it is necessary to protect the feet and legs in the same way as the hands and forearms.* Suitable leather leggings and spats are available and should be used to prevent burns to the legs, feet, and ankles. Further details are given in Section 1.17.

1.14 Screens

People working in the vicinity of a welding arc, including other welders, can be exposed to stray radiation from the arc, and can be caused considerable discomfort. Looking at an unscreened welding arc, even from a distance of several metres and for a few seconds only, can cause 'Arc Eye'. The painful effects of exposure will not be felt until between four and twelve hours later, and it is very likely that the affected worker wakes up at night with the characteristic pains of arc eye. *Persons affected usually complain of a feeling of 'sand in the eyes' which become sore, burn, and water.* Where possible, each arc should be screened in such a way that stray radiation is kept to a minimum. *The walls of welding shops or individual welding booths should be painted with a matt, absorbent type of paint with a very low reflective quality.* The colour of the paint does not have to be black, as experiments have proved that matt pastel shades of grey, blue, or green are equally efficient.

If a person is exposed to a flash, the effects of 'Arc Eye' can be minimized by the immediate use of a SPECIAL EYE LOTION which should be available in the FIRST AID BOX.

One experience of the distressing results which follow exposure to stray radiation is usually sufficient to make the sufferer more careful in the future!

1.15 Fire hazards

Attention must be drawn to one particular hazard which is very often ignored, that is the danger of burns from freshly welded metal. Such pieces should be clearly marked HOT in chalk or special marking crayon. This simple safety precaution should also be adopted when FLAME CUTTING operations are carried out. In addition to the precautions taken to protect the operator himself from burns, it must be realised that *sparks, molten metal, and hot slag can cause a fire to start if flammable material is left in the welding area.* All old rags, cotton waste, sacks, paper, etc., should be removed. Metal bins, surrounded by sand or asbestos sheet, should be provided for spent electrode stubs, for hot electrode stubs can burn and penetrate through the soles of footwear and are a hazard when thrown on to the workshop floor. *Treading on an electrode stub can have the same disastrous effect as, for example, inadvertently stepping on a roller skate.* General cleanliness, not always very apparent in welding workshops, is essential.

Wherever gas or electric welding or cutting operations take place the work area should be screened off with asbestos blankets or metal screens. Wooden flooring should be covered with sand, or overlapping metal sheets; sparks or molten globules of metal must not be allowed to fall into gaps between boards. Very often welders have to work overhead and Fig. 1.22 illustrates the safety precautions which are necessary.

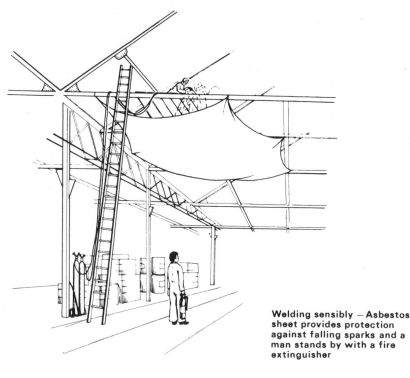

Welding sensibly — Asbestos sheet provides protection against falling sparks and a man stands by with a fire extinguisher

Fig. 1.22 Overhead protection

1.16 Ventilation

When using the majority of types of electrodes *the welder is not likely to suffer any ill effects from welding fumes provided that reasonable ventilation is available.* Localised exhaust ventilation can be provided by a good fume extractor, which not only dilutes and removes fumes but also assists in keeping down the temperature and adds to the comfort and efficiency of the welder. Figure 1.23 illustrates a typical portable fume extractor.

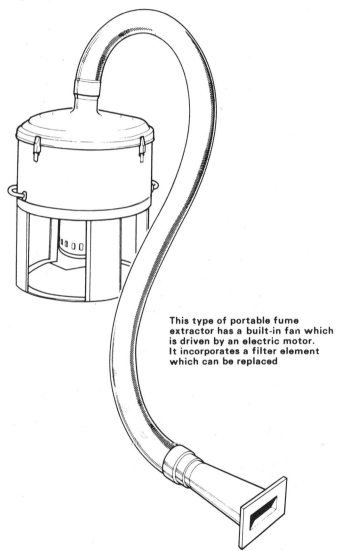

This type of portable fume extractor has a built-in fan which is driven by an electric motor. It incorporates a filter element which can be replaced

Fig.1.23 Portable fume extractor unit

PART C GAS WELDING AND CUTTING

1.17 Personal protection

The need for protective equipment and clothing should be the first consideration before commencing gas welding and cutting operations.

Goggles

These are essential and *must be worn to protect the eyes from heat and glare, and from particles of hot metal and scale.* Goggles used for welding and cutting are fitted with approved filter lenses, as listed in Table 1.4. Figure 1.24 shows the main features of good quality welding goggles.

Table 1.4 Filters for gas welding and cutting

GRADE OF FILTER REQUIRED		RECOMMENDED FOR USE WHEN WELDING
Welding without flux	Welding with flux	
3 / GW	—	Thin sheet steel
—	3 / GWF	Aluminium, magnesium and aluminium alloys. Lead burning, oxy-acetylene cutting
4 / GW	—	Zinc-base die castings. Silver soldering. Braze welding light gauge copper pipes and light gauge steel sheet
—	4 / GWF	Oxygen machine cutting — medium sections. Hand cutting, flame gouging and flame descaling
5 / GW	—	Small steel fabrications. Hard surfacing
—	5 / GWF	Copper and copper alloys. Nickel and nickel alloys. Heavier sections of aluminium and aluminium alloys. Braze welding of un-preheated heavy gauge steel and cast iron
5 / GW	—	Heavy steel sections. Preheated cast iron and cast steel. Building up and reclaiming large areas
—	6 / GWF	Braze welding of preheated cast iron and cast steel

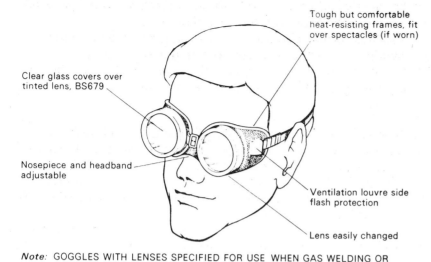

Note: GOGGLES WITH LENSES SPECIFIED FOR USE WHEN GAS WELDING OR CUTTING MUST NOT BE USED FOR ARC WELDING OPERATIONS

Fig.1.24 The essential features of good quality welding goggles

Protective clothing

Garments made of wool are generally considered not to be readily flammable. However, a high percentage of outer clothing, normally worn by workers, is usually made from flammable materials. *Cuffs on overalls, or turn-ups on trousers are potential fire traps; hot slag, sparks, and globules of hot metal can so easily lodge in them.*

The protective clothing worn will depend upon the nature of the work, and the following suggestions are offered as a general guide:
(a) Asbestos or leather gloves should be worn for all cutting operations which involve the handling of hot metal.
(b) Safety boots should be worn to protect the feet from hot slag and, in particular, from falling off-cuts.
(c) The wearing of asbestos spats is strongly advised for most cutting operations.
(d) The wearing of an asbestos or leather apron will help to prevent sparks and hot metal globules reaching and burning the sensitive parts of the operator's body. *Many welders have experienced the folly of working with their shirts open to the waist, thus presenting a ready made receptacle for sparks and hot metal spatter.*

(e) Leather gauntlets should be worn when welding or cutting on overhead or vertical structures.
(f) In situations where welding and cutting operations are carried out overhead it is advisable for those persons working or standing below to protect themselves against falling sparks. For this purpose it is recommended that safety helmets or leather skullcaps should be worn by those in the vicinity. Figure 1.25 shows typical protective clothing worn for cutting operations.

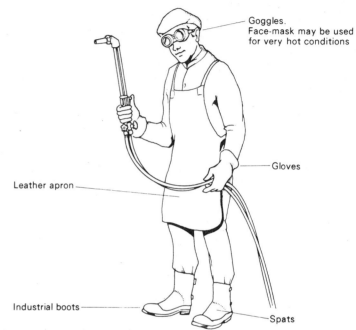

Fig.1.25 Protective clothing for cutting operations

1.18 Fire hazards

The following important precautions should be rigidly observed:
1. Do not position gas cylinders and hoses where sparks or slag can fall on them.
2. Wooden floors should be kept thoroughly wetted with water, or completely covered, for example with sand.
3. Wooden structures should be adequately protected by sheet metal or asbestos.
4. All combustible materials should be removed to a safe position, or, if this is not possible, should be properly protected against flying sparks. *It must be realised that sparks from cutting can travel up to 9 metres along a floor.*

5. Suitable safety measures must be taken in the case of cracks or openings in walls or floors.
6. KEEP FIRE FIGHTING EQUIPMENT READY TO HAND. A responsible person should keep the site under observation for at least half an hour after the completion of the work in order to watch for, and deal with, any outbreak of fire. *There is always the danger of material smouldering for hours before a fire breaks out.*

1.19 Explosion risks

Basically, the heat required for gas welding and cutting operations is generated by the combustion of a suitable fuel gas with oxygen. The vast majority of gas welding processes employ an *Oxygen/Acetylene* gas mixture, and Table 1.5 lists a number of efficient welding gas mixtures in common use.

Table 1.5 Welding gas mixtures

FUEL GAS	MAXIMUM FLAME TEMPERATURE	
	with air (°C)	with oxygen (°C)
Acetylene	1 755	3 200
Butane	1 750	2 730
Coal Gas	1 600	2 000
Hydrogen*	1 700	2 300
Propane	1 750	2 500

*The oxy-hydrogen flame has an important application in under-water cutting processes.

Explosions can occur when ACETYLENE gas is present in AIR in any proportions between 2 per cent and 82 per cent. This gas is also liable to explode when under unduly high pressure, even in the absence of air, therefore THE WORKING PRESSURE OF ACETYLENE SHOULD NOT EXCEED 0·62 BARS.

When using gas welding processes the first essential requirements are:
(a) ENSURE THAT THERE IS ADEQUATE AND PROPER VENTILATION.
(b) EXAMINE THE EQUIPMENT AND SEE THAT IT IS FREE FROM LEAKS.

Explosions in the equipment itself may be caused by FLASH-BACK. *Flash-backs occur because of faulty equipment or incorrect usage.* Approximately 80 per cent of these occur when lighting welding or cutting torches, the other 20 per cent when they are in use. *As long as the flow of gas equals the burning speed a stable flame will be maintained at the torch nozzle, otherwise mixed gases will arise in one of the hoses, resulting in a flash-back.*

Flash-backs may also occur in the following circumstances:
(a) By dipping the nozzle-tip into the molten weld pool.
(b) By putting the nozzle-tip against the work and stopping the flow.
(c) By allowing mud, paint, or scale to cause a stoppage at the nozzle.

In every case the obstruction will cause the OXYGEN to flow back into the ACETYLENE supply pipe and communicate ignition back towards the cylinder or source.

The British Oxygen Company issues many booklets on the safe use and storage of welding gases and these should be consulted before using gas welding equipment.

Figure 1.26 shows a suitable 'hose protector' which will arrest a flash-back; it is built into the hose union.

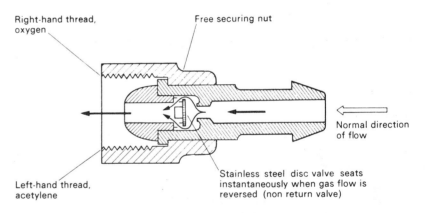

Note: THESE PROTECTORS WILL PREVENT THE FLOW OF GASES TO THE WELDING OR CUTTING TORCH IF THE HOSES ARE INADVERTANTLY REVERSED

Fig.1.26 Hose protector

COPPER tube or fittings made of copper must never be used with ACETYLENE, and alloys used in the construction of pipes, valves, or fittings should not contain more than 70 per cent copper — THE ONLY EXCEPTION IS THE WELDING OR CUTTING NOZZLE. This is because copper, when exposed to the action of acetylene, forms *a highly explosive compound called COPPER ACETYLIDE, which is readily detonated by HEAT or FRICTION.*

Explosions can occur in the regulators on oxygen cylinders as a result of dust, grit, oil, or grease getting into the socket of the cylinder; dust (especially COAL DUST) is highly flammable. The outlet sockets of cylinder valves should be examined for cleanliness before fitting regulators otherwise, if not removed, foreign matter will be projected on to the regulator valve-seating when the cylinder valve is opened. The outlet socket can usually be cleaned by turning on the cylinder valve for a brief moment and closing it soundly. This is commonly known as CRACKING THE CYLINDER, and is shown in Fig. 1.27.

1. Open the control valves on the torch.
2. Release the pressure-adjusting control on the regulators.
3. Open the cylinder valves to turn on the gas.
4. Set the working pressures by adjusting the regulator control.
5. Having established the correct pressure for each gas, close the control valves on the torch.

The system is now ready for leak testing, for which the procedure is clearly indicated in Fig. 1.28.

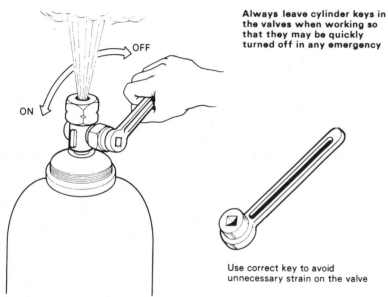

Fig.1.27 Cracking the cylinder

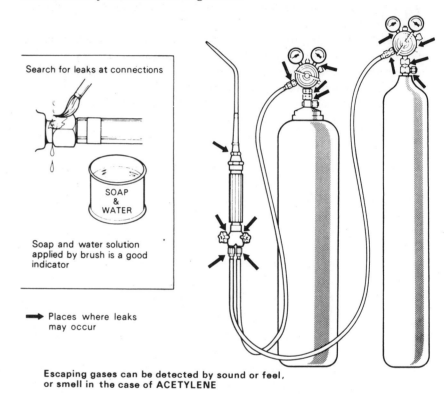

Fig.1.28 Testing for leaks

1.20 Testing for leaks

It is extremely dangerous to search for gas leaks with a naked flame. OXYGEN is *odourless* and, whilst it does not burn, it readily supports and speeds up combustion. ACETYLENE has *an unmistakeable smell*, rather like garlic, and can be instantly ignited by a spark or even a piece of red hot metal.

Before leak testing, it is considered good practice to 'pressurize' the system, and the procedure is as follows:

1.21 Safety in the use of gas cylinders

Cylinders for compressed gases are not themselves dangerous. They must comply with rigid government standards and should be regularly inspected and tested.

The fact that gas cylinders are a familiar sight in factories or on sites where welding and cutting is carried out is often the reason why ordinary safeguards are neglected.

For general identification purposes, narrow cylinders contain gases of HIGH PRESSURE, whilst broad cylinders contain LOW PRESSURE gases.

A great deal of information is available, usually in the form of safety booklets issued by manufacturers, on the use, handling, and storage of gas cylinders. A few of the many safety precautions will now be considered.

1. Cylinders must be protected from mechanical damage during storage, transportation, and use. Acetylene cylinders must *always* be kept upright.
2. Cylinders must be kept cool. On no account should the welding flame, or any other naked light, be allowed to play on the cylinders or regulators. They must also be shielded from direct sunlight, wet, and frost on an open site.
3. Cylinders must always be stored in well ventilated surroundings to prevent the build-up of pockets of explosive mixtures of gases should any leaks occur. **DO NOT SMOKE IN A GAS CYLINDER STORE.**
4. Correct automatic pressure regulators must be fitted to all cylinders prior to use. The cylinder valve must always be closed when the cylinder is not in use or whilst changing cylinders or equipment.
5. Keep cylinders free from contamination. Oils and greases ignite violently in the presence of oxygen; similarly, do not wear dirty or greasy clothes in the presence of compressed gases.

1.22 Welding in confined spaces

Normally AIR contains 21 per cent OXYGEN. Only a small increase in this percentage results in the air becoming 'oxygen enriched'. One danger to a welder working in a confined space, should the air become enriched with oxygen, is that his clothing could spontaneously ignite causing severe burns. ANY SITUATION IN WHICH THE NATURAL VENTILATION IS INADEQUATE SHOULD ALWAYS BE REGARDED AS A 'CONFINED SPACE'.

Safety regulations require specific measures to be taken:
(a) Prevent the use of oxygen to ventilate any confined space; and
(b) Ensure that adequate provision is made for ventilating, and that it is used.

Ventilation is of vital importance when welding or cutting is carried out in confined or enclosed spaces, and neglect in this respect has often resulted in loss of life.

(c) Any worker in a confined space must be kept under constant observation by an assistant outside. This assistant should be competent to control the supply of gases and carry out emergency procedures.
(d) The assistant outside the vessel or confined area should have immediately to hand a pail of water and a suitable fire extinguisher.
(e) When welding or cutting is carried out inside boilers or vessels, gas cylinders should always be kept OUTSIDE.
(f) ALWAYS PASS THE WELDING OR CUTTING TORCH TO THE OPERATOR WORKING IN A CONFINED SPACE ALREADY LIGHTED.
(g) Never leave torches and hoses in a confined space when they are not in use, as, for example, during a meal break or overnight. *A very small leak of either the oxygen or the fuel gas*, over such a period, *can produce a dangerous atmosphere if contained in a confined space.*

PART D MISCELLANEOUS

1.23 Precautions to be observed when forging

The general safety precautions to be observed where forging operations are involved can be summarized as follows:

1. The floor area where a smith's hearth is installed should be of fireproof material.
2. It is essential that workers employed in forging operations should wear adequate protective clothing, such as safety footwear, leather gloves, and leather aprons.
3. Ensure that the fuel used in the hearth is free from foreign material such as stones.
4. A properly designed flue must be fitted to conduct any fumes to the outside atmosphere. DUST AND FUMES ARE A POTENTIAL SOURCE OF DANGER. *Carbon monoxide* is a dangerous gas produced by combustion, especially if the fuel in the hearth is allowed to burn without an adequate air supply. *The effect of this gas, which cannot be detected by smell, is to diminish the oxygen-carrying capacity of the blood. This could render the worker dizzy and weak, and it may lead to a state of unconsciousness.*
5. In respect of item 4, it is essential that the electricity supply to any extractor fan should be arranged independently of any emergency control ('stop button') to other equipment.
6. Anvils and quenching tanks should be situated as close as possible to the hearth. THE DANGERS OF CARRYING HOT METAL AROUND CANNOT BE OVER-EMPHASISED.

7. The smith's anvil should be mounted on a stable base, preferably a cast iron stand, and maintained in good condition. A damaged anvil face can be dangerous and should not be used.
8. Scale should be removed from the workpiece on withdrawal from the fire. This is generally accomplished by means of a simple flat scraper. If this is not done, every hammer blow on the white hot metal will be accompanied by a shower of hot oxide in all directions.
9. Care must be taken when quenching a hollow section, such as a tube. *A jet of scalding steam may shoot out of the open end which is not in the quenching tank.*
10. Always maintain hand-forging tools in undamaged and good condition. After use, tools should be quenched and replaced in a proper storage rack.

1.24 Precautions to be observed when testing containers

Containers have to be tested for leaks either (*a*) after their manufacture; (*b*) during routine maintenance; or (*c*) after a repair has been effected. The type of test employed will depend upon the service conditions under which the container will have to operate. *Oil and fuel tanks, for example, are containers in which the contents are not under pressure — the only pressure involved is that exerted by the weight of the liquid itself.* However, the contents of a great many containers in service are under high pressure, and such containers are called 'pressure vessels'.

Testing by filling

Open-topped containers, such as water tanks or large vats, are usually tested by filling with water and allowed to stand for a period, during which time any leaks will be indicated. *Precautions should be taken when very large containers are being tested because the very large volume of water required will be extremely heavy.*

It is recommended that large open-topped containers should be well supported on a strong base before commencing the test, and the test should preferably be carried out in the open.

Immersion tests

An immersion test can only be applied to small tanks. It must be appreciated that for every 0·028 m³ capacity the water would press upwards with a force of approximately 276 N. *It follows, therefore, that for a container with a capacity of several cubic metres, a considerable force would be required to hold it under the water.* Coupled with that, the container would need to be revolved in order to be sure of locating leaks on the under-side. Thus only relatively small containers can be tested easily by this method. In this test the tank or container is pressurised and completely immersed in water. Any leaks will be indicated by bubbles and their positions marked by an indelible pencil.

Pressure test — pneumatic

The tank should be rendered safe by the recommended methods laid down by the Factories Act. It is then partly filled (about one-tenth of its capacity) with paraffin and rocked so that all internal surfaces become covered with paraffin. The external seams and all other joints (including rivet and bolt heads) must then be coated with a paste prepared from METHYLATED-SPIRIT and WHITENING, applied with a brush. *The spirit will evaporate and the whitening surface remain.* It is general practice to seal off all outlets except one which is provided with a pump connection. Air is pumped into the tank to the correct test pressure, as indicated on a pressure gauge. *Any leakage will at first be indicated by a drop in pressure on the gauge.* The exterior of the container should be visually examined and *any cracks, faulty seams and joints, or faulty bolts and rivets will be clearly indicated by the pressurised paraffin discolouring the whitening.* The leaks are then marked with an indelible pencil, and the whitening removed with a suitable solvent. The container is drained and all volatile fumes expelled. The leaks should then be repaired and the same procedure adopted for any subsequent pressure tests, which are repeated until the container withstands the test.

There is always the risk of an explosion when testing a pressure vessel with compressed air. **Such 'Pneumatic Tests' must be carried out under close supervision by the inspecting authority.** ADEQUATE PRECAUTIONS, SUCH AS BLAST WALLS OR PITS AND MEANS OF REMOTE OBSERVATION ARE ESSENTIAL.

Pressure test — hydraulic

The safest method for testing vessels which operate at medium or high pressures in service is the 'HYDRAULIC TEST'. Any failure due to the internal pressure escaping by leakage will only result in a slight spillage. *Unlike air or gas under pressure, water cannot be compressed and is only a means of transmitting high pressures.*

With the hydraulic test, the vessel being tested is subjected to an internal test pressure of 1·5 times the safe working pressure for not less than 30 minutes. This is considered sufficient minimum length of time to permit a thorough examination to be made of all seams and joints. By comparison, when a vessel is subjected to a pneumatic test, the test pressure should not exceed the design pressure. *In the interest of safety, it is important that any vessel subjected to a hydraulic test should be properly vented so as to prevent the formation of 'air pockets' before the test pressure is applied.* It is recommended that during the test the temperature of the water should not be below 7°C. Figure 1.29 illustrates the correct procedure for hydraulic pressure testing.

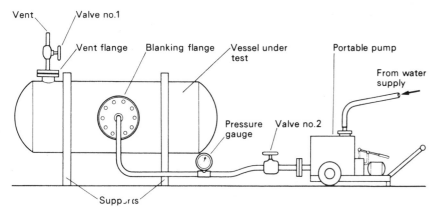

Testing procedure
1. Fill vessel with water by removing vent flange
2. Reconnect vent flange leaving valve No.1 open
3. Open valve No 2 and start pumping
4. When water is flowing out of vent pipe – INDICATING THAT ALL THE AIR HAS BEEN FORCED OUT – close valve No.1; PRESSURE WILL NOW BEGIN TO BUILD UP INSIDE THE VESSEL
5. When the desired pressure has been reached on the pressure gauge, turn off valve No.2 and stop pumping. ANY DROP IN PRESSURE ON THE GAUGE INDICATES LEAKAGE

Fig.1.29 Pressure testing

Repairs to containers

Leaks in containers usually require repairing by some process which involves heat, such as welding, brazing, or soldering, therefore at this stage reference will be made to the requirements of the Factories Act. They specify that no plant, tank, or vessel which has contained any explosive or flammable substance should be subjected to:
(a) Any welding, brazing, or soldering operations;
(b) Any cutting operation which involves the application of heat; or
(c) Any operation involving application of heat for the purpose of taking apart or removing the plant, tank, or vessel or any part of it until all practicable steps have been taken to remove the substances and any fumes arising from them or render them non-explosive or non-flammable.

The risks are fully dealt with in Safety, Health and Welfare Booklet, New Series, No. 32 entitled: *Repair of Drums and Tanks – Explosion and Fire Risk.*

1.25 General precautions to be observed with the use of cranes

The movement of materials by mechanical means has been discussed in Chapter 1 of *Basic Engineering*, and emphasis was placed on the dangers which could arise when lifting heavy loads with the aid of cranes, hoists, and fork-lift trucks. In this section some of the hazards arising from the use of mobile cranes will be outlined in the following precautionary measures:
1. Loads should only be lifted vertically. It is a hazard to swing loads out manually to gain additional radius, for in doing so the effect is to extend the length of the jib and throw stresses on the crane for which it was not designed. This effect is shown in Fig. 1.30.

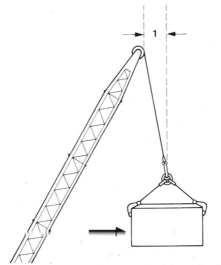

The effect of swinging the load out manually to obtain extra reach is to extend the jib radius (1) and seriously overstress the crane

Fig. 1.30 Effect of over-reaching the jib

2. Loads should always be kept directly and vertically under the lifting point of the jib. Severe overstressing of the jib can be caused by dragging loads inwards or sideways or by moving loads out of the vertical. This effect is indicated in Fig. 1.31.

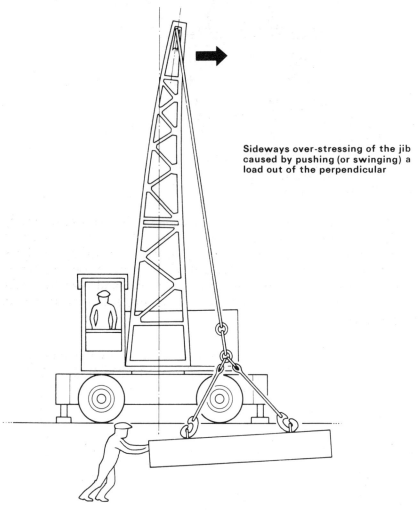

Fig. 1.31 Effect of swinging the load

3. NO PERSON SHOULD STAND UNDERNEATH A LOAD SUSPENDED FROM A LIFTING DEVICE, NEITHER SHOULD A LOAD BE TRAVERSED OVER ANY PERSON.
4. Tyres of mobile cranes need frequent checking. If they are faulty or inflated at the wrong pressure, the result will be crane instability, as shown in Fig. 1.32.

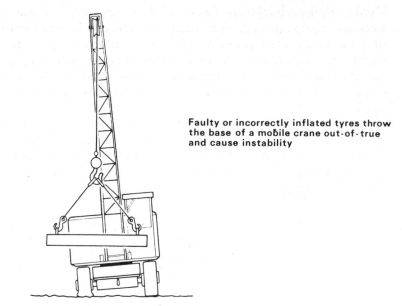

Fig. 1.32 Importance of stability

5. Mobile cranes should not be moved with the jib in near minimum radius. Figure 1.33 illustrates the danger of the jib whipping back and causing the crane to overturn.

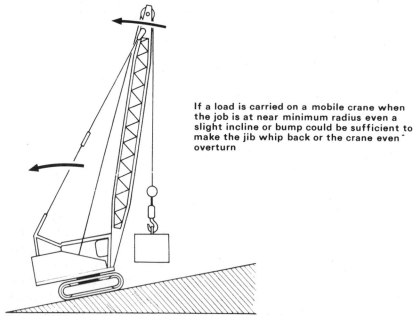

Fig. 1.33 Care when moving the load

6. Mobile cranes should be fitted with *'Outriggers'*. These are auxiliary equipment for extending the effective bases of cranes thus increasing their stability. The crane shown in Fig. 1.31, is fitted with outriggers. On soft ground it is recommended that strong timber baulks (for example, railway sleepers), be placed under the outriggers in order to spread the load and increase stability.

7. Where jibs of mobile cranes have to pass under or operate near to overhead power lines, there is a very real danger that high voltage can arc between the power cable and the jib. Figure 1.34 indicates one of the precautions taken on site to ensure that no part of site equipment can approach too near any live overhead cable.

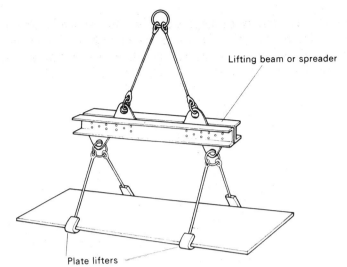

Fig. 1.35 Use of a spreader

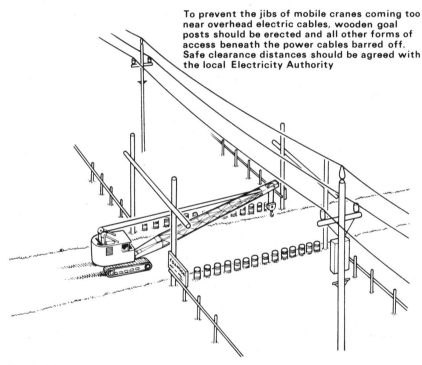

Fig. 1.34 Care when near overhead cables

8. Spreaders should be used where the load is a long one, and where it is necessary to distribute the loading to avoid excessive stresses in the object being lifted. Spreaders are frequently used when lifting long plates or rods which would be liable to buckle. Figure 1.35 shows use of a spreader with two pairs of plate lifters when handling long plates.

9. When a load is raised on a multi-leg sling, the legs should be evenly disposed about the *'Centre Of Gravity'* of the load. Failure to observe this precaution will result in the load tilting until the centre of gravity is vertically below the crane hook, and in an extreme case, one leg could hang vertically and therefore take the full load. This very important precaution is shown in Fig. 1.36.

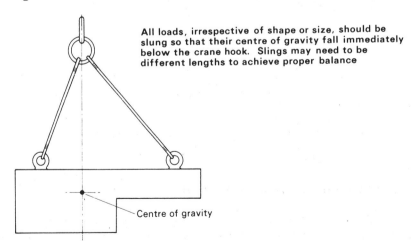

Fig. 1.36 The use of slings

The author is indebted to *Construction Safety* for much of the safety advice offered in this section.

2 Communications

Colour coding saves time

2.1 Orthographic drawing—third-angle projection

In order to interpret orthographic drawings correctly it is very important for the craftsman to be able to distinguish between first- and third-angle projections. Any misinterpretation, even of a relatively simple component, will result in some details being fabricated in the opposite hand. In Fig. 2.1 the same component is shown drawn (a) in first-angle (English) projection, and (b) in third-angle (American) projection, for comparison of the two methods.

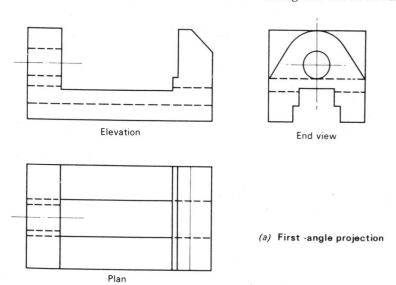

(a) First-angle projection

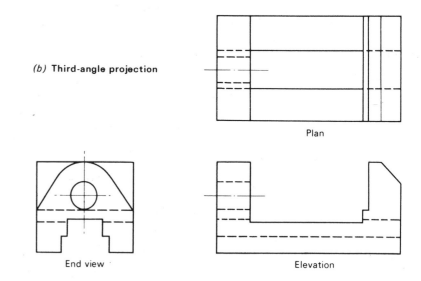

(b) Third-angle projection

Fig. 2.1 First and third-angle projection

The method of orthographic drawing in first-angle projection has been explained in *Basic Engineering*. As a reminder the method of projecting or reading drawings in first-angle projection is summarised in Table 2.1.

Table 2.1 Reading drawings in first-angle projection

1.	DEFINITION: A FIRST-ANGLE PROJECTION is that in which each view is so placed that it represents the side of the object remote from it in the adjacent views.
2.	The FRONT ELEVATION is drawn on a datum line and an END ELEVATION on the right of it (on the same datum line) represents a view on the left of the front elevation.
3.	An end elevation drawn on the left (on the same datum line) will represent a view from the right of the front elevation.
4.	A PLAN VIEW is drawn immediately below the elevation view which shows the most important details. A plan view represents a view looking directly on top of the object.
5.	In each case the view is obtained by turning the front elevation through 90° in a direction towards the viewer.
6.	The ANGLE, the first 90° accounts for this method or orthographic projection being called FIRST-ANGLE PROJECTION.

It will be noticed in Fig. 2.1 that a drawing in third-angle projection has exactly the same views as the first-angle projection, but the views are positioned in a different relationship to each other. Figure 2.2 shows how the views are conventionally positioned in a third-angle drawing. The end view can be at either end or even at both ends at the same time for clarity. Figure 2.3 illustrates the geometrical construction employed to produce the views given in Fig. 2.2. To avoid confusion *the projection used must always be clearly stated on the drawing.*

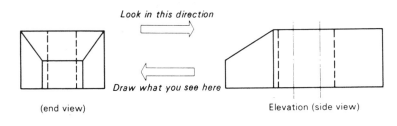

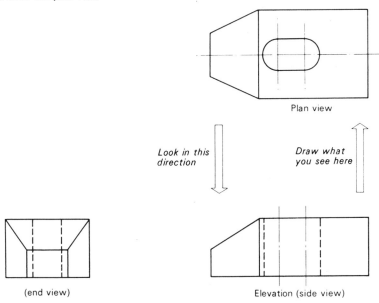

Fig. 2.2 Third-angle orthographic drawing

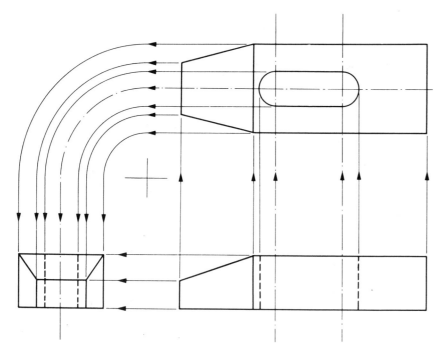

The component is first built up using fine and feint construction lines. The outline of the component is then 'lined in' more heavily so that it stands out. For standard line types see *Basic Engineering* Fig 2.19

Note: Except for a complex development or a constructed curve, full geometric construction is seldom used in the drawing office.

Fig. 2.3 Third-angle — Geometric construction

2.2 Selection of views

In practical work it is important to chose the combination of views that will describe the shape of a component in the best and most economical way. Often only two views are necessary. For example, a cylinder, if on a vertical axis, would require only a front elevation and a plan view. If the cylinder is on a horizontal axis, only a front elevation and an end view is required. Conical and pyramidal shapes can also be described by two views. Figure 2.4 illustrate two-view shapes drawn in third-angle projection.

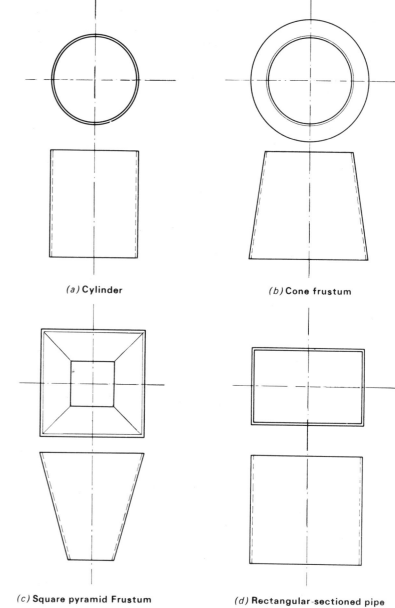

(a) Cylinder (b) Cone frustum

(c) Square pyramid Frustum (d) Rectangular-sectioned pipe

Fig. 2.4 Two-view shapes (third-angle projection)

Sometimes two views are used to describe an object or component on the assumption that the contour in the third direction is of a shape that can naturally be expected. Figure 2.5 explains why in many cases two views do not fully describe the shape of a component.

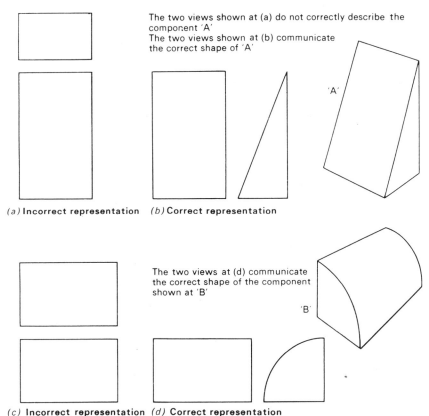

(a) Incorrect representation (b) Correct representation

(c) Incorrect representation (d) Correct representation

Fig. 2.5 Importance of selecting the correct views to describe a component

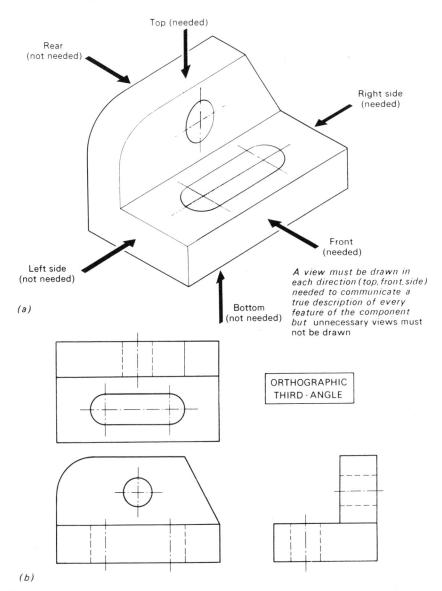

Fig. 2.6 Selection of views

The two views shown at (a) would suggest that the component is rectangular and of square cross-section. But these two views may be the front elevation and plan views of the tapering rectangular component 'A' as shown in the two views at (b). The two views of a component shown at (c) do not describe the component at all. It might be assumed to be of square cross-section, but it could as easily be of round, triangular, quarter-round, or of any other shaped cross-section, *which can only be indicated by a side view*. The figure at (d) illustrates how the component shown at 'B' is clearly described by two views. Figure 2.6 illustrates the six principal views of an object and indicates which views need to be selected in order that the object will be fully represented. In practice, any simple component would be visualised mentally and the necessary views selected without a pictorial sketch as shown in Fig. 2.6(b). However, in more complicated work, a pictorial sketch may be used to advantage. It is never necessary to sketch all possible views in order to make a selection.

2.3 Pictorial drawing—oblique

Examples of pictorial drawing techniques that are applicable to engineering communications are illustrated in Fig. 2.1 in *Basic Engineering*. Two of these techniques, 'oblique' and 'isometric' drawing, will now be considered in detail.

27

In orthographic projections, sufficient views must be drawn in order to show the three principal dimensions of a component, i.e. HEIGHT, WIDTH, and DEPTH. The purpose of a pictorial drawing is to attempt to show these three dimensions of a component in one view only. This method of engineering communication can be more easily understood by a person who is unskilled in the reading of orthographic drawings.

Lines that disappear into the drawing are termed *'receding lines'* or *'receders'*. These may have to be distorted in length and position in order to give the drawing 'realism'. The simplest technique is called OBLIQUE DRAWING. The procedure for making such a drawing is shown in Fig. 2.7.

whilst Fig. 2.9 shows alternative methods of how curved surfaces may be constructed.

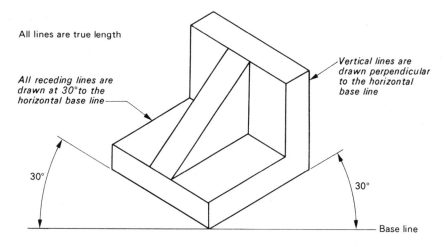

Fig. 2.8 Isometric drawing

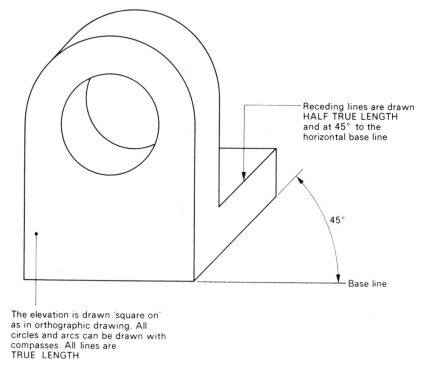

Fig. 2.7 Oblique drawing

2.4 Pictorial drawing—isometric

An ISOMETRIC DRAWING is rather more difficult to produce, but it has the advantage of showing horizontal surfaces more clearly. With this technique all the curved surfaces have to be constructed. Figure 2.8 illustrates the steps in making a simple isometric drawing,

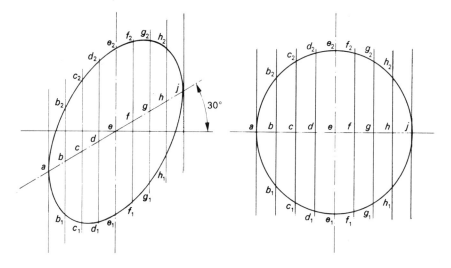

Circle - isometric. Becomes an ellipse   Circle - orthographic. True size and shape

1. *Construct a grid over the true circle by dividing its centre line into an equal number of parts, a b c ……j, and erecting a perpendicular at each point*
2. *Construct a similar grid on the isometric centre line*
3. *Step off distances b_1-b-b_2, c_1-c-c_2 etc. on the isometric grid by transferring the corresponding distances from the true circle*
4. *Draw a fair curve through the points plotted*

(a) **The construction of isometric curves**

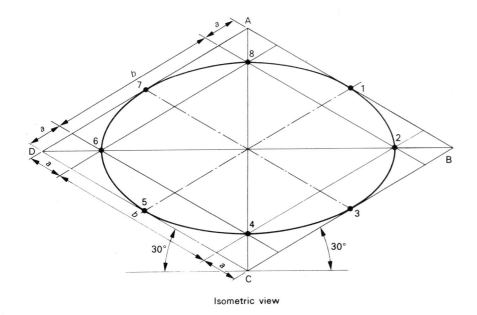

Isometric view

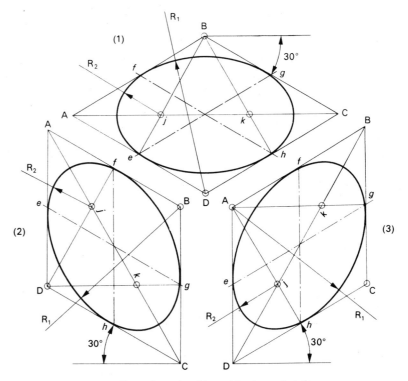

Three-dimensional isometric views of circle

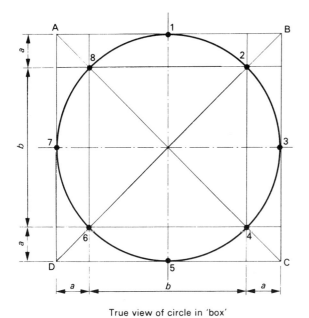

True view of circle in 'box'

(b) Construction of isometric circle – eight-point method

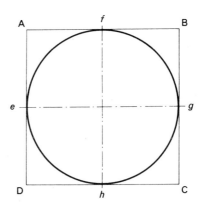

True view of circle in 'box'

(c) Construction of an isometric circle - circular arc method

Fig. 2.9 Pictorial drawing

CURVE-GRID METHOD

Figure 2.9(a) shows how a curve may be constructed using a *grid*. The advantage of this method is that complex, multi-radii curves may be easily plotted.

ISOMETRIC CIRCLE — EIGHT-POINT METHOD

Figure 2.9(b) shows the *'eight-point method'* of constructing an isometric circle.

Method of construction

1. 'Box' the given circle — the circumscribing square touches the circle at TANGENT points 1, 3, 5 and 7.
2. Draw an ISOMETRIC view of the 'box' and mark the mid-points of its sides.
3. Join points 3—7 and 1—5 and draw diagonals A C and D B.
4. Draw the diagonals in the true view, locating points 2, 4, 6 and 8 and draw light lines parallel to the four lines forming the box, thus obtaining dimensions *a* and *b*, which apply to all four sides of the box.
5. Transfer these dimensions to the isometric box, and using a 60°/30° set square obtain points 2, 4, 6 and 8.
6. Sketch a neat elliptical curve through the points 1—8.

ISOMETRIC CIRCLE — 'CIRCULAR ARC' METHOD

Figure 2.9(c) shows the *'circular arc'* method of constructing an isometric circle.

The TRUE VIEW shows the circle in a circumscribing square A B C D, *e g* and *f h* are the intersecting centre lines.

Method of construction

1. Draw the isometric square and transfer the correct location of the centre lines from the true view.
2. The long diagonal A C or B D is drawn on the isometric square.
3. From B or D (in views (1) and (2)) or from A or C (in view (3)) draw lines to the mid-points of the opposite sides. In view (1) the mid-points opposite to B are *e* and *h*. In view (2) the mid-points opposite to D are *f* and *g*. In view (3) the mid-points opposite to A are *g* and *h*. Where these two lines cross the long diagonal at *j* and *k*, are two centres for circular arcs to draw the approximate ellipse; the other two centres are the corners B and D (in views (1) and (2)) and the corners A and C (in view (3)).
4. Using these centres and radii R_1 and R_2 draw the arcs with compasses to complete the ellipse.

In the examples of isometric drawing shown in Fig. 2.8 and 2.9, both the vertical lines and the receders have been drawn 'true length'. To be strictly correct, the receders should be drawn to *isometric scale* and only the vertical lines should be drawn true scale. Figure 2.10 shows the previous example drawn correctly to isometric scale. However, it is common practice to draw all the lines true length.

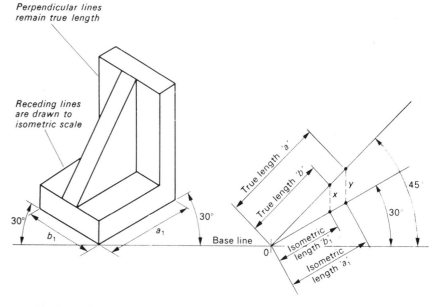

1. *The isometric scale is constructed by drawing two lines at 30° and 45° respectively through the same point 'O' on the base line*
2. *True lengths are stepped off along the 45° line and perpendiculars (x, y) are dropped to cut the 30° line*
3. *The corresponding lengths, cut off the 30° line, represent the isometric scale lengths for the receding lines of the drawing*

Fig. 2.10 Isometric scale

2.5 Use of square and isometric paper

Figure 2.11 shows how an oblique drawing of a simple component may be drawn on squared paper. For convenience, the component shown in Fig. 2.7 has been dimensioned and redrawn on squared paper.

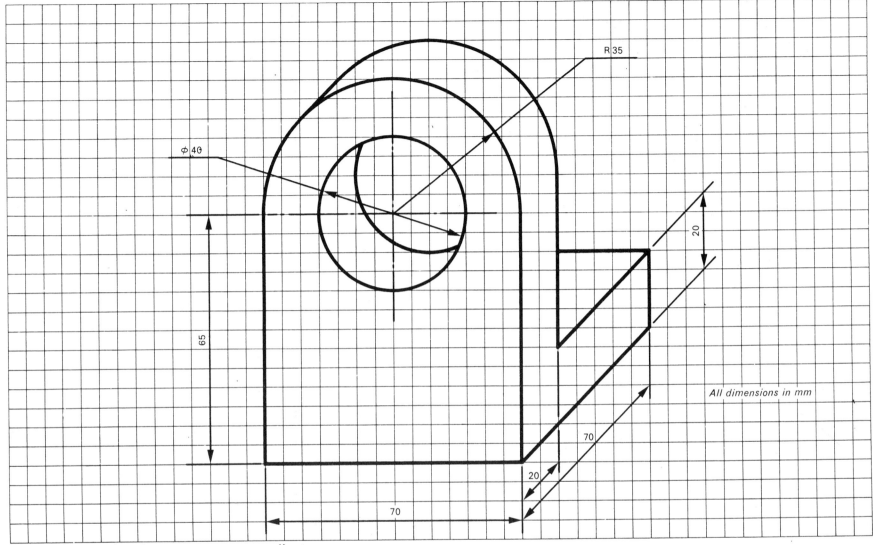

Fig. 2.11 Oblique drawing on square or co-ordinate paper

Squared paper is very useful for making oblique drawings for the following reasons:

1. It assists in the drawing of straight lines.
2. It helps to maintain the oblique proportion. In this example twenty of the 5 mm squares have been chosen for the 100 mm height of the front face of the component. All the other dimensions on the front face are drawn to this scale. The oblique faces are drawn to only one-half of this scale. For example, the 70 mm length is drawn to an equivalent length of 7 squares.
3. The 45° lines for the oblique faces are drawn diagonally through the squares. Squared paper is also an aid to quick and accurate free hand orthographic sketching as shown in Fig. 2.12. Projections are much easier to make than on plain paper. To transfer dimensions from the top view (plan) to the side view, the squares are counted. In this example each square is 5 mm.

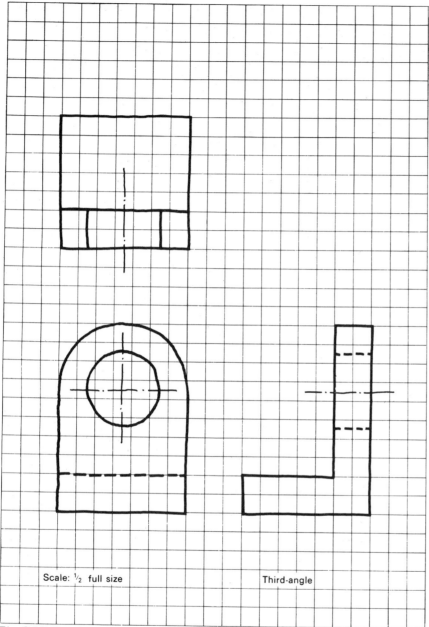

Fig. 2.12 Freehand orthographic sketching on square paper

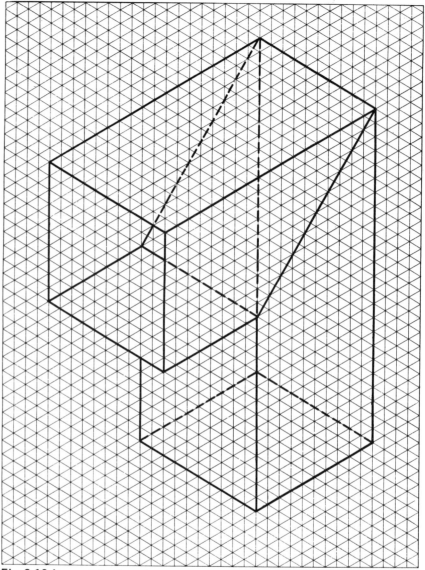

Fig 2.13 Isometric drawing on isometric paper

Figure 2.13 shows an isometric drawing of a square duct elbow produced on isometric paper. *This example shows that all truly isometric lines are either at 30° or 90° to the horizontal.*

When making isometric or oblique drawings, it is good practice to adopt the method of 'boxing' in which an imaginary 'box' is drawn to just contain the component. Additional boxes may be required to simplify the construction of parts of the component. Figure 2.14 illustrates the use of 'boxing' on isometric drawings of (*a*) a square hopper, and (*b*) a right cone. Both these examples have been drawn on isometric paper.

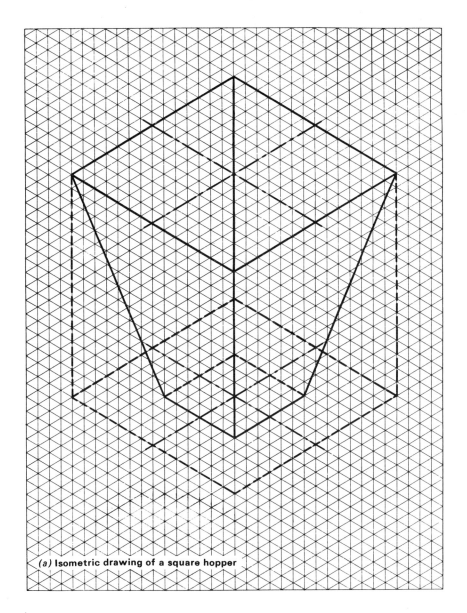

(a) Isometric drawing of a square hopper

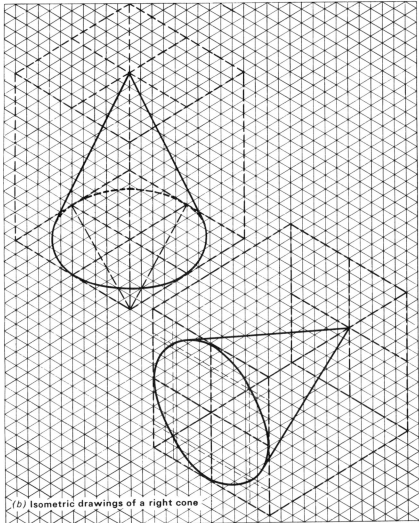

(b) Isometric drawings of a right cone

Fig 2.14 Examples in the use of isometric paper

2.6 Dimensioning

The sections in this chapter so far have dealt basically with orthographic drawing techniques of describing the shape of the object or component. A two-view drawing may communicate the complete shape description of a sheet metal cylinder, as shown at (a) in Fig. 2.14, but *neither communicates to the craftsman any size descriptions nor material specifications*. However, when size and material specifications are added, as shown at (b) in Fig. 2.15, the drawing becomes a 'working drawing', which is suitable for use in the fabrication shop.

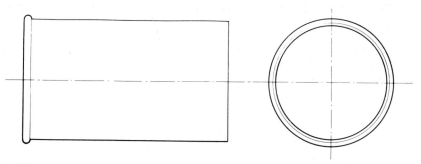

(a) These two views clearly indicate the exact shape of the component but do not communicate any dimensions or material specifications, and therefore cannot be used for a working drawing

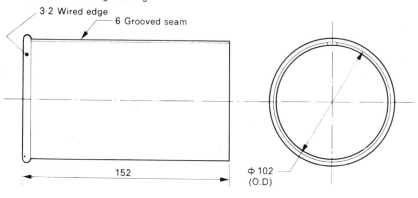

Dimensions in millimetres
Material: 22 swg galvanised steel
6-off

NOT TO SCALE   SLEEVE

(b) Required information added to the two views provides a working drawing which can be used by the sheet metal worker or fabricator

Fig 2.15 The need for dimensioning

In order to be proficient in communicating the required sizes of the drawn shapes by 'dimensioning', three basic skills are required.

1. A knowledge of the techniques used in dimensioning.
2. The ability to select the essential dimensions which are needed in order to make the component.
3. The ability to position the selected dimensions according to accepted Engineering Drawing Practice.

The drawings in this chapter are in accordance with the standards of practice laid down in B.S. 308:1972.

Dimensioning has already been introduced briefly in Chapter 2 of *Basic Engineering*. Figure 2.16 serves as a reminder of some of the basic methods of dimensioning.

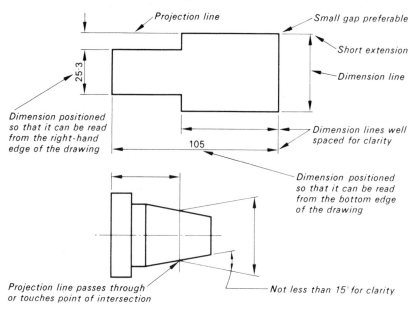

(a) **Projection and dimension lines**

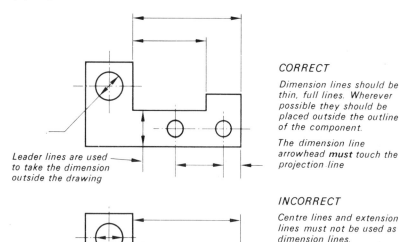

CORRECT

Dimension lines should be thin, full lines. Wherever possible they should be placed outside the outline of the component.

The dimension line arrowhead **must** touch the projection line

INCORRECT

Centre lines and extension lines must not be used as dimension lines.

Wherever possible dimension line arrowheads should not touch the outline direct, but should touch the projection line that extends the outline

(b) **Correct and incorrect dimensioning**

Four types of lines are used in dimensioning an engineering drawing:

1. CENTRE LINES
2. PROJECTION LINES
3. DIMENSION LINES
4. LEADER LINES

CENTRE LINES These are necessary to indicate the existence and location of circular features on drawings. They are thin, chain lines which are a valuable aid in reading a drawing, and must be included for all circular features in all views where appropriate.

PROJECTION LINES These are thin lines, sometimes termed 'Extension Lines', which extend away from the outline of the component in order to remove the dimension from the view. *In this manner the view remains unobstructed and easier to read.* Such dimension extension lines should start just clear of the drawing outline and extend a little beyond the arrowhead on the dimension line.

DIMENSION LINES These should also be thin lines terminating in arrowheads. The arrowheads should be easily readable and normally not less than 3 mm long. *The point of an arrowhead should just touch the projection line.*

Centre lines or any portion of the drawing outline must never be used as a dimension line, as doing so will lead to confusion.

LEADER LINES While size descriptions are communicated by means of dimensions, many features are better described by means of a note. Leader lines are used for this purpose. Such lines are continuous thin lines terminating in arrowheads or dots. Arrowheads are always used where the leader line *touches the outline* of the feature or component. *Dots* are used where the leader line terminates *within the outline* of the feature or component.

Leader lines which touch lines should be nearly normal to the surface. *They should not be parallel to adjacent dimension or projection lines where confusion may arise.* Typical examples of the use of 'leader lines' are shown in Fig. 2.17.

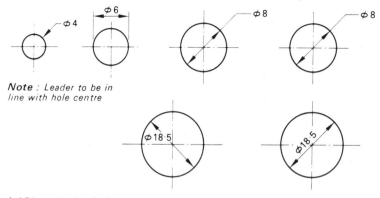

Note: Leader to be in line with hole centre

(c) Dimensioning holes

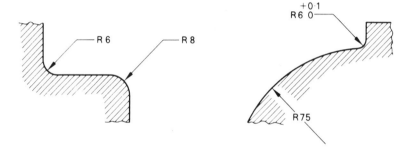

(d) Dimensioning the radii of arcs which need not have their centres located

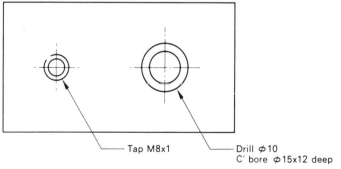

(e) Use of notes to save full dimensioning

Fig. 2.16 Dimensioning

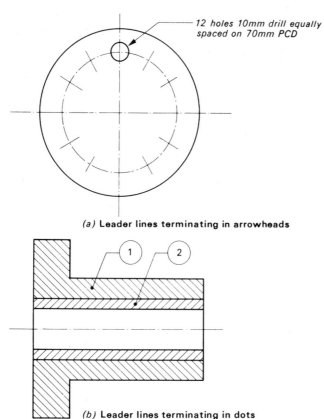

(a) Leader lines terminating in arrowheads

(b) Leader lines terminating in dots

Fig. 2.17 Leader lines

2.7 Selection of dimensions

Certain basic principles apply to the selection of dimensions for various shapes. In the same way as the number of views in a drawing are kept to a minimum, so too must dimensions be selected so that the fabricator will not be confused by unnecessary details. Duplication of dimensions should also be avoided.

In this section a few selected examples will be used to illustrate how the basic principles of dimensioning are applied.

1. Square and rectangular pieces of flat stock have the two basic dimensions of width and length. The thickness is usually given by means of a note or other specification.
2. Flat discs have only one basic dimension which is the diameter. The thickness is specified as in the previous example.
3. If the flat stock is square or rectangular but has rounded corners, the radius of the rounded corner is given. Figure 2.18(a), shows a rectangular sheet metal blank with four rounded corners. In this case the overall dimensions of the blank are given and the corner radius is shown.

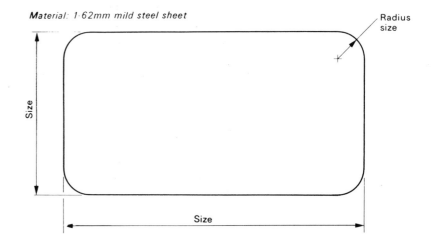

(a) Rectangular sheet metal blank with radiused corners

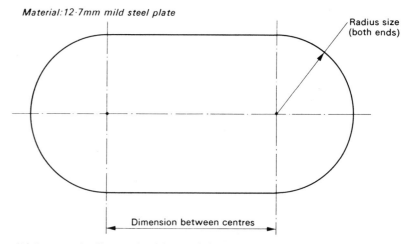

(b) Rectangular flat stock with rounded corners

Fig. 2.18 Examples of the application of dimensions

4. Figure 2.18(b) shows how a metal blank with rounded ends is dimensioned. It will be noticed that only the centre-line distance (centre to centre) and the radius for the ends are given because these are the only dimensions required for marking-out and cutting the blank.

5. Prisms are dimensioned by indicating the width and length measurements on one view and the thickness or height dimension in a related view, as shown in Fig. 2.19(a).
6. Cylinders are dimensioned as shown in Fig. 2.19(b)

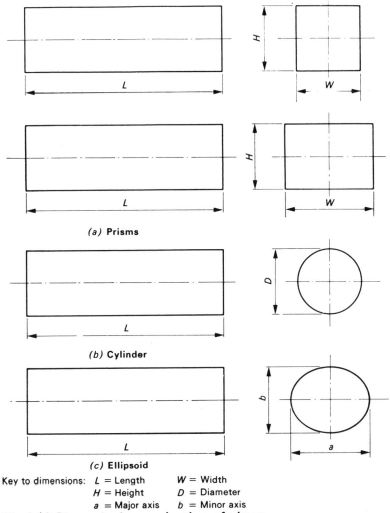

Key to dimensions: L = Length   W = Width
                    H = Height   D = Diameter
                    a = Major axis   b = Minor axis

Fig. 2.19 Dimensioning— selection of views

7. Cones are dimensioned by giving the diameter of the base and the true height in the elevation view, as shown at A in Fig. 2.20. A cone frustum is dimensioned by giving the top and bottom diameters and the vertical distance between them in the elevation view, as shown at B in Fig. 2.20. An alternative method of dimensioning is to give an angular value either between elements, or between the cone base and elements, instead of the top or frustum diameter, as shown at C in Fig. 2.20.

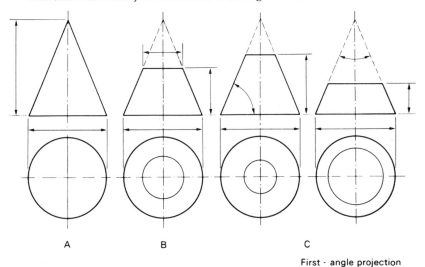

First - angle projection

Fig. 2.20 Dimensioning (cones and cone frustums)

8. Pyramids are dimensioned by indicating the measurements for the base in the plan view, and the vertical height in the elevation view, as shown at A in Fig. 2.21. Pyramid frustums are dimensioned as shown at B in Fig. 2.21.

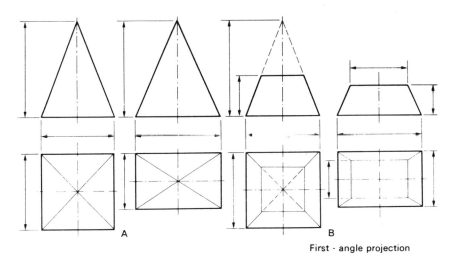

First - angle projection

Fig 2.21 Dimensioning (pyramids and pyramid frustums)

37

2.8 Colour coding - compressed gas cylinders

Table 2.2 is an abstract from the British Standards for the colour coding of compressed gas cylinders in common industrial use. Table 2.3 shows how the use of colour bands denotes the hazard properties of the cylinder contents.

Table 2.2 Compressed gases in common industrial use

NAME OF GAS	CHEMICAL FORMULA	BASE COLOUR OF CONTAINER	COLOUR OF BANDS	REMARKS
Acetylene	C_2H_2	Maroon	None	Only acetylene regulators should be used. Cylinders must be stood upright.
Air	—	French grey	None	—
Ammonia	NH_3	Black	Red and Yellow	The red band is adjacent to the valve fitting.
Argon	Ar	Peacock blue	None	
Carbon dioxide	CO_2	Black	None	Main use for gas-shielded arc welding processes.
Coal gas	—	Signal red	None	Used for heating and cutting. Only coal gas regulator should be used.
Helium	He	Middle brown	None	Used for gas-shielded arc welding processes.
Hydrogen	H_2	Signal red	None	Main use for heating and cutting. Only hydrogen regulator should be used.
Methane	CH_4	Signal red	None	—
Nitrogen	N_2	French grey	Black	—
Oxygen	O_2	Black	None	Only oxygen regulator should be used.
Propane (commercial)	—	Signal red	None	An acetylene regulator may be used with this gas.

Note:
For the purpose of identification the container should be marked with the name of the gas and the chemical formula in accordance with I.S.O. Recommendation R448 — 'Marking of Industrial Gas Cylinders for the identification of their content'.

Table 2.3 Colour bands to denote hazard properties of contents of compressed gas cylinders

NATURE OF GAS OR MIXTURE	COLOUR OF BANDS	CONTAINER NECK
Non-flammable and non-poisonous	None	
Non-flammable and poisonous	Yellow	YELLOW
Flammable and non-poisonous	Red	RED
Flammable and poisonous	Red and Yellow	RED / YELLOW

2.9 Welding symbols and application

Most manually welded joints are a variation of the *butt weld* as shown in Fig. 2.22(a), or the *fillet weld* as shown in Fig. 2.22(b).

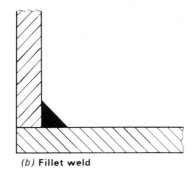

Fig. 2.22 Basic welded joints

Welding drawings must give all the details necessary to specify the type of weld, such as:

1. Edge preparation.
2. Filler material.
3. Type of joint.
4. Length of run.
5. Size of weld.

This information is communicated by use of WELDING SYMBOLS specified in B.S. 499, together with any appropriate notes and dimensions. Figure 2.23 shows a selection of standard welding symbols.

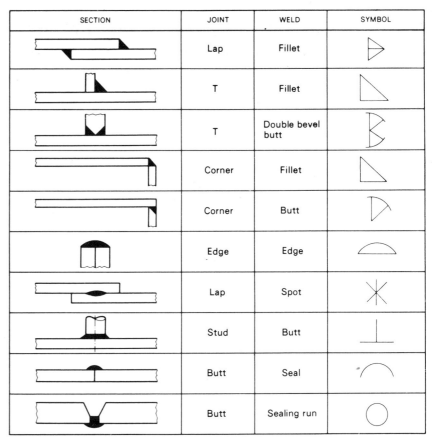

Fig. 2.23 Weld symbols

To apply the weld symbol to a drawing a system of arrows and reference lines is used.

(a) When the weld symbol is *above* the reference line the weld is made on the side of the joint *opposite* the arrowhead, i.e. the OTHER SIDE.

(b) When the weld symbol is *below* the reference line the weld is made on the *same side* of the joint as the arrowhead, i.e. the ARROW SIDE.

(c) When the weld symbol is on *both sides* of the reference line, the welding is to be carried out on *both sides* of the joint.

The reference line which connected to the arrow at an angle (as shown in the examples) is the *datum line* for determining the position of the weld symbol in order that the later may indicate the location of the weld. *The welding symbol is inverted and suspended from this datum line for welds made from the arrow side of the joint. For welds*

39

to be made on the other side of the joint, the weld symbol is based on the datum line. Applications of the use of welding symbols are illustrated in Fig. 2.24.

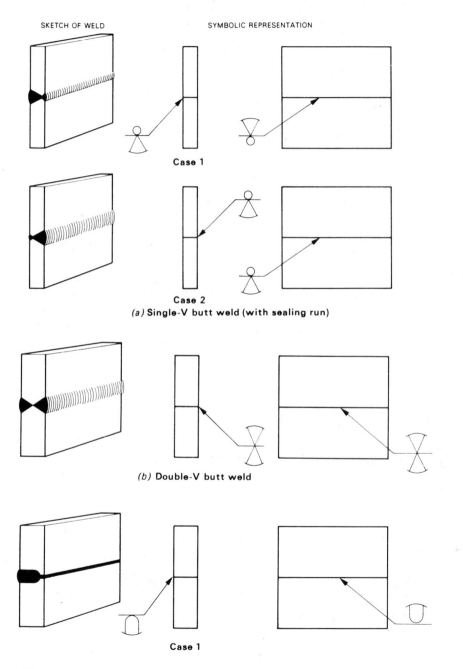

(a) Single-V butt weld (with sealing run)

(b) Double-V butt weld

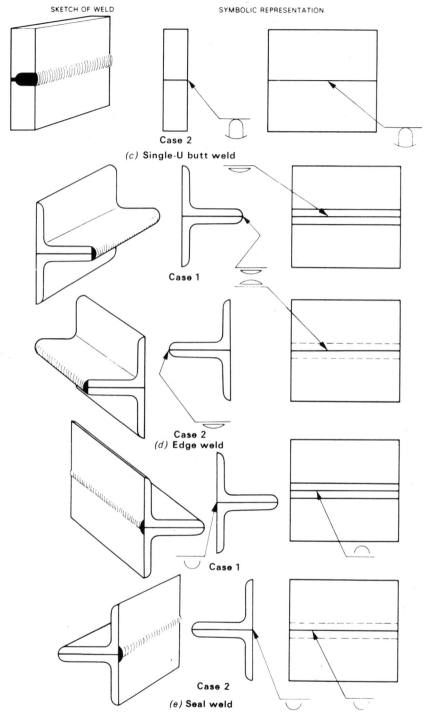

(c) Single-U butt weld

(d) Edge weld

(e) Seal weld

| SKETCH OF WELD | SYMBOLIC REPRESENTATION |

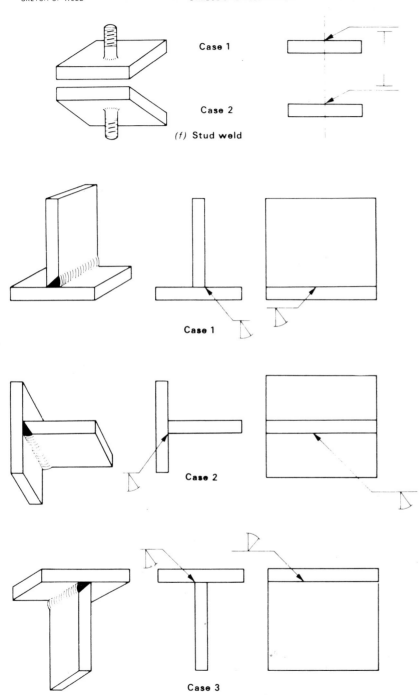

(f) Stud weld

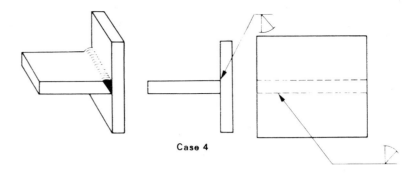

(g) Single-bevel butt weld in four positions

Note how the arrow head is used to indicate which plate is prepared and how the disposition of the symbol about the reference line indicates the side from which the weld is to be made and therefore the wide part of the preparation

Fig. 2.24 Some examples showing significance of the arrow and position of weld symbol in relation to the reference line

In addition to the use of symbols, dimensional information is required to specify the size of the weld and the length of the run. Written notes can also be added to the drawing to specify the filler material and flux used, and also the welding process to be used (i.e. gas, manual metallic-arc, etc.). Some examples are given in Fig. 2.25.

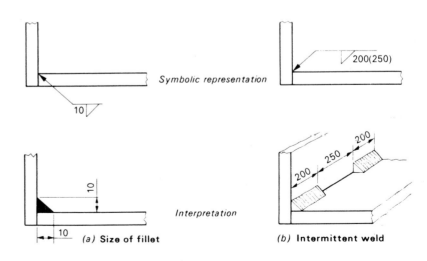

(a) Size of fillet   (b) Intermittent weld

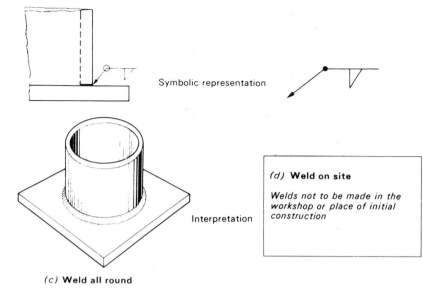

Fig. 2.25 Additional weld information

2.10 Setting out angles—use of set squares

The setting out of angles using a protractor was introduced in Fig. 2.37 of *Basic Engineering*. However, angles that are multiplies of 15° can be set out quickly and accurately using a 45° set square and a 60° set square. Figure 2.26 shows how these set squares can be combined together to produce a range of angles.

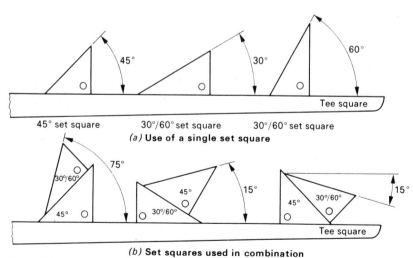

Fig. 2.26 Construction of angles using set squares

2.11 Setting out angles—use of compasses

Angles which are multiples of 7½° can be set out very accurately using compasses alone. The basic angle from which all the others can be derived is the 60° angle shown in Fig. 2.27(a). The constructions for 90° and 45° angles are shown at (b) and (c), respectively, in Fig. 2.27.

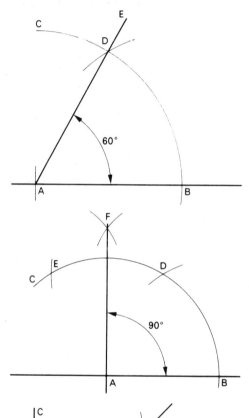

(a) Construction of a 60° angle

1. Let A be the apex of the angle
2. With centre A draw an arc BC of large radius
3. Step off BD equal in radius to AB
4. Draw a line AE through D
5. The angle EAB is 60°

(b) Construction of a 90° angle

1. Let A be the apex of the angle
2. With centre A draw an arc BC of large radius
3. Step off BD and DE equal in radius to AB
4. With centre D draw any arc F
5. With centre E draw an arc equal in radius to DF
6. Join AF with a straight line
7. Angle BAF is 90°

(c) Construction of a 45° angle

1. Draw AB and AC at right angles (90°) to each other as described in Fig. 2.29(b)
2. With centre A, and with large radius, draw an arc to cut AB at D and AC at E
3. With centres E and D draw arcs of equal radius to intersect at F
4. Draw a straight line from A through F
5. Angle BAF is 45° (AF bisects – halves ∠ BAF)

Fig. 2.27 Construction of angles using compasses

The bisection of the 60° angle to produce 30° and the successive bisection of this angle to produce 15° which is bisected to produce 7½° is shown in Fig. 2.28.

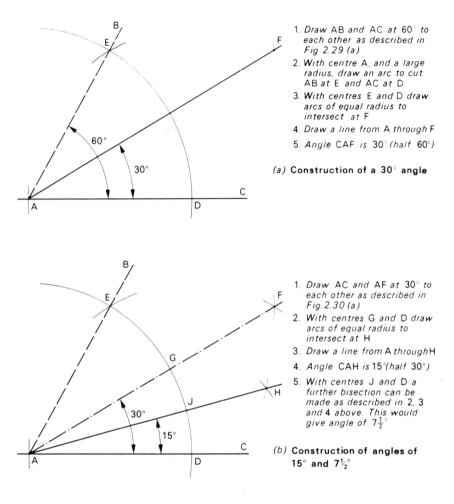

1. Draw AB and AC at 60° to each other as described in Fig 2.29 (a)
2. With centre A, and a large radius, draw an arc to cut AB at E and AC at D
3. With centres E and D draw arcs of equal radius to intersect at F
4. Draw a line from A through F
5. Angle CAF is 30° (half 60°)

(a) Construction of a 30° angle

1. Draw AC and AF at 30° to each other as described in Fig. 2.30 (a)
2. With centres G and D draw arcs of equal radius to intersect at H
3. Draw a line from A through H
4. Angle CAH is 15° (half 30°)
5. With centres J and D a further bisection can be made as described in 2, 3 and 4 above. This would give angle of 7½°

(b) Construction of angles of 15° and 7½°

Fig. 2.28 Bisection of angles

2.12 Construction of plane figures

All the regular plane figures met with in fabrication engineering can be set out by geometrical constructions, using compasses and set squares. Some of these figures, which are to be constructed in this section of the chapter, are described in Fig. 2.29.

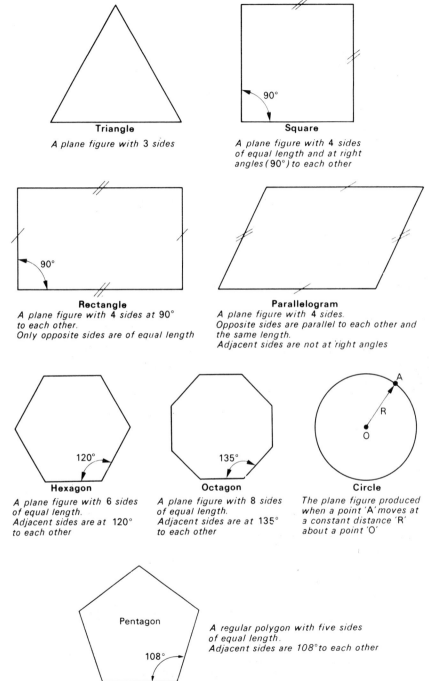

Triangle
A plane figure with 3 sides

Square
A plane figure with 4 sides of equal length and at right angles (90°) to each other

Rectangle
A plane figure with 4 sides at 90° to each other.
Only opposite sides are of equal length

Parallelogram
A plane figure with 4 sides. Opposite sides are parallel to each other and the same length.
Adjacent sides are not at right angles

Hexagon
A plane figure with 6 sides of equal length.
Adjacent sides are at 120° to each other

Octagon
A plane figure with 8 sides of equal length.
Adjacent sides are at 135° to each other

Circle
The plane figure produced when a point 'A' moves at a constant distance 'R' about a point 'O'

Pentagon
A regular polygon with five sides of equal length.
Adjacent sides are 108° to each other

Fig. 2.29 Some regular plane figures

43

The basic constructions associated with these figures are given in the following illustrations:

1. Triangle — Fig. 2.30.
2. Square — Fig. 2.31.
3. Rectangle — Fig. 2.32.
4. Parallelogram — Fig. 2.33.
5. Pentagon — Fig. 2.34.
6. Hexagon — Fig. 2.35.
7. Octagon — Fig. 2.36.
8. Circle — Fig. 2.37.

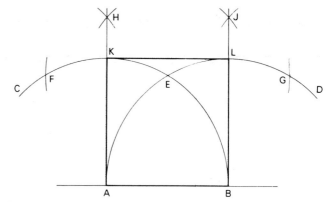

1. *Mark off one side of the square AB on the base line*
2. *With centre 'A' and radius AB draw the arc BC*
3. *With centre 'B' and radius AB draw the arc AD*
4. *With centre 'E' and radius AB step off 'F' and 'G' on arcs BC and AD respectively*
5. *With centres 'E' and 'F' draw arcs of equal radius to intersect at H*
6. *With centres 'E' and 'G' draw arcs of equal radius to intersect at J*
7. *Erect perpendiculars AH and BJ*
8. *The arcs BC and AD cut the perpendiculars AH and BJ at 'K' and 'L' respectively*
9. *To complete the square join 'K' and 'L'*

Fig. 2.31 To construct a square

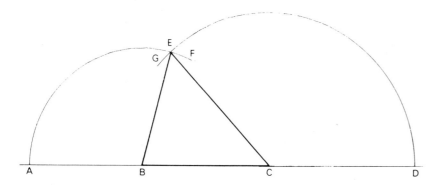

1. *Draw AB, BC and CD equal in length to the sides of the required triangle*
2. *With centre B and radius AB draw the arc AF*
3. *With centre C and radius CD draw the arc DG*
4. *Where the arcs intersect at E is an apex of the triangle*
5. *Join BE and CE with straight lines to form the triangle BCE*

Fig. 2.30 To construct a triangle

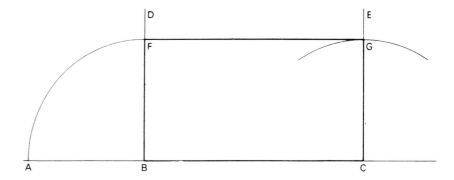

1. *Mark off the lengths of a short side AB and a long side BC on the base line*
2. *Using the constructions demonstrated in the previous examples, or using a set square for simplicity, erect perpendiculars BD and CE at 'B' and 'C'*
3. *With centre 'B' and radius AB draw an arc to cut BD at 'F'*
4. *With centre 'C' and radius AB draw an arc to cut CE at 'G'*
5. *To complete the rectangle join F and G with a straight line*

Fig. 2.32 To construct a rectangle

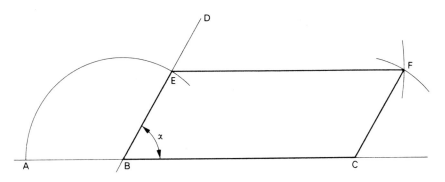

1. Mark off one short side AB and one long side BC on the base line
2. Draw BD at the required angle α to the base line using an adjustable set square or protractor
3. With centre 'B' and radius AB draw an arc to cut BD at 'E'
4. With centre 'C' and radius AB draw an arc
5. With centre 'E' and radius BC draw an arc to cut the arc drawn in (4) at F
6. To complete the parallelogram join EF and CF with straight lines

Fig. 2.33 To construct a parallelogram

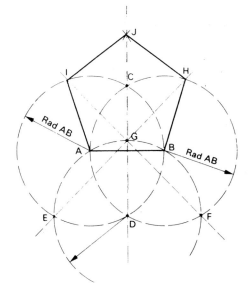

Method of construction

1. Draw the given side AB
2. With centres A and B and compasses set to radius AB, draw two circles to intersect at C and D
3. Join CD
4. With centre D and radius AB, draw an arc to cut the two previously drawn circles at E and F, and CD at G
5. From E and F draw lines through point G to cut the two circles at H and I
6. With radius AB and centres H and I draw arcs to interest at J on DC produced
7. Join AIJHB to complete the pentagon

(b) To construct a pentagon given the length of one side

Fig 2.34 Two methods of constructing a pentagon

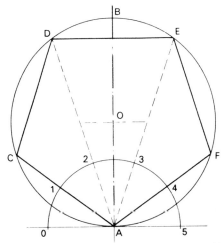

Method of construction

1. With centre O and compasses set to required radius draw the given circle
2. Draw a vertical diameter AB and a tangent at A.
3. With centre A and any suitable radius draw a semi-circle
4. With the compass divide the semi-circle into five equal parts (ie. same number as sides in the pentagon) and number 0–5
5. From tangent point A draw lines through the points 1, 2, 3 and 4 on the semi-circle to cut the given circle at C, D E and F
6. B Complete the pentagon by joining CDEF

Note: This method of construction may be used to construct any regular within a circle. The semi-circle is divided into the same number of equal parts as there are sides in the polygon

(a) To construct a pentagon in a given circle

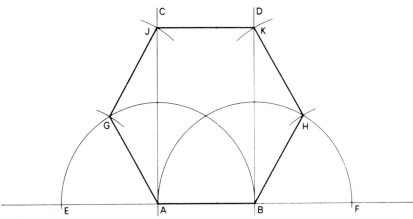

1. Mark off one side AB on the base line
2. Erect perpendiculars AC and BD using one of the previously demonstrated constructions or set square
3. With centre 'A' and radius AB draw an arc to cut the base line at 'E'
4. With centre 'E' and radius AB draw an arc to cut the previous arc (3) at 'G'
5. With centre 'B' and radius AB draw an arc to cut the base line at 'F'
6. With centre 'F' and radius AB draw an arc to cut the previous arc (5) at 'H'
7. With centre 'G' and radius AB draw an arc to cut the perpendicular AC at 'J'
8. With centre 'H' and radius AB draw an arc to cut the perpendicular BD at 'K'
9. To complete the hexagon join AG, GJ, JK, KH and HB with straight lines

Fig. 2.35 To construct a hexagon

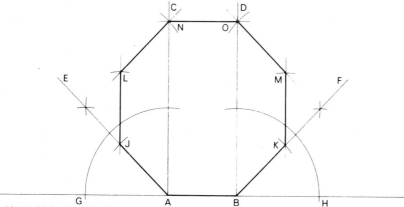

1. Mark off the length of one side AB on the base line
2. Erect perpendiculars AC and BD from 'A' and 'B' respectively
3. Bisect the angle CAG so that EA lies at 45° to GA
4. Similarly, bisect the angle DBH so that BF lies at 45° to BH
5. With centre 'A' and radius AB draw an arc to cut AE at J
6. With centre 'B' and radius AB draw an arc to cut BF at K
7. Erect perpendiculars at 'J' and 'K'
8. With centres 'J' and 'K' and radius AB draw arcs to cut the perpendiculars at L and M respectively
9. With centres 'L' and 'M' and radius AB draw arcs to cut AC and BD at 'N' and 'O' respectively
10. To complete the octagon join AJ, JL, LN, NO, OM, MK and KB with straight lines

Fig. 2.36 To construct an octagon

2.13 The construction of ovals and ellipses

Ovals and ellipses are common shapes encountered in sheet metal work. The essential differences between these two figures will now be explained.

The oval differs from the true ellipse in the fact that *the oval is an approximate ellipse drawn with compasses*, and is therefore a construction made up of arcs of circles. *Because no part of a true ellipse is part of a circle it cannot be drawn with compasses.*

The shape of an ellipse can be traced out by a point which moves so that the sum of its distances from two fixed points on the major axis is always constant. These fixed points are called the 'foci' or 'focal points' of the ellipse. This fundamental principle is illustrated in Fig. 2.38.

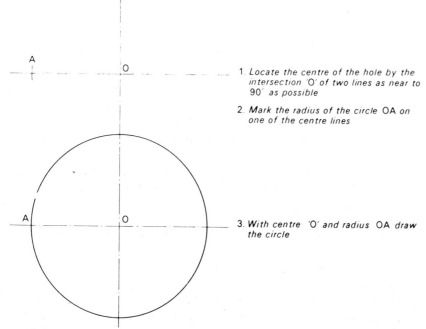

1. Locate the centre of the hole by the intersection 'O' of two lines as near to 90° as possible
2. Mark the radius of the circle OA on one of the centre lines
3. With centre 'O' and radius OA draw the circle

Fig. 2.37 To construct a circle

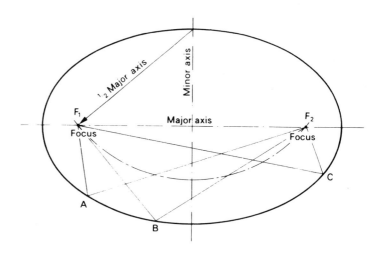

Note: Lines F_1AF_2, F_1BF_2 and F_1CF_2 are all the same length

Fig 2.38 Principle of the ellipse

The ELLIPSE is symmetrical about two AXES which bisect each other at 90°. The longer 'axis' is termed the MAJOR AXIS (or major diameter), and the shorter 'axis' is termed the MINOR AXIS (or minor diameter).

The path of an ellipse may be traced out by a point which moves from two fixed points on the major axis (termed FOCI) in such a way that the sum of its distances from the foci is always constant.

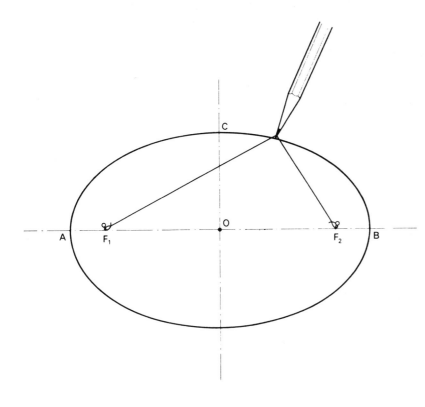

Fig 2.39 Practical method of drawing an ellipse

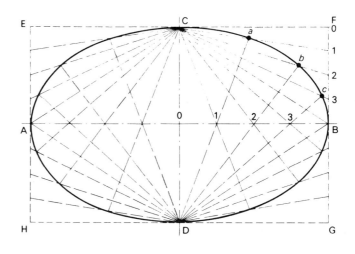

Fig 2.40 Construction of an ellipse - rectangle method

Method of construction

1. Draw the major and minor axes.
2. With centre C and radius equal to A O (½ major axis) draw arcs to locate the focal points F_1 and F_2 on the major axis A B.
3. Fix pins in the focus points. Make two loops in a piece of cotton such that its length (including loops) is exactly equal to the major axis. Slip the loops over the locating pins.
4. Place the pencil point so that the cotton is stretched taut, as shown in the diagram. The path of the ellipse may now be traced by moving the pencil point whilst keeping the cotton taut.

Method of construction

1. Draw the major and minor axis A B and C D.
2. Draw the rectangle E F G H whose sides are equal to the length of the axis.
3. Divide the major axis A B into any equal number of parts, and divide the ends of the rectangle (E H and F G) into the same number of equal parts.

Note: For clarity only one quarter of the rectangle has been numbered.

4. From each end of the minor axis C D draw radial lines (as shown opposite), into points 1, 2 and 3.
5. Draw radial lines from C and D through the points 1, 2 and 3 on the major axis A B these will intersect the prevision radial lines at points *a*, *b* and *c*.

47

6. The outline of the ellipse may now be drawn by joining the location points between A C, C B, B D and D A.

ELLIPSE CONSTRUCTION — CONCENTRIC CIRCLE METHOD

Figure 2.41 shows how an ellipse may be constructed by the concentric circle method.

7. The outline of the ellipse may now be drawn by joining the location points between A C, C B, B D and D A.

PLOTTING AN ELLIPSE USING TRAMMELS

Figure 2.42 shows how an ellipse may be plotted using trammels.

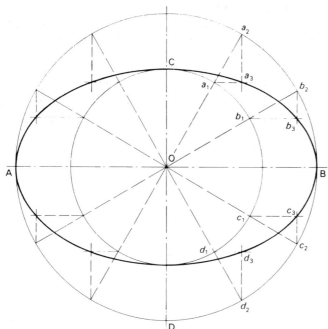

Fig 2.41 Construction of an ellipses - concentric circle method

Method of construction

1. Draw the major and minor axis, A B and C D. These intersect at O.
2. Using each axis as a diameter, with centre O, draw two CONCENTRIC CIRCLES as shown.
3. Using a 60°/30° set square in conjunction with a Tee square draw four diameters as indicated by a, b, c and d, on the minor circle and a_2, b_2, c_2 and d_2, on the major circle.
   From a_2, b_2, c_2 and d_2, draw vertical lines towards the major axis A B.
5. From a, b, c and d, draw horizontal lines towards the minor axis C D.
6. These lines will intersect and locate points on the ellipse at a_3, b_3, c_3 and d_3.

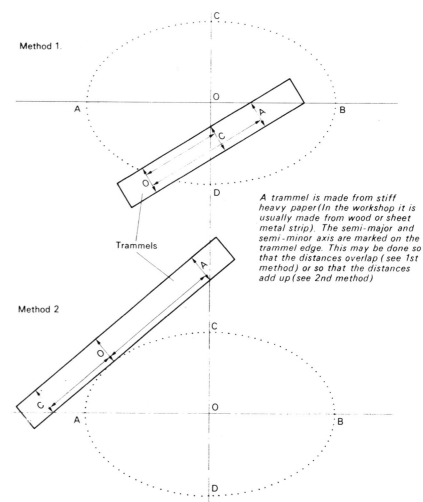

Fig. 2.42 The plotting of an ellipse using trammels

1st method

Mark the paper trammel such that the distance A O = ½ major axis A B and the distance C O = ½ minor axis C D, these distances overlapping having a common starting point O.

1. Draw the major and minor axes A B and C D.
2. The semi-major point A is placed on the major axis and the semi-minor point C is placed on the minor axis. The point O on the trammel will locate one point on the ellipse (as shown). Additional points are located by moving the trammel keeping A and C on the axis.

2nd method

Mark the paper trammel such that the distances A O and C O representing ½ major and ½ minor axis add up.
1. Draw the major and minor axis A B and C D and extend them.
2. The points A and C are positioned on the axis as shown above, and the point O on the trammel locates one point on the ellipse. Additional points for plotting the ellipse are obtained by moving the trammel keeping A and C on the axis.

ELLIPSE CONSTRUCTION – PROJECTION

Figure 2.43 shows how an ellipse may be projected.

Method of projecting the ellipse

1. Draw a semicircle on the diameter C D and divide it into six equal parts.
2. From the points on the semicircle project lines square to C D and through it back to the base line A B.
3. In any convenient position below A B draw the major axis of the ellipse $A^1 B^1$ parallel to it.
4. Project lines from the points on the base line A B and square to it through the major axis for plotting the ellipse as shown.
5. Mark off the various widths at 1, 2, 3, 4 and 5 on the semicircle and transfer them above and below the major axis as points for plotting the ellipse.
6. The curve for the ellipse may now be drawn through points $A_1, 1_1, 2_1, 3_1, 4_1, 5_1, B_1, 5_2, 4_2, 3_2, 2_2, 1_2$ and A_1.

This method of constructing a true ellipse is used in pattern development.

OVAL CONSTRUCTION

Figure 2.44 shows how an oval (approximate ellipse) may be constructed.

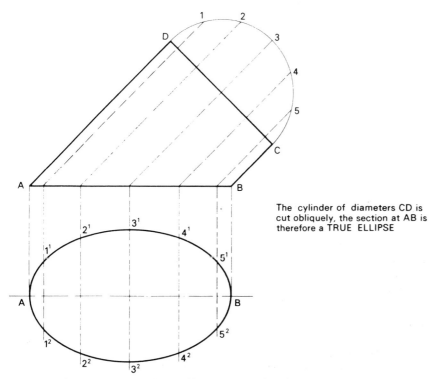

The cylinder of diameters CD is cut obliquely, the section at AB is therefore a TRUE ELLIPSE

Fig 2.43 Construction of an ellipse by projection

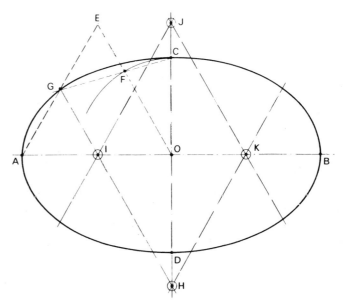

The curves for the shape of the oval are drawn with compasses

Fig 2.44 Construction of an oval (approximate ellipse)

49

Method of construction

1. Draw the major and minor axis A B and C D. These intersect at O.
2. Using a 60°/30° set square in conjunction with a Tee square draw an EQUILATERAL TRIANGLE with sides equal to ½ the major axis, as shown at A E O.
3. With compasses set to ½ the minor axis and centre O, draw an arc from C to cut side E O of the triangle at F.
4. Draw a line from C through F to cut side E A of the triangle at G.
5. Using the 60°/30° set square draw a line from G parallel to E O to cut the major axis A B at I and the minor axis C D produced at H.
6. H and I are the centres for drawing arcs to produce the approximate ellipse (oval). For corresponding centres, mark B K = A I and O J = O H using compasses. Curve G C is drawn with H as centre. Curve G A is drawn with I as centre.

2.14 Further useful geometric constructions

It will be explained in Chapter 5 that geometric constructions can be used to simplify the marking out of geometric components. In addition to the plane figures just described (2.12 and 2.13) the following constructions are also of great use to the craftsman.

1 PARALLEL LINES

Figure 2.45 shows how parallel lines may be drawn.

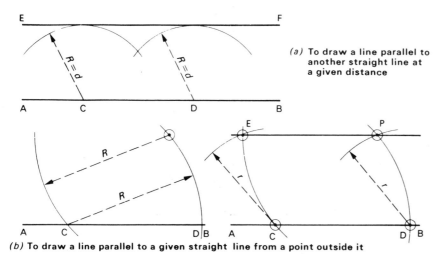

(b) To draw a line parallel to a given straight line from a point outside it

Fig 2.45 Construction of parallel lines

It will be seen in Fig. 2.45(*a*) that to draw a line parallel to another straight line at a given distance the following construction should be used.

With radius R (= to required distance) and any two points C and D on the given line A B, draw arcs. **Draw the line E F tangential to the two arcs.**

It will be seen in Fig. 2.45(*b*) that to draw a line parallel to a given straight line from a point outside it the following construction should be used.

With the centre at the point P, and any radius R describe an arc to locate a point C on the given line A B. With centre C and the same radius R draw an arc to locate a point D on the given line A B. With centre C and radius r (= D P) draw an arc to intersect the arc radius R at E. **Draw a line through P and E.**

2 TO LOCATE THE CENTRE OF AN ARC OR A CIRCLE

This construction is given in Fig. 2.46. Select three points C, D and E on the arc A B. Bisect the arc lengths between these points, using C, D and E as centres. **The point O where the bisectors intersect will be the centre of the given arc or circle.**

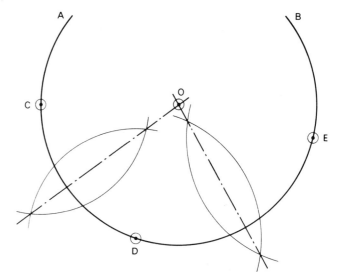

Fig 2.46 To locate the centre of a given arc or circle

3 DIVISION OF A STRAIGHT LINE

The construction for dividing a straight line into a number of equal or unequal parts was given in Chapter 2 of *Basic Engineering*.

4 TO CONSTRUCT A CIRCLE INSIDE OR OUTSIDE A GIVEN TRIANGLE

Figure 2.47(a) shows how a circle is drawn inside (*inscribed*) a triangle. The method is to bisect any two angles (Fig. 2.28). The point O where the bisectors intersect is the centre of the circle which is tangential to all three sides of the given triangle.

Figure 2.47(b) shows how a circle is drawn outside (*circumscribed*) a triangle. The method is to bisect any two sides of the given triangle. The point O where the two bisectors intersect is the centre of the circle which may be drawn to pass the points A, B and C.

5 ARCS OF CIRCLES

Figure 2.48 shows various constructions associated with arcs of circles. Figure 2.48(a) shows how to divide a semi-circle into a number of equal parts.

Bisect the DIAMETER A C. With a radius equal to HALF THE DIAMETER, and centres A, D and C draw arcs to cut the semi-circular circumference at points 1, 2, 3 and 4. The semi-circle is now divided into SIX EQUAL PARTS.

Note: This construction is a basic method used in pattern development for cylindrical and conical fabricated articles in sheet metal and plate work. It is also used for pyramids with HEXAGONAL bases.

Figure 2.48(b) shows how to construct a right-angle in a semi-circle. Select any point C on the circumference of the given semi-circle and draw lines from the DIAMETER to termination points A and B to C. *Then the angle* ACB *is* 90°. Similarly the angles at E and D are right-angles.

Figure 2.48(c) shows how to construct an OGEE curve. The termination points of the curve A B are normally given dimensionally. Join A and B by a straight line and bisect it at C. With radius *R* equal to the required radius of the curve, and with points A, C and B as centres draw arcs to locate D and E. *With the same radius and centres D and E complete the curve.*

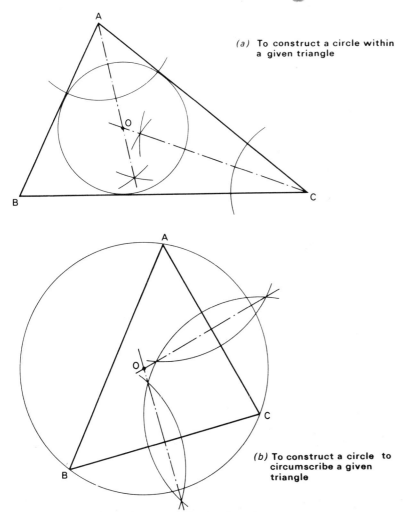

(a) To construct a circle within a given triangle

(b) To construct a circle to circumscribe a given triangle

Fig. 2.47 Inscribed and circumscribed triangles

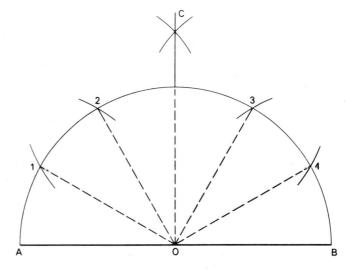

(a) To divide a semi-circle into equal parts

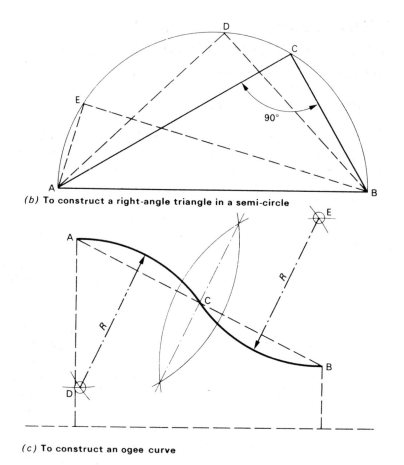

(b) To construct a right-angle triangle in a semi-circle

(c) To construct an ogee curve

Fig. 2.48 Constructions associated with arcs of circles

6 TANGENCY (STRAIGHT LINE)

Figure 2.49 shows how an arc may be drawn to touch two straight lines. Figure 2.49(a) shows the construction for drawing an arc tangential to two perpendicular lines. With the given radius r and centre B locate points E and D. With D and E as centres, and the same radius draw arcs to intersect at point O. *With the same radius and centre O draw the given arc to touch the two straight lines.*

Figure 2.49(c) shows the construction for drawing an arc tangential to two lines forming an obtuse angle.

At a distance equal to the given radius r, construct parallel lines to A B and B C to intersect at a point O.

With centre O and the given radius r, draw the required arc to touch the two straight lines.

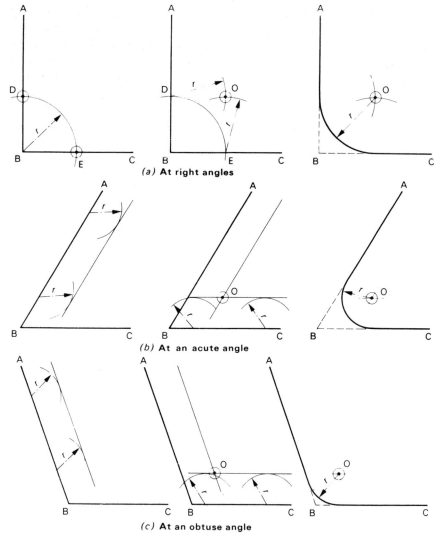

Fig. 2.49 To draw an arc of given radius to touch two straight lines

7 TANGENTIAL ARCS

Figure 2.50 shows the constructions associated with tangential arcs. Figure 2.50(a) shows the construction for an arc that is tangential to a circle and a straight line. Construct a parallel line, at a distance equal to the given radius R_1, to the given straight line A B. Add R_1 to the radius R of the given circle and from the centre of the circle draw an arc to locate point O. *With centre O and radius R_1 draw the required arc to touch the straight line and the circle.*

Figure 2.50(*b*) shows the construction of an external, tangential arc. Let *R* represent the radius of the given arc, and R_1 and R_2 the radii of the circles it is required to touch. Subtract the circle radii from the given radius of the arc, in turn, and from the respective centres of the circles draw arcs to intersect at point O. *With O as centre and radius R draw the required arc to touch the circles externally.*

Figure 2.50(*c*) shows the construction of an internal, tangential arc.

Add the circle radii to the given radius of the arc *R*. From the respective centres of the circles draw arcs to intersect at a point O. *With O as centre and radius Ř draw the required arc to touch the circles internally.*

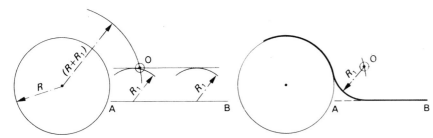

(a) To draw an arc of given radius to touch a circle and a straight line

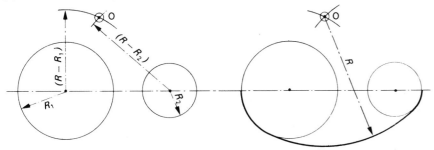

(b) To draw an arc of given radius to touch two circles externally

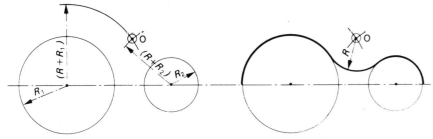

(c) To draw an arc of given radius to touch two circles internally

Fig. 2.50 Tangential arcs

2.15 Development of surfaces

The three basic methods used in pattern development are:

1. The PARALLEL LINE method.
2. The RADIAL LINE method.
3. The TRIANGULATION method.

In this book the principles of these methods will be explained, and the examples used will be limited to right prisms, right pyramids, the right cylinder, and right cone, and their frustums between parallel planes. *The development of a surface is the unrolling or unfolding of that surface so that it lies in one plane.*

The faces of *Prisms* are planes with their edges parallel. The unfolding of these faces will produce a development which takes the form of a simple rectangle.

A cylinder is developed by unrolling its surface, thus producing a *rectangle* having *one side equal in length to the 'circumference' of the cylinder,* the other side being equal to the *length or height of the cylinder.*

A pyramid, when its surface is unfolded, forms a development which consists basically of a number of triangles. *The base of each triangle is equal to the length of the base of the pyramid. The sides of each triangle are equal in length to the 'slant' edges of the pyramid.*

A cone, is developed by unrolling its surface. The circular base of the cone unrolls around a point, which is the apex of the cone, for a distance equal to its circumference. *The radius of the arc producing the base of the development is equal to the 'slant height' of the cone.*

In practice a complete cone is rarely required, except perhaps 'flat cones' which are used as 'caps', for example, on stove pipes. However, in the fabrication industry conical sections are constantly required to be manufactured. These components are part cones, often referred to as TRUNCATED CONES. When the cone is cut off parallel with its base, i.e. the top portion removed, the remaining portion is called the FRUSTUM.

2.16 Parallel line developments

The 'parallel line method' of pattern development depends upon a principle of locating the shape of the pattern on a series of parallel lines. All articles or components which belong to the class of PRISMS, *which have a constant 'cross-section' throughout their length, may be developed by the parallel line method.*

Elementary examples of parallel line development are illustrated in Fig. 2.51 to 2.53 inclusive.

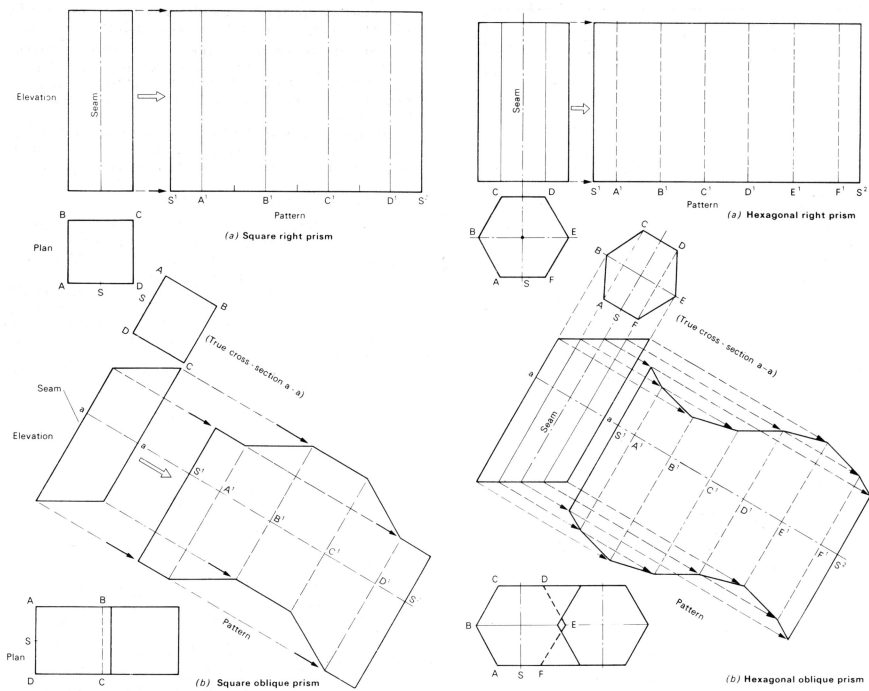

Fig. 2.51 Developments of square prisms (parallel line)

Fig. 2.52 Developments of hexagonal prisms (parallel line)

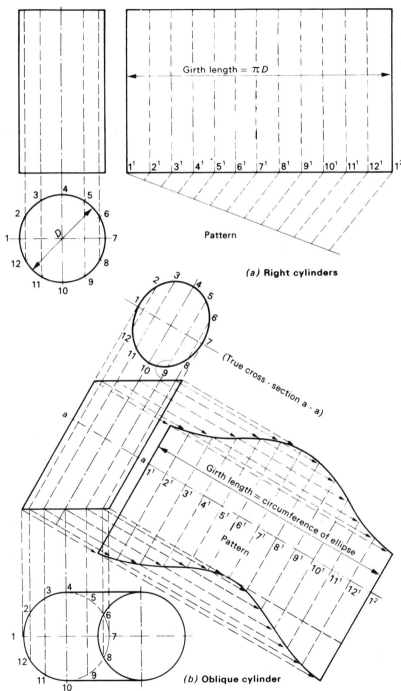

Fig. 2.53 Developments of cylinders (parallel line)

In order to distinguish the basic difference between a 'right prism' and an 'oblique prism' it is essential to apply the basic rule:

IF THE CROSS-SECTION OF THE ENDS BETWEEN PARALLEL PLANES IS AT 90° TO THE AXIS, THE COMPONENT IS A RIGHT PRISM.

IF THE CROSS-SECTION OF THE ENDS BETWEEN PARALLEL PLANES IS NOT NORMAL TO THE AXIS, THE COMPONENT IS AN OBLIQUE PRISM.

Therefore in order to develop the correct 'stretch-out' for the length of the pattern the distance around the TRUE CROSS-SECTION *must be used.* The examples illustrated should be self-explanatory.

2.17 Radial line development

The 'radial line method' may be applied for developing the pattern of any article or component which *tapers to an Apex*. This method is also adaptable to the development of 'frustums' which would normally *taper to an apex if the sides are produced.*

The principle of radial line development is based on the location of a series of lines which radiate down from the apex along the surface of the component to a base, or an assumed base, from which a curve may be drawn whose perimeter is equal in length to the perimeter of the base.

Elementary examples of radial line development are illustrated in Figs 2.54 to 2.57 inclusive.

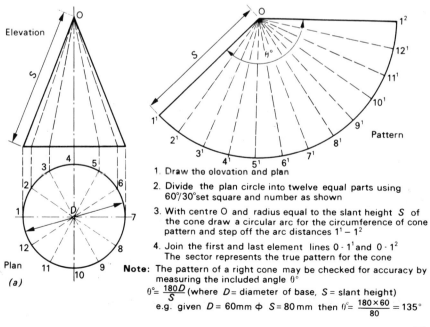

1. Draw the elevation and plan
2. Divide the plan circle into twelve equal parts using 60°/30° set square and number as shown
3. With centre O and radius equal to the slant height S of the cone draw a circular arc for the circumference of cone pattern and step off the arc distances $1^1 - 1^2$
4. Join the first and last element lines $0 - 1^1$ and $0 - 1^2$. The sector represents the true pattern for the cone

Note: The pattern of a right cone may be checked for accuracy by measuring the included angle $\theta°$

$\theta° = \dfrac{180 D}{S}$ (where D = diameter of base, S = slant height)

e.g. given D = 60mm φ S = 80 mm then $\theta° = \dfrac{180 \times 60}{80} = 135°$

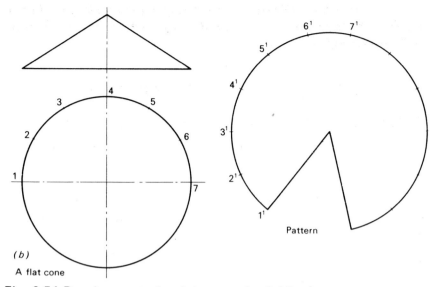

Fig. 2.54 Development of a right cone (radial line)

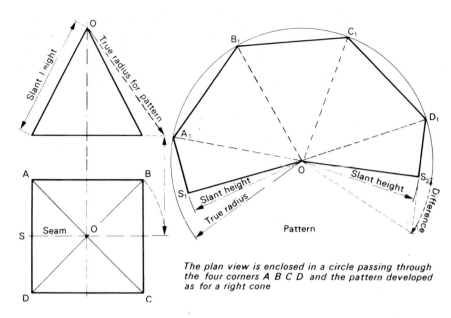

Fig. 2.56 Development of a square-based pyramid (radial line)

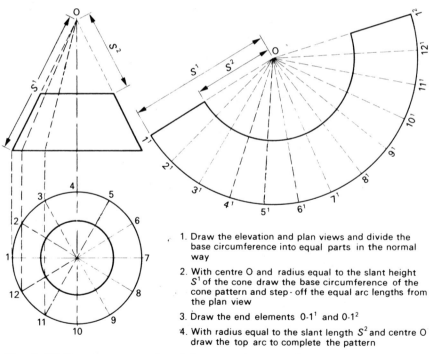

1. Draw the elevation and plan views and divide the base circumference into equal parts in the normal way
2. With centre O and radius equal to the slant height S^1 of the cone draw the base circumference of the cone pattern and step-off the equal arc lengths from the plan view
3. Draw the end elements $0\text{-}1^1$ and $0\text{-}1^2$
4. With radius equal to the slant length S^2 and centre O draw the top arc to complete the pattern

Fig. 2.55 Development of a right cone frustum (radial line)

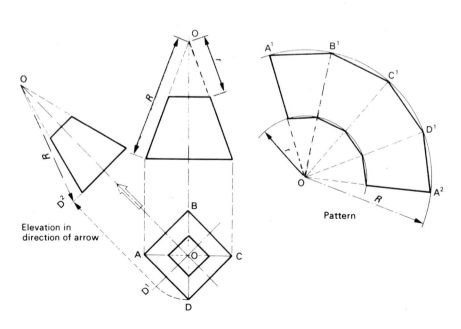

Fig. 2.57 Development of a square-based pyramid frustum (radial line)

Cones and pyramids are very closely related geometrical shapes. *A pyramid may be considered as 'a cone with a limited number of sides'.* Similarly a cone may be considered as 'a pyramid of an infinite number of sides'. *In practice, many large conical shapes in heavy gauge metal are often formed on the press brake as if they were many-sided pyramids.*

Although cones and pyramids have very similar characteristics, care must be taken when developing patterns for pyramids. It is very important to recognise one specific difference between a cone and a pyramid in order to avoid mistakes in development.

Figure 2.56 shows two views of a right pyramid which completely describe the object. The elevation shows the true slant height of faces of each triangular face which are square to the plan view. However, *the slant corners of the pyramid in the plan view are not normal to the elevation.* In order to establish their true length for the pattern, the plan view would have to be rotated until one slant corner was square to the elevation. This is not possible on the drawing board, but an arc may be drawn in plan as shown in the figure (radius O B) to the centre line and the point projected up to the base of the elevation view. The distance from the apex to this point will provide the true radius for swinging the arc for the basis of the pattern. It will also be noticed, in this example, that the seam is to be along the centre of one face of the pyramid. Therefore the true length of the joint line is equal to the slant height shown in the elevation. The three full sides are marked off along the basis curve in the pattern, in the same manner as for a right cone, an arc is swung each end using centre O and radius equal to the slant height, and the last two triangles are completed by swinging arcs from A and D with a radius equal to half the length of one side in the plan. The fundamental difference between the cone and pyramid is clarified in Fig. 2.57.

2.18 Development by triangulation

Triangulation is by far the most important method of pattern development since a great number of fabricated components transform from one cross-section to another. A typical 'square-to-round' transformer is illustrated in Chapter 5. The basic principle of triangulation is to develop a pattern by dividing the surface of the component into a number of triangles, determine the true size and shape of each, and then lay them down side by side in the correct order to produce a pattern.

To obtain the true size of each triangle, the true length of each side must be determined and then placed in the correct relationship to the other sides.

THE GOLDEN RULE OF TRIANGULATION:

'PLACE THE PLAN LENGTH OF A LINE AT RIGHT ANGLES TO ITS VERTICAL HEIGHT, THE DIAGONAL WILL REPRESENT ITS TRUE LENGTH.'

An elementary example of the method of triangulation is shown in Fig. 2.58.

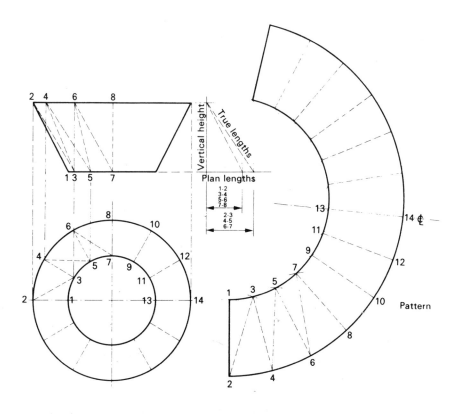

Fig 2.58 Development of a truncated cone (triangulaution)

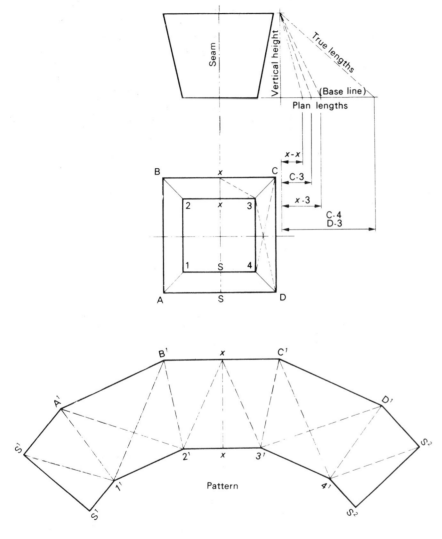

Fig 2.59 Development of a square hopper (triangulation)

2. It will be seen that the lengths A B, B C, C D and D A of the large square, and lengths 1,2, 2,3, 3,4 and 4,1 of the small square are TRUE LENGTHS in plan since they lie in the same horizontal plane, and therefore have no vertical height.
3. For the first triangle in the pattern take the true length distance B C (in plan) and mark it in the pattern. Draw a vertical centre line x,x. Mark the plan length x,x along the base line at 90° to the vertical height, and obtain its TRUE LENGTH and mark it on the pattern. Obtain the true length of diagonal $x,3$ in plan and swing arcs from x (on B C) in the pattern. Complete the triangle in the pattern by taking true length $x,3$ (in plan) and swing arcs each side of the centre line to locate 2 and 3.
4. Join B,2 and C,3 in the pattern (this represents one side of the hopper) check these two sides by plotting plan length C—3 against the vertical height.
5. For the next triangle mark true length arc B A. Obtain the true lengths of diagonal 2,A in plan by plotting it against the vertical height and swing an arc from 2 in the pattern to locate point A. Join B,A in the pattern.
6. For the next triangle swing true length arc 2,1 and true length arc B,1 these will intersect to locate point 1 in the pattern. Join A,1 and 2,1 to complete a second side of the hopper.
7. Take 3,1 in plan and swing an arc from 1 in the pattern. Take true length A,S from the plan view and swing an arc from A in the pattern to obtain true length 1,S by plotting its plan length against the vertical height and swing an arc from 1 in the pattern to locate points S. The last triangle S,1,S is completed by swinging an arc from S equal to the true length of the font line. Join A,S, 1,S and S,S.

Note: By commencing the pattern in the middle at $x-x$ (i.e. opposite the seam). The whole pattern to be obtained by repeating the marking out procedure each side of the centre line.
Check the pattern for symmetry — if drawn correctly the last two triangles are right-angle triangles.

A more complex example of triangulation is shown in Fig. 2.59.
1. Draw the elevation and plan views. The corner points in plan are lettered A B C D and numbered 1, 2, 3, 4 and 5,5 denotes the seam.

2.19 Comparison of right and oblique cones

Although the development of oblique cones and their frustums is beyond the scope of this book, it is important to be able to recognize the essential differences between these very similar geometrical shapes.

The essential differences between right and oblique cones is explained in Fig. 2.60.

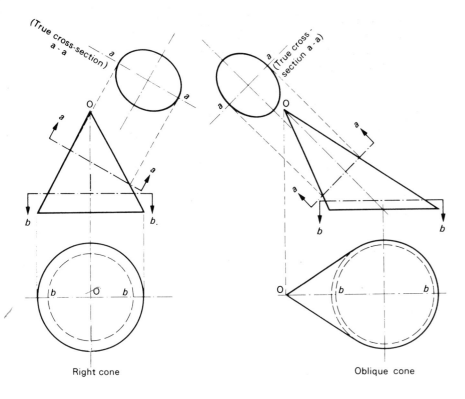

Note:

When an oblique cone is cut by a plane normal to its axis the cross-section produced is a true ellipse

Whereas, in the case of the right cone, the distance from the apex to the base is the same at all positions round the surface, the oblique cone varies from point to point

Fig. 2.60 Comparison of right and oblique cones

A RIGHT CONE has a circular base and its apex lies perpendicularly over the centre of its base.

AN OBLIQUE CONE has a circular base, but its apex does not lie perpendicularly over the centre of its base. The axis of an oblique cone leans to one side of the perpendicular. Both cones when cut by a plane parallel to the base (as shown at b,b) present a true circle at the plane of cutting. In the case of the right cone a cutting plane parallel to its base is normal to its axis. A right cone cut obliquely presents an elliptical cross-section (as shown at a,a).

Figure 2.61 shows typical frustums of oblique cones.

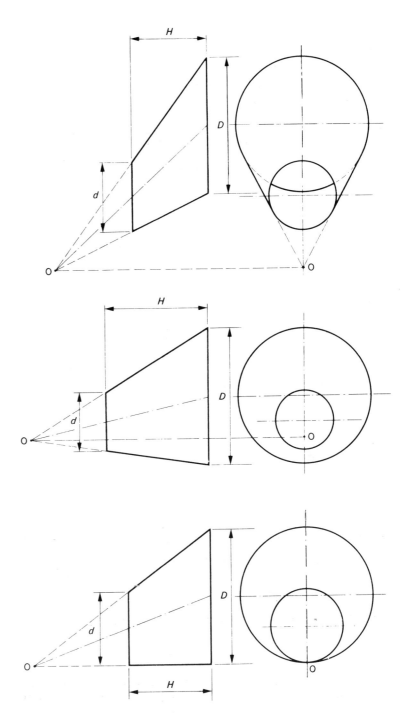

Fig 2.61 Typical oblique cone frustums

3 Calculations

Trigonometry not 'Trigger-nometry' should be used to solve triangle problems, Fred

3.1 Mass and weight

The concept of *mass* and *weight* has already been introduced in sections 7.7 and 7.8 of *Basic Engineering*.

MASS is the quantity of 'matter' or material in a component and is a constant quantity. The SI Unit for mass is the kilogramme (kg).

1 000 gramme = 1 kilogramme

1 000 kilogramme = 1 tonne

WEIGHT is the name given to the effect of the force of gravity acting on the mass of the component.

A *mass* of 1 kg equals a *weight* of 9·81 N approximately at sea level.

Table 3.1 shows some calculations of mass and weight. For most *workshop calculations* it is normal to work on the conversion of:

1 kg mass = 10 N weight

This not only makes the problem easier to work out, it also gives a 'margin of safety'.

Table 3.1 Mass/weight conversion

MASS TO WEIGHT		
1. *A steel casting has a mass of 3 810 kg. Calculate its weight in newtons (N).*		
(1 kg mass	=	10 N weight [9·81 N to be more precise])
3 810 kg mass	=	3 810 x 10
	=	38 100 N weight
	=	38·1 kN (kilo-newton)
2. *A brass screw has a mass of 27 g. Calculate its weight in newtons (N).*		
27 gramme	= 27 ÷ 1 000	= 0·027 kg.
0·027 kg mass	= 0·027 x 10	= 0·27 N weight.

Note how the mass in grammes had to be converted to kilo-grammes before the calculation could proceed.

WEIGHT TO MASS

3. *An aluminium casting weighs 2 850 N. Calculate its mass in kilogramme (**kg**).*

   (1 N weight = 0·1 kg mass)

   2 850 N weight = 2 850 ÷ 10

   = <u>285 kg mass</u>

4. *A large machine weighs 50 kN. Calculate its mass in tonnes.*

   (1 kilo-newton = 1 000 newton)

   50 kN = 50 000 N

   50 000 N = 50 000 ÷ 10

   = 5 000 kg

   1 000 kg = 1 tonne

   5 000 kg = <u>5 tonne</u>

It is useful to be able to calculate the mass and weight of components in the workshop. For example:

1. To check if lifting tackle is adequate to lift the component safety.
2. To calculate the material cost, as most raw materials are sold by 'weight'.
3. To calculate the charge of a furnace to see if there is sufficient metal to make the casting.

3.2 Density

Density is defined as *mass per unit volume*.

That is, density = $\dfrac{\text{mass}}{\text{volume}}$

For example, the density of copper is 0·008 9 g/mm^3 or, each cubic millimetre of copper has a mass of 0·008 9 gramme.

$$\text{MASS} = \text{VOLUME} \times \text{DENSITY}$$

EXAMPLE: Calculate the mass of a component of volume 0·3 metre^3. The density of copper is 0·008 9 g/mm^3.

Mass = volume × density

= 0·3 × 10^6 × 0·008 9 (convert volume to mm^3)

= 2 670 gramme

= <u>2·67 kilo-gramme</u>

The corresponding weight of the component would be:

weight = mass × acceleration due to gravity

= 2·67 × 9·81 (mass must be in kg)

= <u>26·2 newton (N)</u>

For workshop purposes it would be accurate enough to calculate the weight as:

weight = 2·67 × 10

= <u>26·7 N</u>

The 'error' of 0·5 N is on the right side to give a margin of safety.

Table 3.2 gives the densities for some typical engineering materials.

Figure 3.1 shows how the mass and weight of the solids given in *Basic Engineering*, table 3.23, can be calculated.

Table 3.2 Densities of common engineering materials

MATERIAL	DENSITY		
	g/mm^3	g/cm^3	kg/m^3
Aluminium	0·002 56	2·56	2 560
Brass (70/30)	0·008 21	8·21	8 210
Bronze	0·008 52	8·52	8 520
Copper	0·008 65	8·65	8 650
Lead	0·011 4	11·4	11 400
Steel	0·007 73	7·73	7 730
Tin	0·007 3	7·3	7 300
Zinc	0·007	7·0	7 000

SOLID	DATA	CALCULATIONS	
		MASS	WEIGHT
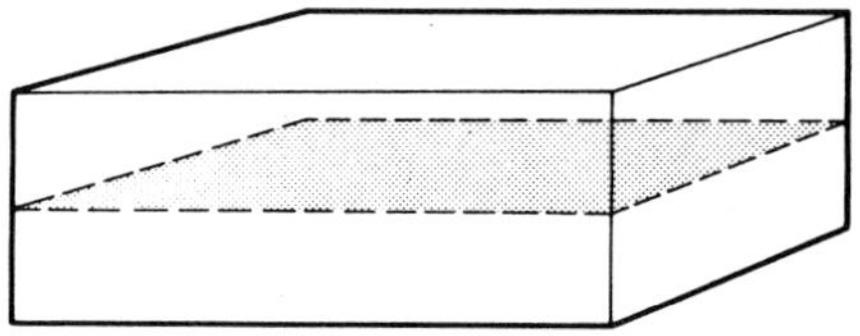 Rectangular prism (copper)	Volume = 72 000 mm^3 Density = 0·008 65 g/mm^3	Mass = volume × density = 72 000 × 0·008 65 = <u>622·8 g</u>	622·8 g = 0·622 8 kg 0·622 8 kg = 0·622 8 × 10 = <u>6·228 N</u>
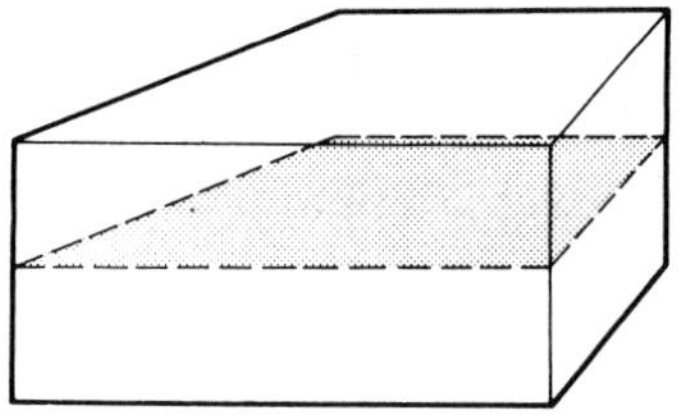 Trapezoidal prism (steel)	Volume = 27 000 mm^3 Density = 0·007 3 g/mm^3	Mass = volume × density = 27 000 × 0·007 3 = <u>197·1 g</u>	197·1 g = 0·197 1 kg 0·197 1 kg = 0·197 1 × 10 = <u>1·971 N</u>
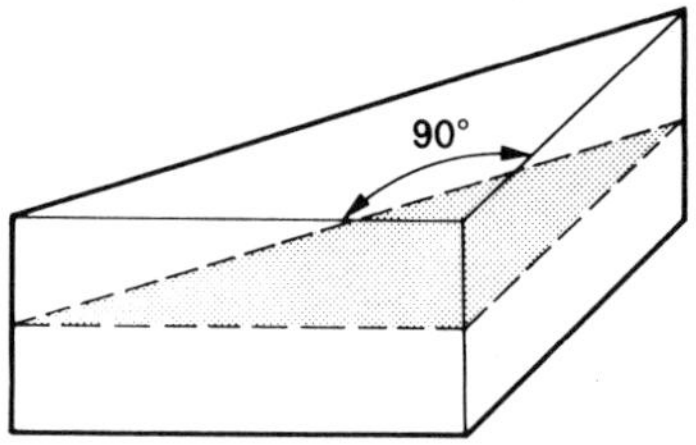 Triangular prism (aluminium)	Volume = 300 000 mm^3 Density = 2·56 g/cm^3	300 000 mm^3 = 300 cm^3 Mass = volume × density = 300 × 2·56 = <u>768 g</u>	768 g = 0·768 kg 0·768 kg = 0·768 × 10 = <u>7·68 N</u>
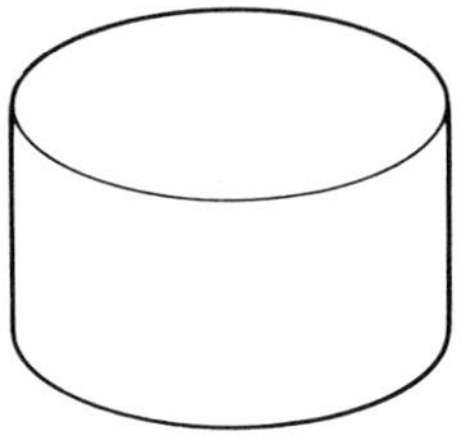 Cylinder (brass)	Volume = 4 713 000 mm^3 Density = 8 210 kg/m^3	4 713 000 mm^3 = 0·004 713 m^3 Mass = volume × density = 0·004 713 × 8 210 = <u>38·7 kg</u>	38·7 kg = 38·7 × 10 = <u>387 N</u>

Note: The volumes of these solids were calculated in Basic Engineering, table 3.23

Fig. 3.1 Calculations of volume, mass and weight

Most engineering components can be broken down into the basic shapes shown in Fig. 3.1. Figures 3.2 and 3.4 inclusive give some examples of calculations of the mass and weight of simple engineering components.

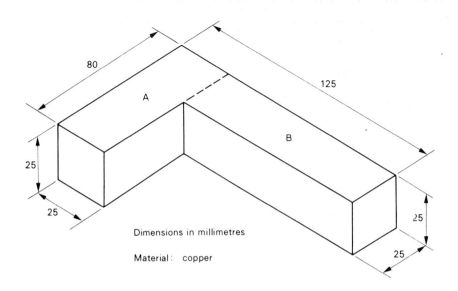

Dimensions in millimetres

Material: copper

The figure can be broken down into two rectangular prisms A and B

1. *Volume of Prism* A

   Volume = length × breadth × thickness

   = 80 × 25 × 25

   = 50 000 mm³

2. *Volume of Prism* B

   Volume = 100 × 25 × 25

   = 62 500 mm³

3. *Total volume* = 50 000 + 62 500 = 112 500 mm³

   Density of copper from table 3.2 = 0.008 65 g/mm³

4. *Mass* = volume × density

   = 112 500 × 0.008 65

   = 973.125 g

5. *Weight* = mass (kg) × 10

   = 0.973 1 × 10

   = 9.731 N

Fig. 3.2 Example — **volume, mass and weight (1)**

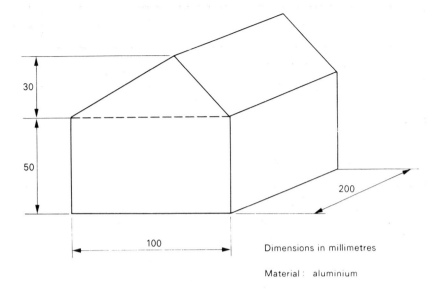

Dimensions in millimetres

Material: aluminium

The figure can be broken down into a rectangular prism and a triangular prism

1. *Volume of rectangular prism*

   Volume = length × breadth × thickness

   = 100 × 50 × 200

   = 1 000 000 mm³

2. *Volume of triangular prism*

   Volume = ½ base × height × thickness

   = ½ × 100 × 30 × 200

   = 300 000 mm³

3. *Total volume* = 1 000 000 + 300 000 = 1 300 000 mm³ = 1 300 cm³

   Density of aluminium from table 3.2 = 2.56 g/cm³

4. *Mass* = volume × density

   = 1 300 × 2.56

   = 3 328 g

5. *Weight* = mass (kg) × 10

   = 3.328 × 10

   = 33.28 N

Fig. 3.3 Example — **volume, mass and weight (2)**

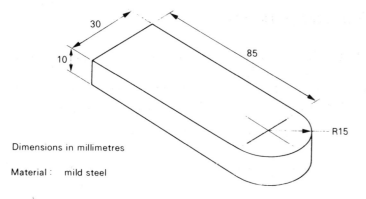

Dimensions in millimetres

Material: mild steel

The figure can be broken down into a rectangular prism and half a right cylinder

1. *Volume of rectangular prism*

   Volume = length × breadth × thickness

   = 85 × 30 × 10

   = 25 500 mm³

2. *Volume of half cylinder*

   Volume = $\frac{1}{2}(\pi R^2) \times 10$

   = $\frac{1}{2}$ × 3·14 × 15 × 15 × 10

   = 3 532·5 mm³

3. *Total volume* = 25 500 + 3 532·5 = 29 032·5 mm³

   Density from table 3.2 = 0·007 73 g/mm³

4. *Mass* = volume × density

   = 29 032·5 × 0·007 73

   = 224·4 g

5. *Weight* = mass (kg) × 10

   = 0·224 4 × 10

   = 2·244 N

Fig. 3.4 Example — volume, mass and weight (3)

3.3 Sheet metal components

Sheet metal is often sold by *mass per unit area* for a given thickness, rather than by using volume and density. Table 3.3 gives some examples for sheets of typical engineering materials. Figure 3.5 shows how blank weights can be calculated using the data in table 3.3.

Table 3.3 Mass/unit area for sheet metal
(The values given are for mild steel. For other metals see notes.)

THICKNESS (mm) ISO R388		MASS/UNIT AREA		NOTES ON USE OF TABLE
1st choice	2nd choice	g/cm²	kg/m²	
0·020		0·015 5	0·155	1. The mass/unit area given in the tables can be converted to the values for metals other than steel by use of the following multiplying factors.
	0·022	0·017 0	0·170	
0·025		0·019 3	0·193	
	0·028	0·021 6	0·216	
0·032		0·024 7	0·247	
	0·036	0·027 8	0·278	
0·040		0·030 9	0·309	
	0·045	0·034 8	0·348	
0·050		0·038 7	0·387	Aluminium × 0·331 1
	0·056	0·043 3	0·433	Brass × 1·062 (70/20)
0·063		0·048 7	0·487	
	0·071	0·054 9	0·549	Bronze × 1·102
0·080		0·061 8	0·618	Copper × 1·119
	0·090	0·069 6	0·696	
0·100		0·077 3	0·773	Lead × 1·475
	0·112	0·086 5	0·865	Tin × 0·944
0·125		0·096 6	0·966	
	0·140	0·108 2	1·082	Zinc × 0·906
0·160		0·123 7	1·237	
	0·180	0·139 1	1·391	2. To calculate the mass/unit area of 1 mm thick copper sheet.
0·200		0·154 6	1·546	
	0·224	0·173 2	1·732	
0·250		0·193 3	1·933	
	0·280	0·216 4	2·164	Mass/unit area for 1 mm thick steel is 7·73 kg/m² multiplying factor for copper is 1·119. Therefore mass/unit area for copper 1 mm thick will be.
0·315		0·243 5	2·435	
	0·355	0·274 4	2·744	
0·400		0·309 2	3·092	
	0·450	0·347 9	3·479	
0·500		0·386 5	3·865	
	0·560	0·432 9	4·329	
0·630		0·487 0	4·870	7·73 × 1·119
	0·710	0·548 8	5·488	
0·800		0·618 4	6·184	= 8·65 kg/m²
	0·900	0·695 7	6·957	
1·00		0·773 0	7·730	
	1·120	0·865 0	8·650	
1·25		0·996 2	9·962	
	1·40	1·082 2	10·822	
1·60		1·236 8	12·368	
	1·80	1·391 4	13·914	

2·00		1·546 0	15·460
	2·24	1·731 5	17·315
2·50		1·932 5	19·325
	2·80	2·164 4	21·644
3·15		2·434 9	24·349
	3·5	2·744 1	27·441
4·00		3·092 0	30·920

3.4 Bar components

Bar is often sold by mass per unit length for a given cross-section, rather than by using volume and density. Table 3.4 gives some examples for bars of typical engineering materials. Figure 3.6 shows how blank weights can be calculated using the data in table 3.4.

Table 3.4 Mass/metre run for mild steel bars

CIRCULAR SECTION		SQUARE SECTION		HEXAGONAL SECTION	
Diameter mm	Mass kg	Size mm	Mass kg	Size (A/F) mm	Mass kg
4	0·098	4	0·125	4	0·107
5	0·152	5	0·194	5	0·168
6	0·200	6	0·280	6	0·240
7	0·296	7	0·380	7	0·330
8	0·390	8	0·495	8	0·430
9	0·495	9	0·630	9	0·545
10	0·610	10	0·774	10	0·670
12	0·875	12	1·13	12	0·965
14	1·190	14	1·52	14	1·34
16	1·558	16	1·99	16	1·70
18	1·97	18	2·52	18	2·00
20	2·43	20	3·10	20	2·68
25	3·80	22	3·74	22	3·24
30	5·47	24	4·47	24	3·57
35	7·45	26	5·25	26	4·53
40	9·75	28	6·08	28	5·75
45	12·3	30	6·97	30	6·05
50	15·3	32	7·91	32	6·88
60	22·0	34	8·96	34	7·79
70	29·8	36	10·3	36	8·70
80	39·0	38	11·4	38	9·70
90	49·1	40	12·4	40	10·75
100	61·0	42	13·65	42	11·85
125	95·0	44	15·00	44	13·00
150	137·0	46	16·4	46	14·4
175	186·0	48	17·9	48	15·5
200	242·0	50	19·3	50	16·8
225	308·0	55	23·4	55	20·2
250	380·0	60	28·0	60	24·1
300	549·0	65	32·6	65	28·3

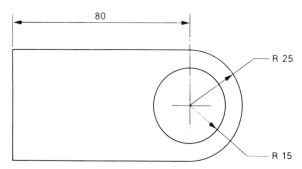

Dimensions in millimetres
Material: 1·25mm Thick mild steel

1. Area of rectangle = 80 × 50 = **4 000mm²**

   Area of semi-circle = ½ × 3·14 × 25 × 25 = **981 mm²**

   Area of hole = 3·14 × 15 × 15 = **706 mm²**

   Total area of blank = 4000 + 981 − 706 = **4 275 mm²**

   Mass/unit area for 1·25 mm thick mild steel = 0·996 g/cm² (Table 3.3)

   Area of blank in cm² = 4 275 ÷ 100 = **42·75 cm²**

   Therefore, mass of blank = 42·75 × 0·996
   = 42·6 g (mass)
   = **0·426 N (weight)**

2. If aluminium sheet was used instead of mild steel, the multiplying factor × 0·3311 would have to be used (Table 3.3)

   Mass/unit area for 1·25mm thick aluminium = 0·996 × 0·3311
   = 0·329 7 g/cm²

   Therefore, mass of blank in aluminium = 42·75 × 0·329 7
   = 14·1 g (mass)
   = **0·141 N (weight)**

Fig. 3.5 Calculation of sheet metal blank weights

Note: To use the above tables for materials other than mild steel, use the multiplying factors given with table 3.3.

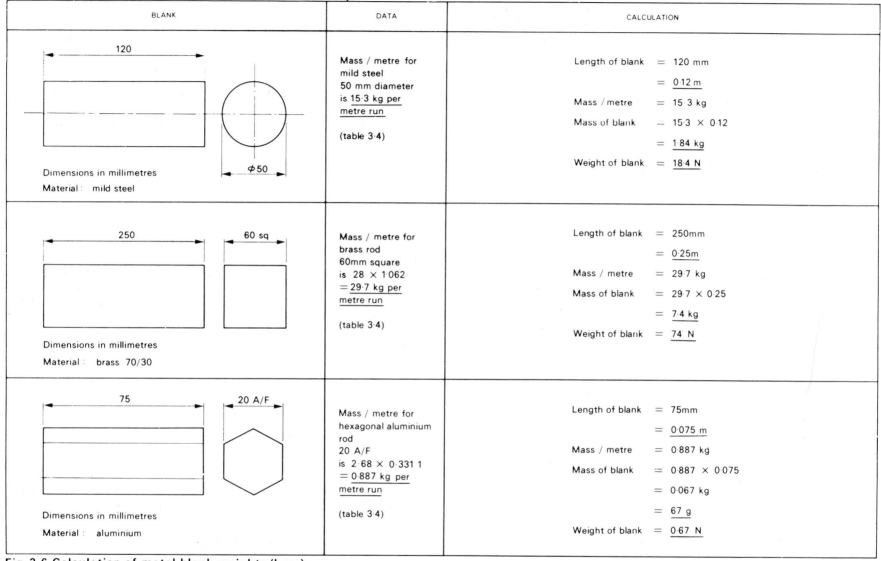

Fig. 3.6 Calculation of metal blank weights (bars)

3.5 Mass of the contents of a container

If the volume of a container is known and the density of the fluid or powder stored is known then the mass of the fluid or powder can be calculated as shown in Fig. 3.7.

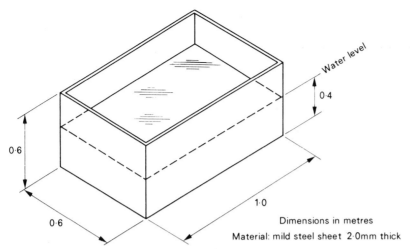

To find the mass of the open-topped container shown together with water filled to a depth of 0·4 m

Area of sheet metal:
 ends: 2 × 0·6 × 0·6 = 0·72 m²
 sides: 2 × 1·0 × 0·6 = 1·20 m²
 bottom: 1·0 × 0·6 = 0·60 m²

Total area   = 2·52 m²

Mass per unit area for 2·0 mm thick mild steel = 15·46 kg/m² (table 3.3)

Mass of tank = 2·52 × 15·46 = 38·96 kg

Volume of water (allowing for thickness of sheet metal)

Volume = 0·398 × 0·596 × 0·996
       = 0·236 2 m³

Density of water = 1·0 g/cc

Mass of water = 1 × 0·2362 × 1 000 000
              = 236 200 g
              = 236·2 kg

Total mass of tank and water = 38·96 + 236·2
                             ≃ 275 kg

no	log
0·398	1̄·5999
0·596	1̄·7752
0·996	1̄·9983
0·236 2	1̄·3734

Fig. 3.7 Mass of contents of container

3.6 Introduction to trigonometry

Figure 3.8 shows two triangles whose angles are the same but the lengths of whose sides vary. In order that the angles remain the same the lengths of the corresponding sides must vary in the same ratio as shown.

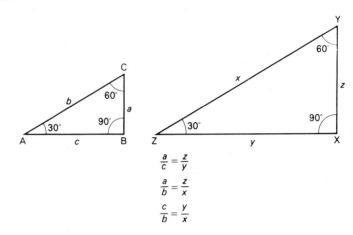

$$\frac{a}{c} = \frac{z}{y}$$
$$\frac{a}{b} = \frac{z}{x}$$
$$\frac{c}{b} = \frac{y}{x}$$

Fig. 3.8 Similar triangles

It can be seen, therefore, that the size of the acute angles in any right-angled triangle can be stated as the ratio of any two of the sides. Figure 3.9 shows a right-angled triangle and shows how the sides of the triangle are named.

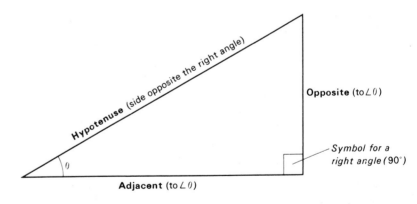

Fig. 3.9 The sides of the right-angled triangle

The ratios of these sides for a given angle are given special names as shown in Fig. 3.10. They are called the *trigonometrical ratios*. To keep things simple only the right-angled triangle will be considered in this chapter, and for most workshop purposes this is sufficient. At a more advanced level, trigonometry can be applied to any sort of triangle and angles of any magnitude.

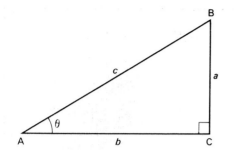

Tangent (tan) θ = $\dfrac{\text{opposite}}{\text{adjacent}}$ = $\dfrac{a}{b}$

Cosine (cos) θ = $\dfrac{\text{adjacent}}{\text{hypotenuse}}$ = $\dfrac{b}{c}$

Sine (sin) θ = $\dfrac{\text{opposite}}{\text{hypotenuse}}$ = $\dfrac{a}{c}$

Fig. 3.10 The trigonometrical ratios

3.7 Trigonometrical tables

These are used to evaluate problems involving the sides and angles of triangles (trigonometry). Tables of natural tangents, natural sines, and natural cosines are included at the end of this book. Trigonometrical tables are used in a similar manner to the tables of logarithms introduced in *Basic Engineering*. Figures 3.11 to 3.13 inclusive show how the tables should be read. Figures 3.14 to 3.16 give examples involving the use of trigonometry.

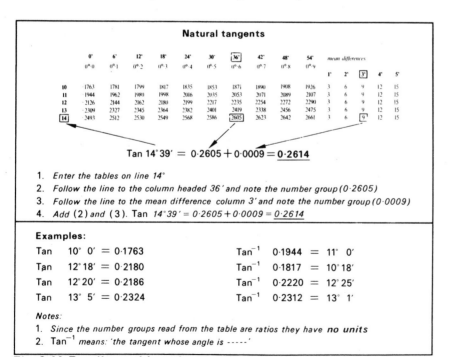

Fig. 3.11 Reading tables of natural tangents

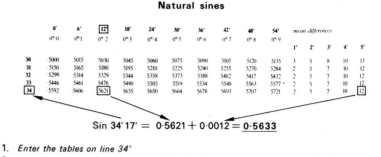

Fig. 3.12 Reading tables of natural sines

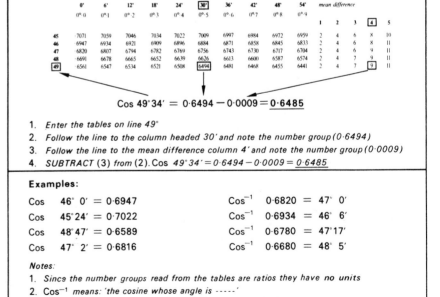

Fig. 3.13 Reading tables of natural cosines

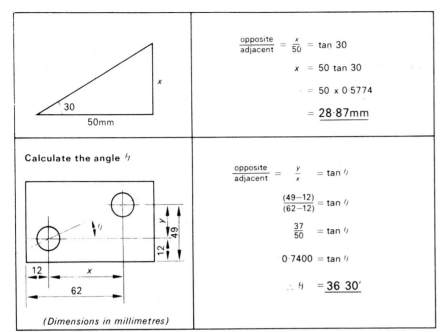

Fig 3.14 Use of trigonometry - tangents

$\dfrac{\text{opposite}}{\text{adjacent}} = \dfrac{x}{50} = \tan 30$

$x = 50 \tan 30$

$= 50 \times 0.5774$

$= \underline{28.87\text{mm}}$

Calculate the angle θ

$\dfrac{\text{opposite}}{\text{adjacent}} = \dfrac{y}{x} = \tan \theta$

$\dfrac{(49-12)}{(62-12)} = \tan \theta$

$\dfrac{37}{50} = \tan \theta$

$0.7400 = \tan \theta$

$\therefore \theta = \underline{36°\,30'}$

(Dimensions in millimetres)

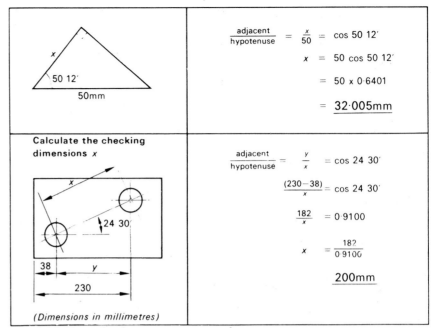

Fig 3.16 Use of trigonometry - cosines

$\dfrac{\text{adjacent}}{\text{hypotenuse}} = \dfrac{x}{50} = \cos 50°\,12'$

$x = 50 \cos 50°\,12'$

$= 50 \times 0.6401$

$= \underline{32.005\text{mm}}$

Calculate the checking dimensions x

$\dfrac{\text{adjacent}}{\text{hypotenuse}} = \dfrac{y}{x} = \cos 24°\,30'$

$\dfrac{(230-38)}{x} = \cos 24°\,30'$

$\dfrac{182}{x} = 0.9100$

$x = \dfrac{182}{0.9100}$

$\underline{200\text{mm}}$

(Dimensions in millimetres)

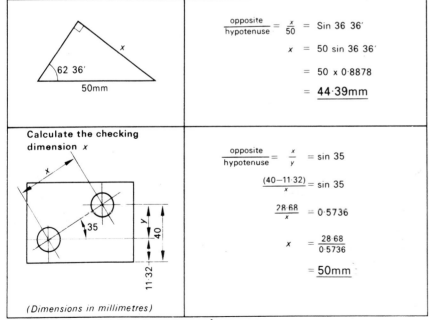

Fig 3.15 Use of trigonometry - sines

$\dfrac{\text{opposite}}{\text{hypotenuse}} = \dfrac{x}{50} = \sin 36°\,36'$

$x = 50 \sin 36°\,36'$

$= 50 \times 0.8878$

$= \underline{44.39\text{mm}}$

Calculate the checking dimension x

$\dfrac{\text{opposite}}{\text{hypotenuse}} = \dfrac{x}{y} = \sin 35$

$\dfrac{(40-11.32)}{x} = \sin 35$

$\dfrac{28.68}{x} = 0.5736$

$x = \dfrac{28.68}{0.5736}$

$= \underline{50\text{mm}}$

(Dimensions in millimetres)

4 Science

Some chemical changes in metals can be desirable

4.1 Chemical changes in metals

Chemical changes in metals occur as a result of using different oxy-acetylene *flame settings*. These basic flame settings are fully explained in Part A of Chapter 11.

Metal which has become heated to a very high temperature, as in the case of welding, very rapidly becomes oxidised by the oxygen which is present in the atmosphere. Other chemical reactions also take place when metals are joined by thermal processes.

4.2 Chemical reactions during welding

Whether in the solid or molten state, heated materials tend to absorb and/or react with gases in the surrounding atmosphere. *The higher the temperature is raised, the more rapid becomes the rate of absorption and reaction, and these usually show a marked increase as the material becomes molten.*

The fuel gases used for gas-welding flames disassociate when ignited in the presence of oxygen or air. They break down into CARBON MONOXIDE, CARBON DIOXIDE, HYDROGEN, and WATER VAPOUR.

It is these new gases which react with materials rather than the original fuel gases.

If metal is welded without protection from the atmosphere, the OXYGEN and NITROGEN in the air will chemically combine with the molten metal to form OXIDES and NITRIDES. If these are allowed to dissolve in the WELD POOL the result will be the production of poor and brittle welds.

OXYGEN is generally harmful and a metal which oxidises readily is likely to be difficult to weld. Oxidised welds are undesirable and can easily be identified by the appearance of their surface which is irregular and pitted. Such welds lack STRENGTH and DUCTILITY.

NITROGEN dissolves in many liquid materials and may react with some of their constituents. If, when welding certain steels, nitrogen is allowed to enter into the molten weld pool, the resultant weld will be brittle, porous, and low in ductility.

HYDROGEN can cause undesirable problems with many molten metals, in particular steel, aluminium, and 'tough-pitch' copper. The chief cause of GAS POROSITY is the presence of hydrogen in weld metal, or the reaction of hydrogen with oxides present in the molten parent metal which results in the production of STEAM.

There are numerous sources of hydrogen in welding, the chief one being the WELDING FLAME in gas welding or the FLUX COVERING in manual metal-arc welding.

WATER VAPOUR tends to disassociate when it is in contact with the surface of molten metal. Some of the freed HYDROGEN will *dissolve* in the metal. The freed OXYGEN tends to *react* with it.

4.3 Oxidation of welds

Certain metals have such a high affinity for oxygen that the oxides form on the surface almost as rapidly as they are removed.

In gas welding, soft soldering, or brazing operations, these oxides are usually removed and eliminated by the use of suitable fluxes.

The affinity for oxygen, which is a characteristic of some metals, can be used to great advantage in certain welding operations. For example, manganese and silicon, elements common to plain carbon steel, are important in oxy-acetylene welding because they readily react with oxygen when the steel is in the molten condition. The reaction produces a very thin SLAG COVERING which tends to prevent any oxygen from contacting the weld metal. It also prevents the formation of gas pockets (cavities) in the weld.

The action of these elements in gas-welding is the same as in steel-making, where they are used in 'open hearth' or 'electric' furnaces to produce clean deoxidised metal. Manufacturers of steel welding rods ensure that the filler material contains the correct percentage of manganese and silicon.

Steel welding wire and rod is usually supplied 'chemically cleaned and copper coated'. *The copper coating acts as a reducing agent and prevents the filler metal from becoming oxidised.*

4.4 Protecting the 'weld pool' from the atmosphere

The type of flame used in welding various materials plays an important role in securing the most desirable weld deposit.

The proper type of flame condition with the correct welding technique will provide a shielding medium which will protect the molten weld metal from the harmful effects of oxygen and nitrogen in the atmosphere. When welding plain carbon steels, use of the correct flame condition has the effect of stabilising the molten weld metal and preventing the 'burning-out' of CARBON, MANGANESE, and other alloying elements.

Figure 4.1 illustrates how the molten metal or 'weld pool' is protected from the harmful gases of the surrounding atmosphere by the REDUCING GASES present in the oxy-acetylene flame.

In manual metal-arc welding the flux covering on the electrode protects the weld from atmospheric contamination, as shown in Fig. 4.2.

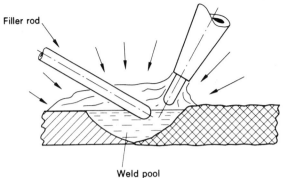

Weld shielded from the surrounding air by the reducing gases of the flame. Filler rod tip must be kept in the reducing zone

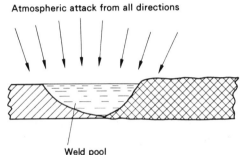

Flame removed, weld pool now prone to atmospheric attack

Fig. 4.1 Protection of the weld by the oxy-acetylene flame

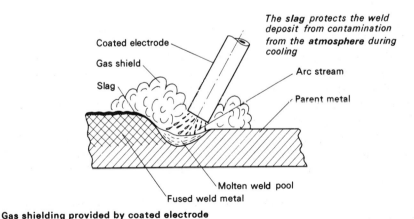

Gas shielding provided by coated electrode

Fig. 4.2 Protection of the weld – manual metal-arc

4.5 Shielding gases in arc welding (specific)

Two arc-welding processes which are rapidly gaining popularity in most modern fabrication and welding workshops, and in shipyards, are the TUNGSTEN INERT GAS METAL-ARC WELDING PROCESS, commonly referred to as TIG, and the METAL INERT GAS WELDING PROCESS, known also as MIG. Although the mechanics of these two important arc-welding processes are beyond the scope of this book, a brief description of the functions of the shielding gases is given below.

Tungsten inert gas welding

The TIG welding process employs a tungsten electrode which is non-consumable. Tungsten has a very high melting point (in excess of 3 000 °C), and therefore is not consumed by the heat of the arc. The atmosphere is excluded from the weld pool by a supply of inert gas. *An inert gas is one which does not chemically combine with any other element, and therefore cannot affect the weld metal.* ARGON and HELIUM are inert gases which are used with this welding process to shield the molten weld pool and parent metal from the atmosphere. When using this process, care must be taken to see that the gas flow is correct. Should it be too low, or should a draught interfere with the shielding of weld, atmospheric contamination will result. With the TIG welding process the arc is maintained between the tip of the tungsten electrode and the parent metal, and unlike other metal-arc welding processes, the filler material is added as in gas welding. Figure 4.3 shows how the atmosphere is excluded from the weld pool.

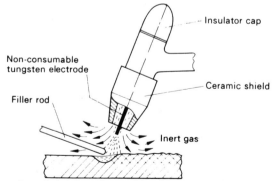

The atmosphere is excluded from the weld by shielding with an inert gas. No chemical reactions take place

Fig. 4.3 Protection of the weld — TIG process

Metal inert gas welding

The MIG welding processes are basically automatic or semi-automatic. The electrode is a bare wire which is continuously fed into the welding gun by means of an electrode wire drive unit as the metal is transferred across the arc. Some electrode holders are termed 'welding torches' because the design of the curved neckpiece resembles an oxy-acetylene welding torch.

The shielding gases used are ARGON, HELIUM, and CARBON DIOXIDE. When carbon dioxide is used the process is referred to as 'CO_2 welding'. Carbon dioxide is not an inert gas, and when it passes through the arc it tends to break down into carbon monoxide and oxygen. To ensure that the liberated oxygen does not contaminate the weld metal, 'DEOXIDISERS' are included in the welding wire which is specially manufactured for these welding processes. These deoxidisers combine with the oxygen to form a very thin slag on the surface of the completed weld. Figure 4.4 illustrates the method of shielding the weld in CO_2 welding.

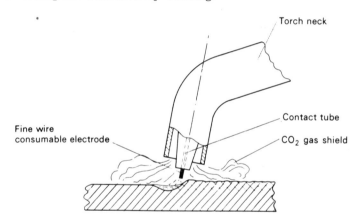

Fig. 4.4 Protection of the weld — MIG process

4.6 The problem of oxides in welding

The removal and dispersion of the oxide film which forms when metals are heated is one of the traditional problems associated with welding.

Practical difficulties in welding occur:
1. When the surface oxide forms a tenacious film.
2. When the oxide has a MELTING POINT very much higher than that of the parent metal.
3. When the oxide forms very rapidly.

One of the most important factors in weld quality is the removal of oxides from the surface of the metal to be welded. Unless the oxides are removed, the following undesirable conditions may result:
1. Fusion may be difficult.
2. Inclusions may be present in the weld metal.
3. The joint will be weakened.
4. The oxides will not flow from the welding zone but remain to become entrapped in the solidifying metal, interfering with the addition of filler material. *This condition generally occurs when the oxide has a higher melting point than the parent metal.*

4.7 The use of fluxes

Fluxes are CHEMICAL COMPOUNDS used to prevent oxidation and unwanted chemical reactions. Table 4.1 lists some common applications of fluxes for gas-welding.

Table 4.1   Common applications of fluxes for gas-welding

METAL or ALLOY	FLUX	REMARKS
Mild steel	—	No flux is required because the oxide produced has a lower melting point than the parent metal. Being less dense it floats to the surface of the molten weld metal as scale which is easily removed after welding. Use a neutral flame.
Copper	Borax base with other compounds	If Borax alone is used, a hard scale of copper borate is formed on the surface of the weld which is difficult to remove. Use a neutral flame.
Aluminium and aluminium alloys	Contains chlorides of lithium and potassium	Aluminium fluxes absorb moisture from the atmosphere — i.e., they are hygroscopic. Always replace the lid firmly on the container when the flux is not in use. The flux residue is very corrosive. On completion of the weld it is essential to remove all traces of this residue. This can be accomplished by scrubbing the joint area with a 5% nitric acid solution or hot soapy water. Use carburizing flame.
Brasses and bronzes	Borax type containing sodium borate with other chemicals	The flux residue is a hard glass-like compound which can be removed by chipping and wire brushing. Use an oxidising flame.
Cast iron	Contains borates, carbonates, and bicarbonates plus other slag-forming compounds	Oxidation is rapid at red heat, and melting point of the oxide is higher than that of the parent metal. For this reason it is important that the flux combines with the oxides to form a slag which floats to the surface of the weld pool and prevents further oxidation.

FLUX REMOVAL: — Many types of fluxes are corrosive to the metals or alloys with which they are used. Therefore it is important that residual flux be removed from the surface immediately after the welding operation. *Methods of removal generally employed include mechanical methods such as chipping and scratch brushing, rinsing or scrubbing with water, and use of acids or other chemicals.*

THE ESSENTIAL FUNCTIONS OF A GOOD FLUX

It is important that the basic functions and characteristics of fluxes used in gas-welding and allied processes be clearly understood. A flux should:
1. Assist in removing the oxide film present on the surface of the metal by attacking and dissolving it.
2. Assist in removing any oxides which may occur during welding by forming 'fusible slags' which float to the surface of the molten weld pool, and not interfere with the deposition and fusion of filler material.
3. Protect the weld pool from atmospheric oxygen and *prevent the absorption and reaction of other gases in the welding flame*, without obscuring the welder's vision, or hampering his manipulation of the molten pool.
4. Have a LOWER MELTING POINT than the metal being welded and the filler material used.

Electrode flux coverings will be fully discussed in Chapter 11.

Most methods of preventing absorption of OXYGEN, HYDROGEN, and NITROGEN by the molten weld metal, aim at *providing a barrier between the molten metal and the surrounding atmosphere.* These can be summarised as:
1. The addition of *Reducing Agents* ('Deoxidisers') to the filler material.
2. Providing an *Inert Atmosphere*, such as *Argon* or *Helium*.
3. Providing a *Shielding Gas*, such as *Carbon Dioxide*.
4. Covering with an *Inert Layer Of Slag*.
5. Covering with a *Slag Capable Of Removing Oxides*.

4.8 Chemical changes in metal affected by the oxy-acetylene flame

The NEUTRAL flame condition

A NEUTRAL flame condition is produced when equal volumes of oxygen and acetylene are supplied to the mixing chamber and the chemical reactions in the flame take place in two stages. The primary reaction produces CARBON MONOXIDE and HYDROGEN which are both REDUCING GASES. The secondary reaction for complete combustion takes extra oxygen from the surrounding atmosphere which means that there is little chance of oxygen combining with the weld metal. With complete combustion there is no free CARBON to be picked up by the weld metal, and, therefore, there is no change in the structure of the weld metal when using a neutral flame, as shown in Fig. 4.5.

The weld metal is in the normal 'cast' condition, consisting of uniform equi-axed crystal formation surrounded by long columnar crystals caused by a faster rate of cooling. No oxides are present, and no addition of carbon to the weld occurs because of complete combustion of the acetylene. The effect on the structure of the parent metal is explained in Chapter 6.

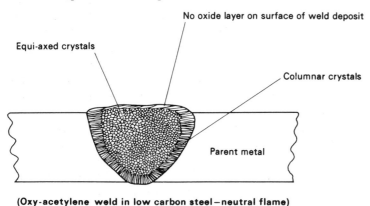

Fig. 4.5 Structure of weld metal when using a neutral flame

The CARBURISING flame condition

When the ratio of oxygen to acetylene becomes less than 1, the flame condition is said to be 'carburising'. *Excess acetylene results in incomplete combustion which produces free* CARBON. The weld metal becomes heavily carburised which results in a major change in the HARDNESS of the weld structure. Figure 4.6 illustrates the effect of the carbon from the flame:

1. The carbon content of molten weld metal is increased by DIFFUSION, as shown in Fig. 4.6(a). *This means that there is a higher carbon content in the weld metal than in the parent metal.* This gives rise to the possibility of HARDENING OF THE WELD AREA during cooling if the carbon content exceeds 0.4 per cent (see: 'Heat treatment', Chapter 6).

2. Iron and steel will dissolve small amounts of CARBON into the structure in the same way as sugar is dissolved in water.

The effect of solutions of metals and gases or non-metals (e.g., carbon) is to alter the MELTING TEMPERATURE 'UP' or 'DOWN'.

When CARBON *is dissolved in* IRON *the melting temperature is always lowered, as shown in Fig. 4.6(b).* This effect is advantageous when welding low-carbon steel; as the welding flame becomes increasingly carburising the amount of 'CARBON PICK-UP' by the surface layers of the weld pool increases, *thus reducing the melting point and speeding up the welding action.*

When low-carbon steels are being hard surfaced the excess acetylene is kept to a minimum. The surface of the steel is made to melt at a lower temperature by the excess carbon in the flame. This is known as surface sweating which allows a layer of hard surfacing to be 'semi-fused' on to it, as shown in Fig. 4.6(c).

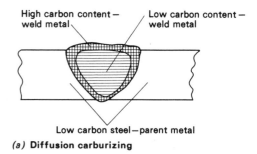

(a) **Diffusion carburizing**

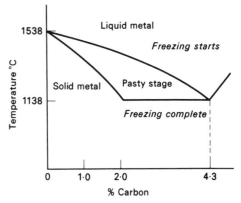

(b) **Effect on melting point**

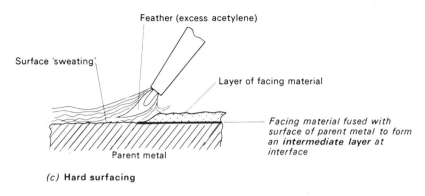

(c) Hard surfacing

Fig. 4.6 Effects of using a carburising flame

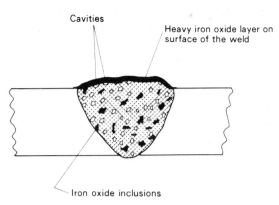

Fig. 4.7 Effects of using an oxidising flame

The OXIDISING flame condition

For complete combustion, 1 volume of acetylene requires 2½ volumes of oxygen, thus with the neutral flame the extra 1½ volumes is obtained from the surrounding atmosphere. When an oxidising flame is used the oxygen is not taken from the atmosphere since there is an excess of oxygen supplied from the cylinder. This means that the weld metal is in contact with an atmosphere which is rich in oxygen. OXYGEN IS A VERY REACTIVE GAS WHICH WILL CHEMICALLY COMBINE WITH MOST METALS TO FORM OXIDES, AND THE CHEMICAL REACTION IS SPEEDED UP BY THE APPLICATION OF HEAT. The effects of using an oxidising flame condition are illustrated in Fig. 4.7.

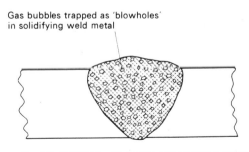

When welding low-carbon steels with an OXIDISING FLAME condition the excess OXYGEN tends to combine with the CARBON in the steel which is a 'REDUCING AGENT'. This results in CARBON MONOXIDE bubbles forming within the liquid weld metal. As the liquid weld metal commences to freeze, some of these bubbles will become trapped as 'blowholes'. The structure of the weld metal will be very porous, and it will have less carbon content near the surface than in the body of the weld structure.

Once the carbon has been oxidised, the *oxygen* will tend to chemically combine with the IRON in the steel to form IRON OXIDES. Iron oxides will be present in the weld structure as undesirable 'INCLUSIONS'. They are harmful to the weld structure because they tend to break up an otherwise normal homogenous weld; *they weaken the cohesive strength of the weld.*

THE HIGHEST FLAME TEMPERATURE IS ATTAINED WITH A HIGHLY OXIDISED FLAME CONDITION. This means that the molten weld pool will tend to boil vigorously. *When steel is heated to a high temperature, this may result in a condition which cannot be remedied by HEAT TREATMENT, and the steel is said to be 'BURNT'.* This condition is due to the fact that the grain boundaries become OXIDISED as a result of absorption of OXYGEN at high temperature, and hence the steel is weakened. The presence of brittle IRON OXIDE films at the grain boundaries renders the steel unfit for service, except as scrap for remelting. *Therefore, an oxidised weld can only be removed (by gouging or machining) and remade with a NEUTRAL FLAME condition.* OXIDES IN WELDS CAN ALSO LEAD TO CORROSION PROBLEMS. Table 4.2 indicates the effect of varying flame conditions when welding low-carbon steel.

Table 4.2 Effects of varying flame conditions when welding low-carbon steel (oxy-acetylene process)

BEFORE TEST		GAS RATIO (R) = OXYGEN/ACETYLENE					
		OXIDIZING FLAMES				NEUTRAL FLAME	CARBURIZING FLAME
ELEMENT	PER CENT	$R = 1.14$	$R = 1.33$	$R = 2$	$R = 2.37$	$R = 1$	$R = 0.82$
Carbon	0.15	0.054	0.054	0.058	0.048	0.15	1.56
Oxygen	—	0.04	0.07	0.09	—	0.02	0.01
Nitrogen	—	0.015	0.023	0.03	—	0.012	0.023

The above are the results of an experiment carried out with various flame conditions when making welds on low carbon steel.

Both the filler rod and the parent metal used had the following analysis:

Carbon	0.15%
Manganese	0.56%
Silicon	0.03%
Sulphur	0.03%
Phosphorus	0.018%

The results clearly show that with a NEUTRAL FLAME there is no decarburation of the weld when $R = 1$.

With an increase of 14 per cent OXYGEN ($R = 1.14$) the carbon content of the weld drops to 0.05 per cent — SLIGHTLY OXIDISING FLAME.

Use of a CARBURISING FLAME, where $R = 0.8$, causes the weld to absorb free CARBON from the flame, and the final carbon content in the weld is 1.56.

These results clearly indicate the effects of the three basic oxy-acetylene flame conditions on the weld metal.

4.9 Linear expansion

If a bar of steel is HEATED UNIFORMLY in a furnace it will EXPAND naturally in ALL DIRECTIONS if it is NOT RESTRAINED in any way. If allowed to COOL EVENLY and WITHOUT RESTRAINT it CONTRACTS to its original shape and size WITHOUT DISTORTION, as shown in Fig. 4.8.

If the original dimensions, the rise in temperature, and the expansion of the metal block (Fig. 4.8) are measured, it will be found that they are related to each other. Thus, for a given metal:
(i) The expansion is proportional to the rise in temperature.
(ii) The expansion is proportional to the original size of the component.

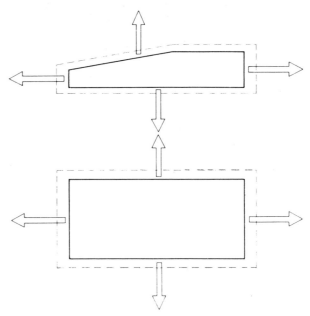

Fig. 4.8 Linear expansion

Expressed mathematically this gives the formula:
$$x = l \alpha t$$
where

x = increase in size
l = original length
α = coefficient of linear expansion
t = rise in temperature

Table 4.3 gives the values of α for some common engineering materials (α is the Greek letter 'alpha'). Figure 4.9 gives some examples of expansion calculations.

Table 4.3 Coefficients of linear expansion

MATERIAL	COEFFICIENT OF LINEAR EXPANSION
Aluminium	0.000 023/°C
Brass	0.000 02/°C
Copper	0.000 017/°C
Mild steel	0.000 011/°C
Wrought iron	0.000 011/°C

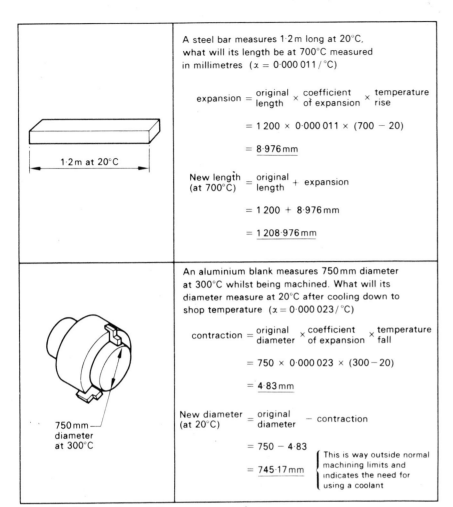

Fig. 4.9 Examples of linear expansion

4.10 Expansion at high temperatures

Care must be taken in applying the formula, used in 4.9 above, at high temperatures. The atoms in the crystals of a metal are arranged in a definite pattern. In some metals, this pattern changes at specific temperatures (critical temperatures) and the atoms become more- or less-tightly packed. This causes a corresponding shrinkage or expansion of the metal. An example is given in Fig. 4.10.

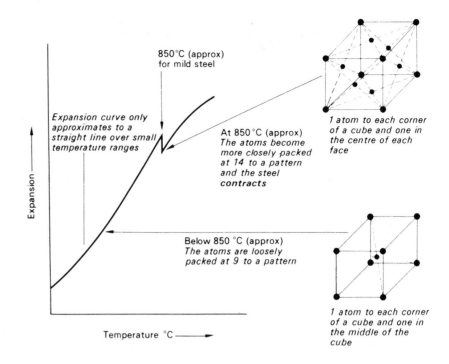

Fig. 4.10 Effect of temperature on the structure of steel

Therefore, when using the formula:
$$x = l\,\alpha\,t$$
care must be taken to observe the following rules:
1. Restrict the temperature range so that non-linearity is unimportant.
2. Avoid areas of change in the crystal structure of the metal.

For example, for ferrous metals, only use the formula for calculations *below red heat*.

4.11 Expansion and contraction in welding and cutting processes

It has just been explained how metals become larger when heated, and become smaller upon cooling. During welding and cutting operations the flame or the arc heats the metal causing it to become larger, or *expand*. When the heat source is removed, the surrounding metal and air exerts a cooling effect upon the heated area, resulting in the metal becoming smaller, or *contracting*. *If this expansion and contraction is not controlled excessive DISTORTION is likely to occur.* There is another problem associated with welding, for if expansion and contraction is *restrained* or controlled too rigidly, severe *stresses* may occur and seriously impair the strength of the weld. *Summarizing: metals EXPAND WHEN HEATED and CONTRACT ON COOLING, it is this effect which is responsible for the introduction of undesirable stresses in the weld and distortion in the work.*

4.12 Theory of distortion

Two factors are necessary to avoid distortion, which in this case are:
1. UNIFORM HEATING AND COOLING OF THE ENTIRE METAL BAR.
2. FREEDOM FROM RESTRAINT, so that EXPANSION and CONTRACTION can take place unhindered.

Unfortunately, when making a weld, unhindered expansion and contraction in all directions is not possible. Thus distortion will result unless suitable precautions are taken.

If a metal bar is restrained in any way during heating, it will not be able to expand in the direction of the restraint. The effect of heating a metal bar under restraint is shown in Fig. 4.11.

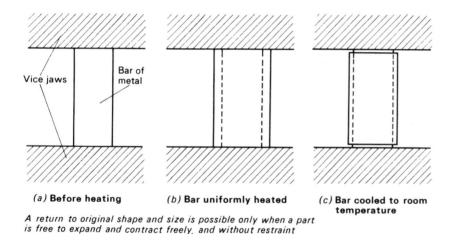

Fig. 4.11 Effect of heating a steel bar under restraint

A metal bar is placed in a vice so that the jaws close against the two ends, as shown in Fig. 4.11(*a*). The bar is in restraint and therefore cannot expand in the direction of the restraint.

When the bar is heated uniformly it can expand in all directions except in the direction where it is held by the vice jaws. The expansion is indicated by the full lines. The dotted lines represent the original shape and size (Fig. 4.11(*b*).

When it contracts upon cooling, however, there is no restraint in any direction and the bar is free to contract in all directions. It does not return to its original shape and size, as shown opposite, but becomes shorter in length, thicker in cross-section, and drops out of the vice jaws. The dotted lines represent the original shape and size (Fig. 4.11(*c*)).

Should a bar be heated over a small area, the expansion will be local and uneven, as is the case in welding. The mass of the surrounding and relatively cool metal will not expand and tends to restrain the expansion of the heated area in all directions except upon the surface. Figure 4.12 illustrates the effects of heating a spot on a steel plate.

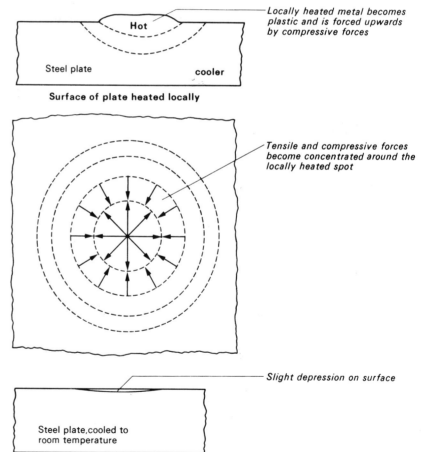

Fig. 4.12 Effects of local heating on a steel plate

When a spot is heated on a smooth steel plate, the metal which is locally heated expands and exerts a force around it. The relatively cooler metal which surrounds the heated spot acts as a restraint and causes the hot plastic metal to be forced upwards into the air.

During the cooling period CONTRACTION takes place and the local area of metal continues to shrink until room temperature is reached. While the metal is shrinking, TENSILE stresses develop around the spot. These tensile stresses in the vertical direction tend to draw the metal downwards, resulting in a slight depression on the surface of the plate where it was locally heated.

Two effects of heating on a steel bar are shown in Fig. 4.13, one of which can be used to great advantage as explained in **4.20**—Use of heat triangles.

When a steel rod, 1 metre long, is heated from 0°C to 600°C, it expands approximately 9 mm in length (LINEAR EXPANSION). When a weld bead is laid on the surface of a flat steel bar, the weld bead itself can be likened to the steel rod heated well above 600°C. The heated rod contracts and returns to its original length upon cooling (Fig. 4.13(a)). The weld bead (heated rod) also contracts in its length upon cooling, but is restrained by the steel plate. The flat steel bar bends as the weld bead cools. This is because the weld bead is not free to move in the direction of its length when contracting.

The locally heated spot of metal expands in all directions. In expanding, some of the force of expansion is expended by bending the bar. Some of the expansion force is relieved by expanding away from the surface of the metal. But some of the expansion force remains in the hot metal as a compressive force. The heated metal has expanded beyond its ELASTIC LIMIT in compression. When the metal cools, the spot shrinks and causes the bar to bend back. But since the metal has taken a permanent set in compression it contracts (shrinks) to a length shorter than the original. The final result is that the bar is no longer straight (Fig. 4.13(b)).

If a bar of metal has a bow in it, application of local heating on the convex side (i.e., the outside of the bend) will straighten it.

4.13 Manufacturing stresses

Internal stresses may have been introduced during previous manufacturing processes such as rolling or forging. Any HOT or COLD ROLLED metal or any CAST metal, unless it has been HEAT TREATED ('ANNEALED'), contains TENSILE and COMPRESSION STRESSES. This is evident, for example, when a large sheet of steel which has straight edges is sheared into strips on a guillotine. *The strips will not be straight.* If a steel beam is flame cut longitudinally through the web section, the two 'Tee' sections produced will be bowed outwards indicating that the web was in tension.

Local heating of sheet metal or plate has the following effects:
1. Sheet metal or thin plate becomes buckled.
2. Thick plate acquires stresses upon cooling.
3. A sheet or plate which is heated at the edge warps upon cooling.

4.14 The problem of distortion in welding

During all welding operations the weld metal and the heated parent metal undergo considerable contraction on being cooled to room temperature.

In oxy-acetylene or manual metal-arc welding, the heating is

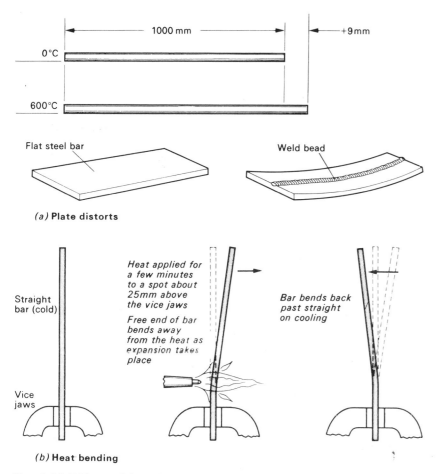

Fig. 4.13 Effect of heating on a steel bar

not uniform because the deposited weld metal is far hotter than the adjacent parent metal in the joint, which in turn is hotter than the parent metal remote from the weld area. This means that the weld will contract more than the parent metal unless prevented by restraint. The volume of the weld metal will contract about 1 per cent in all dimensions. This contraction results in considerable stress in the weld itself. In addition, because of uneven heating, stress is imposed in the parent metal, and these stresses combine in causing the structure or welded fabrication to distort and shrink, unless the component parts are prevented from moving.

Under such conditions, RESIDUAL STRESSES up to the YIELD POINT of the metal may be expected. Should the plastic flow required exceed the metal's capacity to flow, CRACKING may result.

4.15 Types of distortion

Figure 4.14 illustrates the spread of heat from the weld through the parent metal (*a*) and the effects of single-side welding of an unrestrained butt joint (*b*).

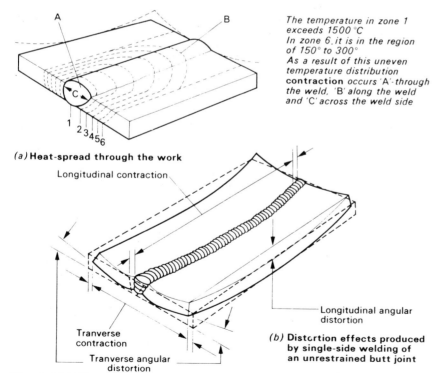

Fig. 4.14 Effects of welding on a metal plate

Longitudinal and transverse distortion *Longitudinal distortion* refers to bending which occurs along the length of the joint, and *transverse distortion* to that in the direction at right angles to the joint.

Transverse contraction If two plates are spaced slightly apart and not 'tack-welded', the gap will close up in advance of the flame or the arc as welding proceeds from one end to the other. This is referred to as *transverse contraction*, and is illustrated in Fig. 4.15.

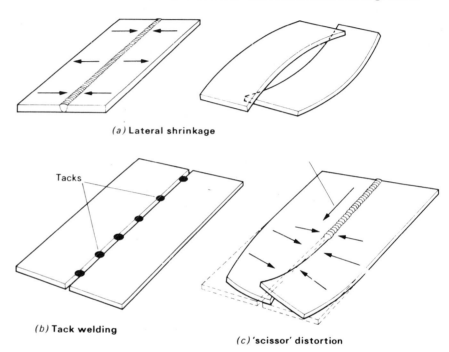

Fig. 4.15 Tranveverse contraction

When two plates are butt-joined together by welding there are COMPRESSION STRESSES at the ends and TENSILE STRESSES at the middle of the joint. *This results in lateral shrinkage*, as shown in Fig. 4.15(*a*). If the same amount of heat which was applied to the joint during welding were applied to one edge of each of the plates (not held together) they would acquire the shape shown.

Sheet metal or plates to be welded can be held in restraint by tack-welding as shown in Fig. 4.15(*b*). This prevents the pre-set gap between the butt edges from closing up in advance of the welding arc or flame due to transverse contraction of the weld metal.

If the plates are spaced slightly apart and not tack-welded, and welding commenced at one end, the effect shown in Fig. 4.15(*c*) will occur. The gap between the plates has narrowed and closed up. *If the plates are thin and long enough they eventually 'scissor' or overlap each other, so great is the contractional force.*

EFFECTIVE TACK-WELDING IS ESSENTIAL TO MAINTAIN THE ROOT GAP IN A JOINT SO THAT ADEQUATE PENETRATION CAN BE OBTAINED.

Angular distortion When two pieces of sheet metal are butt welded together with an oxy-acetylene flame, it will be noticed the completed joint has 'bird-winged' because of greater contractional forces on the top surface of the weld. This is known as ANGULAR DISTORTION, and is illustrated in Fig. 4.16.

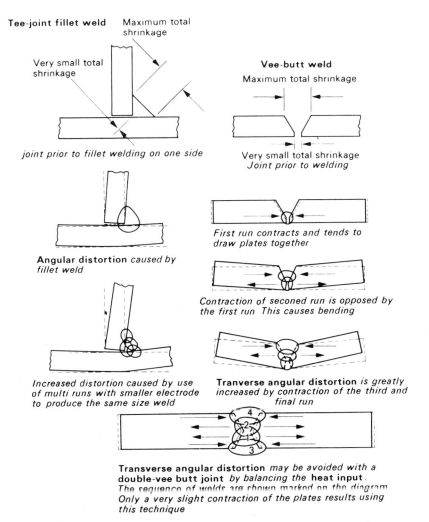

Fig. 4.16 Angular distortion

4.16 Methods of minimising distortion

In spite of the hazards of internal stressing, restrained welding is the generally-employed process; the parts may be held in fixtures (jigs), secured by clamps, or merely tack-welded.

The alternative to the restrained method of assembly is to allow almost complete freedom of the parts during welding, this having the advantage of ensuring the joints are practically free from locked-up stresses.

Every effort must be made to anticipate distortion and make arrangements to correct it before it happens. It may be possible to 'pre-set' the parts so that they distort to their correct positions.

The stresses can often be reduced by using a carefully planned welding sequence, or by pre-heating.

It is not always necessary to prevent movement in all directions, sometimes it is an advantage to allow movement in one direction.

Large assemblies generally require a combination of methods. The main unit is broken down into smaller units or sub-assemblies which are generally welded without restraint. These are then joined to form the complete assembly, and it is at this stage of welding that restraint is often necessary.

The following are the most commonly-used methods for minimising distortion:

1. PRE-SETTING If the shape permits, the parts to be joined may be set up so that when it distorts the surface is pulled level. Pre-setting is illustrated in Fig. 4.17.

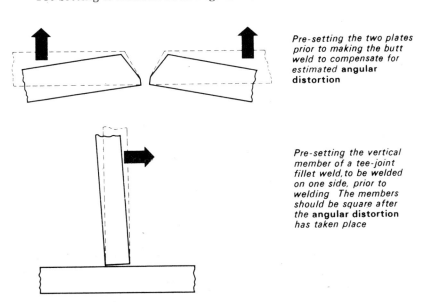

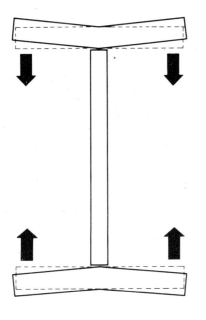

*A 'set' may be put in the horizontal members of the 'I' section, shown opposite prior to fillet welding on both sides of the vertical member. On contraction, the **angular distortion** which results will tend to pull the members square after welding*

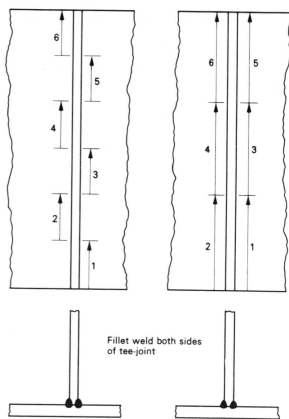

The arrows indicate the sequence of welding; on very long welds, the length of each run is the length of metal deposited by each electrode

Fig. 4.17 Pre-setting to compensate for angular distortion

Fig. 4.18 Principle of balancing contractional stresses

2. COUNTER DISTORTION The distortion in long parts can be considerable if precautions are not taken. If welding is required on both sides, distortion may be practically eliminated if the welding is done from both sides simultaneously, with two welders working opposite to each other. *By this means the distortion in one direction is neutralized by that of the other side.*

   The same effect can be obtained by 'pre-heating' with a welding torch one side of the work so that the distortion caused by welding on the other is cancelled out.

3. BALANCING THE STRESSES It is possible to plan the welding sequence in such a way that the shrinkage forces of one weld are balanced by those of another weld. The basic principle of balancing the stresses is shown in Fig. 4.18.

4. BACK-STEPPING AND SKIP-WELDING Long joints are particularly liable to distortion if welded continuously from end to end, and for this reason one or both of the following methods should be adopted:

   (a) Whenever possible, the weld should start at the middle of the joint, working towards the free ends.

   (b) Either STEP-BACK or SKIP-WELDING should be employed in order to distribute the welding heat over a large area.

The *step-back* method entails making a series of short welds in the opposite direction to the general direction of progression, as illustrated in Fig. 4.19. *By this means one section of the weld is fairly cool before the next is joined to it.*

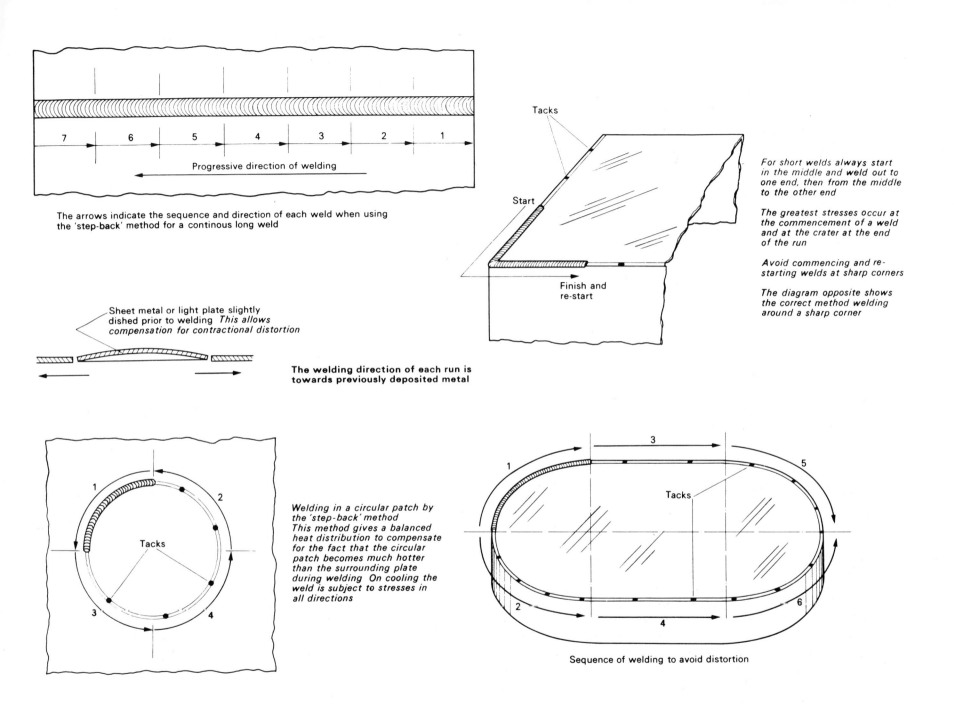

Fig. 4.19 Principles of step 'step back' welding

Skip-welding or *planned wandering* is a method employed to distribute the heat by making a series of short welds in different places along the joint. The first run is made in the centre and thereafter, portions on each side of the centre are welded in turn, run by run, as shown in Fig. 4.20. The runs are usually deposited in the same direction as the general direction of progression. This technique distributes the welding heat over a much wider area than the *step-back* method.

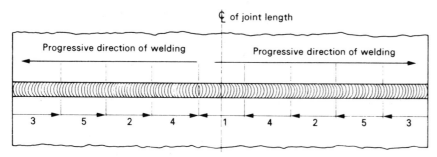

'*Down-hand*' or '*flat*' positional welding on butt-joint

The diagram shows the sequence of welding a long butt-weld with two operators welding simultaneously after the first run has been made.
The arrows indicate the direction and sequence of each weld relative to the centre line of the joint.

Fig. 4.20 Skip-welding (planned wandering)

5. LOCAL HEATING   Localised heat causes distortion, and this effect may often be put to good use to correct distortion arising from welding heat. If carefully chosen areas are heated with a welding torch, the contraction stresses that occur during cooling can sometimes be used to correct distortion due to welding. Angular distortion on a long welded joint may often be corrected by applying heat in straight lines on the reverse side, following the path of the joint. This is particularly useful in the case of distortion arising from fillet welds.

IN ALL WELDING OPERATIONS WHERE DISTORTION IS A FACTOR TO BE AVOIDED, THE DIRECTION OF WELDING SHOULD BE AWAY FROM THE POINT OF RESTRAINT AND TOWARDS THE POINT OF MAXIMUM FREEDOM. The importance of this statement is illustrated in Fig. 4.21.

Consider three cast iron plates as shown in Fig. 4.21(*a*):

1. If the first weld is made along the seam A to B, *no cracking* takes place because the plates either side of the welded joint are free to expand and contract.

2. If the next weld is made from D to C, *cracking* will take place because of the direction of welding — from a free end to a fixed (restrained) one.

3. Welding from C to D, *no cracking* takes place, as this is from a fixed end to a free (unrestrained) one.

(a) **Welding three cast iron plates**

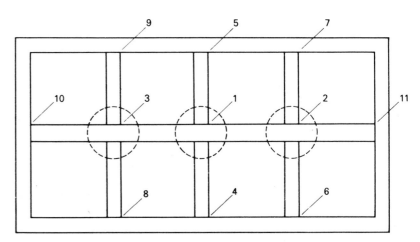

(b) **A suitable welding sequence for a steel frame structure**

Fig. 4.21 The importance of correct welding sequence

4.17 Distortion caused by flame cutting

Figure 4.22 illustrates the possible types of plate deformation (distortion) which may result through flame cutting.

When a narrow strip is flame cut from a plate, the cut strip, on cooling will bow inwards towards the cut edge, as shown in Fig. 4.22(a). This effect results from the hot edge contracting. The other edge of the strip is relatively cold and acts as a restraint to thermal expansion. Figure 4.22(b) shows the remaining plate deformed. The effect of flame cutting has released *residual stresses* (locked-up stresses) which were left in as a result of hot rolling during manufacture.

When flame cutting on a profile machine, the cutting head is adjustable up and down. This is to allow the operator to maintain the correct distance between the cutting flame and the top surface of the plate which tends to buckle during cutting.

When a hole is flame cut out of a plate, the cut edge will contract. Other areas of the plate, being longer, will bend and distort, as shown in Fig. 4.22(c).

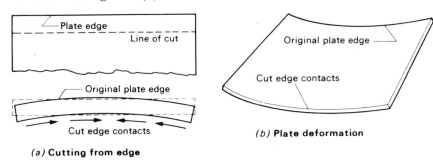

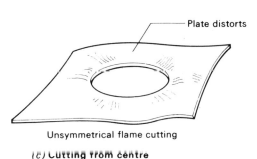

When flame cutting on a profile machine, the cutting head is adjustable up and down This to to allow the operator to maintain the correct distance be between the cutting flame and the top surface of the plate which tends to buckle during cutting

When a hole is flame cut out of a plate, the cut edge will contract Other areas of the plate, being longer will bend and distort, as shown opposite

4.22 The effect of flame cutting

4.18 The use of heat for straightening plate and sections

It has been seen that the application of heat can produce distortion. Heat can be used to advantage, for those same forces of expansion and contraction can be harnessed to remove distortion in plates or to straighten sections. Figure 4.23 illustrates the principle of shrinking a thin plate at the places that are stretched.

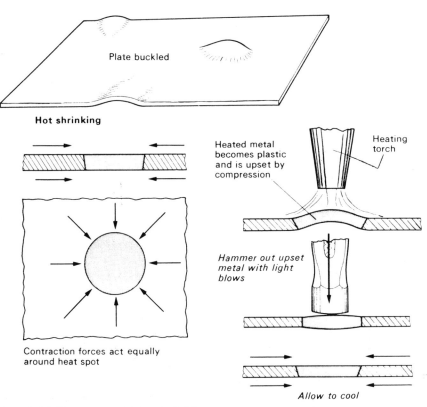

Fig. 4.23 Principle of hot shrinking

A buckled or deformed plate may be straightened by the relatively simple process of 'hot shrinking'. A number of spots in the area of stretched (buckled) metal are heated to a cherry-red (approximately 750°C) and allowed to cool in turn. *The metal which is locally heated becomes plastic, but the surrounding cold metal plate prevents thermal expansion.* The plastic area becomes upset by compressive forces. When a heated spot is allowed to cool, the metal will tend to contract, and it is during this shrinkage that contractional stresses will occur.

The process is repeated until the stretched areas of metal are compressed, and the plate is restored to a straight and flat condition. THIS PROCESS IS WIDELY USED IN LIGHT VEHICLE CRASH REPAIR AND PANEL-BEATING WORKSHOPS.

4.19 Use of heat strips

Figure 4.24 shows the use of *heat strips* for the 'hot straightening' and 'hot shrinking' of plate and wide sections.

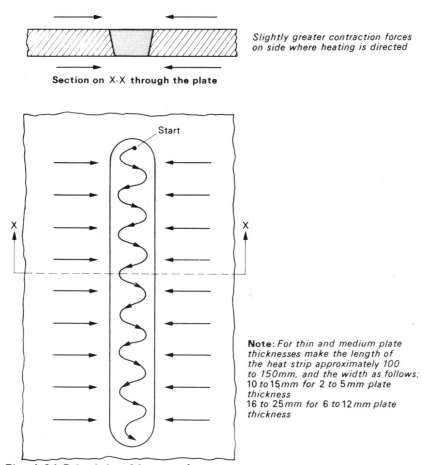

Fig. 4.24 Principle of heat strips

The shrinking forces will be approximately equal for both sides of the plate. Figure 4.24 shows the application of a heat strip which, upon cooling, causes the metal to become compressed, because the contraction forces come in at right angles to the strip.

The length and width of a particular heat strip can be determined by the thickness of the plate. As a general guide: for thicknesses from about 10 mm to 30 mm, the width of the heat strip should be between 20 and 30 mm, the length of the heat strip between 130 and 200 mm.

Heating is commenced at one end of the strip, making sure that the correct heat goes right through the plate (cherry red 750°). The whole heating operation is a continuous one, employing a zig-zag movement of the heating torch towards the opposite end. On cooling the plate will be shorter in length in the locally heated area.

4.20 Use of heat triangles

The use of *heat triangles* for straightening thin angle and flat sections, and the use of 'triangles' of heat strips for the bending and straightening of plate and wide sections are shown in Fig. 4.25.

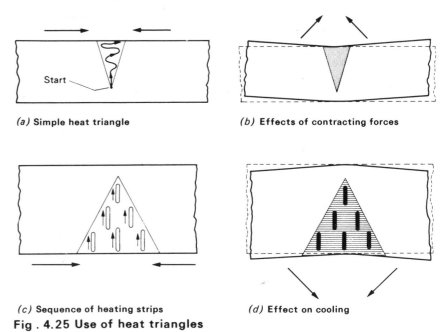

Fig. 4.25 Use of heat triangles

Simple heat triangles may be used as shown in Fig. 4.25(*a*). This entails starting with the heating torch at the apex of the triangle and working towards the base with a gradually widening zig-zag movement. When allowed to cool, the base of the heat triangle will start to contract the most, and the contracting forces tend to cause the plate to bend, as shown in Fig. 4.25(*b*).

The resultant effects of using triangles of heat strips are exactly the same as for the simple heat triangles. Simple heat triangles are used for straightening of thin plate and light sections. Triangles of heat strips are preferred when bending or straightening thick plate and heavy sections.

The order in which the heat strips are applied, in the triangle, is shown in Fig. 4.25(c). Heating with the torch is commenced a short distance in from the edge of the plate, progressively heating from the outside inwards.

4.21 Straightening simple sections

The basic principles of straightening simple sections are explained in Fig. 4.26.

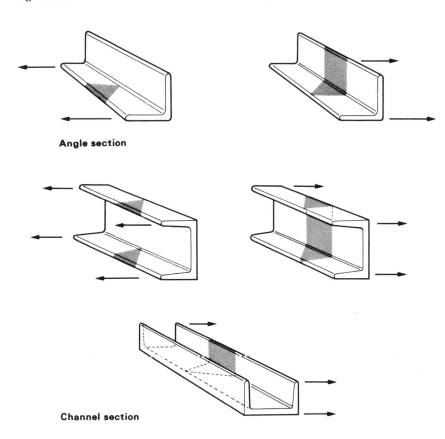

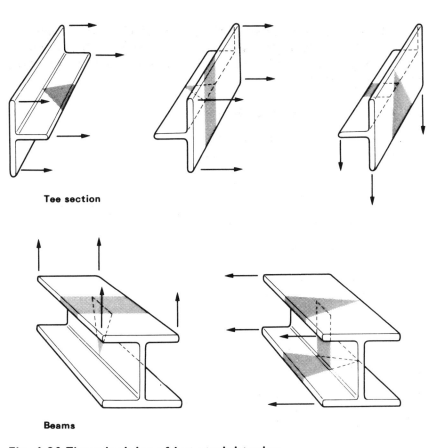

Fig. 4.26 The principles of hot straightening

The arrows indicate the direction in which the ends of the section will move on cooling, after the application of localized heating—'heat triangles'

5 Measurement

Limit gauging ?

The basic principles of measurement and marking out have been introduced in Chapter 5 of *Basic Engineering*. In this chapter these basic principles will be applied to general fabrication shop practice.

5.1 Direct eye measurement

The ENGINEER'S RULE, used for making direct measurements, depends upon VISUAL ALIGNMENT of a mark or surface on the work to be measured against the nearest division on its scale. This may appear to be a relatively simple exercise, but in practice errors can very easily occur, as shown in Fig. 5.1.

These errors can be minimised by using a rule whose thickness is as small as possible — this emphasises the importance of using a thin steel rule.

It is important when making measurements with an Engineer's rule to have the eye directly opposite and at 90° to the mark on the work, otherwise there will be an error — known as 'PARALLAX' — which is the result of any sideways positioning of the direction of sighting.

Reference to Fig. 5.1 will show that:
1. 'M' represents the mark on the work whose position is required to be measured by means of a rule laid alongside it. The graduations of measurement are on the upper face of the rule, as indicated.

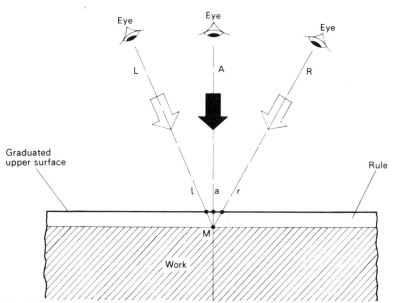

Fig. 5.1 Possible errors with direct eye measurement

2. If the eye is placed along the sighting line A–M, which is at 90° to the work surface, A TRUE READING will be obtained at 'a', for it is then directly opposite 'M'.

3. If, however, the eye is not on this sighting line, but displaced to the right, as at 'R', the division 'r' on the graduated scale will appear to be opposite 'M' and an INCORRECT READING will be obtained. Similarly, if the eye is displaced to the left, as at 'L', an incorrect reading on the opposite side, as at 'l' will result.

5.2   Possible error when using scribers

Care must be taken marking a straight line or marking around a template with a scriber. To reduce the possibility of an error, the scriber must be held against the straight edge or the periphery (in the case of a template) as illustrated in Figure 5.2(*b*).

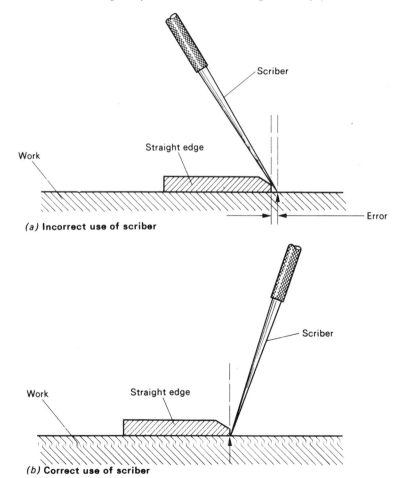

Fig. 5.2 Possible error when using a scriber

5.3   The use of Geometric constructions

Components marked out on sheet metal and plate are often large, by normal engineering standards. This makes the use of try squares, straight edges, protractors, etc., difficult and sometimes impossible. It is often easier to construct the shape required geometrically, using dividers and trammels (beam compasses). Section 2. gave a wide variety of geometrical constructions. These can equally well be set out on sheet metal as on the drawing board.

5.4   Marking-off a large template or workpiece (general)

Large sheet metal fabrications and platework jobs have to be marked out 'in the flat'. Often it is convenient to cut a sheet or plate to the correct shape and dimensions before marking the position of rivet or bolt holes if these are around the edge only, otherwise the entire job must be laid out.

If the required sheet or plate is not too large a 'DATUM LINE' may be scribed adjacent to one edge with the aid of a straight edge and scriber.

When marking out large steel plates, a CHALKLINE is employed for producing long straight lines. The use of a chalkline is illustrated in Fig. 5.3.

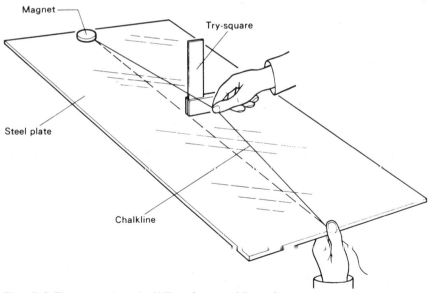

Fig. 5.3 The use of a chalkline for marking plate

1. The chalkline may be positioned at one end by means of a magnet;
2. The line is then thoroughly 'chalked', either with ordinary chalk or French chalk;
3. The line is stretched and firmly held against the work in line with 'witness marks' previously made with a centre punch;
4. The line is then 'flicked' by lifting it vertically with finger and thumb and releasing it;
5. On release the line strikes the surface of the steel plate producing a truly straight chalk line in the required position.

NOTE: *When marking out very long lines, it is good practice to place a try-square against the line near its centre. The chalkline is then pulled straight up against the edge of the square and released.*

BEAM TRAMMELS and TAPE MEASURES are used for striking lines at 90° to each other, and for measuring distances accurately. It is common practice for the craftsman to use a pair of trammel heads or 'trams' and any convenient beam such as a length of wooden batten. Figure 5.4 shows the arrangement of the trammel and the simple means of fine adjustment for accurate marking out.

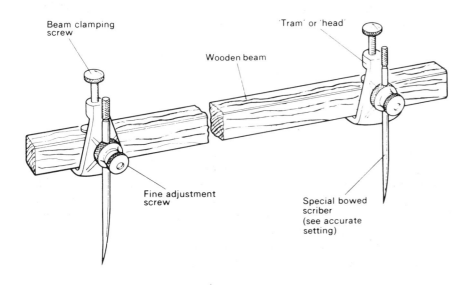

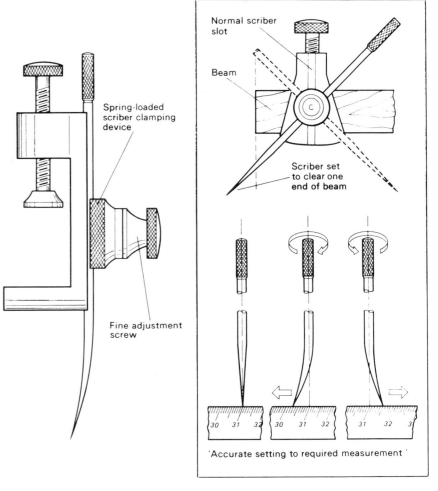

Fig. 5.4 Beam trammels

Lines making angles of 90°, i.e. lines square with each other, may be set out with the aid of beam trammels or a steel tape, as shown in Fig. 5.5(*a*).

A KNOWLEDGE OF GEOMETRIC CONSTRUCTIONS AND ARITHMETIC IS ESSENTIAL FOR MARKING OUT.

NOTE: Where possible existing straight edges and square corners on the plate to be marked should be used.

The normal accuracy obtainable when marking out with DIVIDERS, ODD-LEGS and TRAMMELS is within 0·15mm of the TRUE DIMENSION.

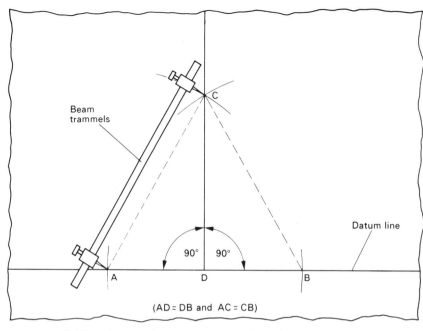

(a) Use of trammels to construct a right angle

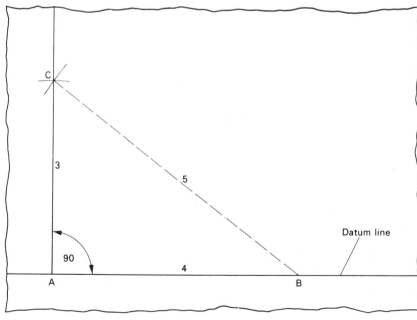

(b) Use of trammels and steel tape to construct a right angle

Fig. 5.5 Applications of beam trammels and steel tape (marking out)

Figure 5.5(*b*) shows how the properties of a right-angled triangle can be used to set out perpendicular lines (*Basic Engineering*, section 3.16).

If the ratio of the sides of a triangle is 3:4:5 then it is a RIGHT ANGLE TRIANGLE.

Scribe a datum line on the plate and mark off a length A B representing 4 units of measurement with a steel tape or beam trammels. Centre-punch mark points A and B. With centre A and the trammels set at a distance equal to 3 units, scribe an arc. With centre B and the trammels set at a distance equal to 5 units scribe another arc. The two arcs intersect at a point C which is marked with a centre punch. With a chalkline mark a line passing through the witness marks A and C.

5.5 Basic methods used for marking-off large-size plates

For economic reasons advantage should always be taken of as many good and straight edges as possible before commencing marking off large-size plates for cutting. *Unnecessary shearing or flame-cutting can be avoided if the edges of the steel plates are examined before marking-off.*

Any one or a combination of THREE BASIC METHODS may be used for obtaining parallel and squared lines to enable a plate to be cut to the required dimensions:

Method 1 Use of square and steel tape

A FLAT SQUARE is used for marking out on large flat surfaces. The flat square differs from an Engineer's try-square in that it is laid on the flat surface of the sheet metal or plate to be marked out. It is larger than the try-square and is made in one piece, consisting of a long arm termed the 'body' and a short arm termed the 'tongue'. The body and tongue are of uniform thickness and form a 90° angle, as illustrated in Fig. 5.6. In many fabrication workshops use is made of a simple made-up square of either wood or light guage steel.

A suitable steel tape is used in conjunction with the flat square.

Select one straight edge on the plate, and with the aid of a flat square and a stick of French chalk, mark a line at 90° to this datum. Extend this line using a chalkline. From these two datums the required dimensions are marked off with French chalk. A steel tape is

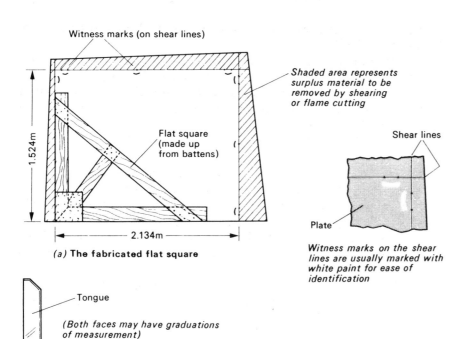

(a) The fabricated flat square

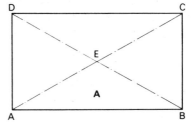

Witness marks on the shear lines are usually marked with white paint for ease of identification

(b) The steel flat square (one piece)

Fig. 5.6 The flat square

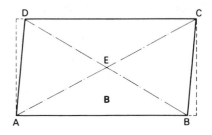

Checking long rectangular outlines

TRUE SQUARENESS of marking may be easily *checked by measuring diagonal corner distances*

A — In an OBLONG or RECTANGULAR figure, the diagonals bisect each other such that
A-E = C-E = B-E = D-E.
The diagonals A-C and B-D are equal, therefore the angles in a rectangle, or square are each 90°

B — In a PARALLELOGRAM the diagonals bisect each other such that:
A-E = E-C and B-E = E-D
The diagonals A-C and B-D are NOT EQUAL therefore none of the angles are 90°

This principle is used to check large rectangular outlines for squareness when marking out

Fig. 5.7 Checking large rectangular outlines

Figure 5.8 illustrates the use of a steel tape for marking-off a plate to measure 1·65m by 1·23m.

used for measuring all dimensions. The shear lines are completed with the aid of a chalkline, and witness marks are made on them with a centre punch.

Before commencing to mark out a large plate:

1. Always check for squareness
2. Where possible, select one straight edge and use as a base datum.

Figure 5.6 showed how a square and steel tape are used for marking-off a steel plate for cutting. Figure 5.7 shows how squareness may be checked.

Method 2   Use of steel tape

A plate of any size may be marked-off with square corners by measuring with a steel tape, units of length in the proportion of 3, 4 and 5 to produce the datum lines at right angles to each other.

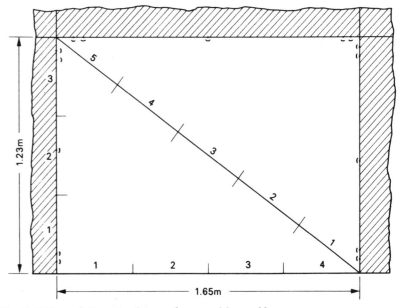

Fig. 5.8 Use of the steel tape for marking off

Select one straight edge on the plate for straightness and use as a base line, otherwise mark a datum line with the aid of a chalkline.

In this example the plate is required to be marked out 1·23m by 1·65m, using a steel tape only.

The method employed has been explained in Fig. 5.6.

In this case a most suitable measurement to be used for the 3:4:5 ratio of the sides of a 90° triangle will be 410 mm, giving the following dimensions to be used for the steel tape:

1 230 mm (3 x 410) : 1 640 mm (4 x 410) : 2 050 mm (5 x 410)

Once a line has been constructed at 90° to the base datum, the dimensions of the sides are measured with the steel tape, the outlines made with a chalkline and witness marked.

The outline is checked for true squareness as explained in Fig. 5.7.

Arcs may be swung with a steel tape by holding the French chalk in the hook at the zero end of the tape.

Method 3   Use of steel tape and trammels

Figure 5.9 illustrates the method of marking-off a steel plate which is required to be 1·58 m x 1·58 m with square corners, using a steel tape and trammels.

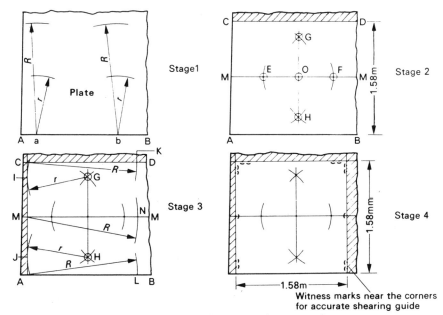

Fig. 5.9 Marking-off with a steel tape and trammels

Stage 1   A suitable straight edge is selected and used as a base line, as shown at A—B.

The trammels are set to the full width of the plate (R = 1·58 m) and with any two points 'a' and 'b' (on the base line A—B) as centres, arcs are struck. With the same centres and the trammels set to approximately half this dimension (radius r) two other arcs are shown struck as in Fig. 5.9.

THE STEEL TAPE IS USED FOR ALL MEASUREMENTS

Stage 2   Parallel lines, C—D and M—M are marked with the chalk-line held tangential to each pair of equal arcs, in turn.

A light centre punch mark is made at O which is approximately half the width M—M.

From the point O on M—M construct a perpendicular G—H, and mark with the chalkline. Lightly centre-punch mark the points G and H.

The points, G, H and O are used to check whether the edges of the plate are straight and parallel to this line of points, to enable use to be made of them.

Stage 3   If both edges prove unsuitable for use, the trammels are set to radius r, and with centres G and H, arcs are struck to provide a suitable shearing margin at points I and J.

The end shear line is made with the chalk line held at a tangent to these arcs.

The plate edge measurements for the length of the plate are made from this line (through I and J). The trammels are set to R = 1·58 m, and a chalkline is made at a tangent to the arcs at points K, N and L, as shown in Fig. 5.9.

Stage 4   The shear lines are witness marked with a centre punch, and white paint marks are made near them.

The finished outline is checked for SQUARENESS by measuring the diagonal lengths.

5.6   Method of marking out bolt holes for flanges

Many fabrications such as boilers, chemical plant and pressure vessels incorporate the use of flanged inlets and outlets. Manholes and inspection covers are also bolted to flanges. Pipes of various diameters are connected by means of flanges.

The flanges are welded to the fabrications and on the ends of pipes, and the connections are made by bolting. Figure 5.10 illustrates the method of laying out the bolt holes on flanges. In practice the standard size of the required flange, the PITCH CIRCLE DIAMETER and the number and size of the bolts is specified in B.S. 1560.

Table 5.1 Data for marking out pipe flanges

FLANGE SIZE DESIGNATION (Nominal bore of pipe) (mm)	APPROXIMATE O.D. OF STEEL PIPE (mm)	DIAMETER OF FLANGE (mm)	BOLT CIRCLE DIAMETER (mm)	NUMBER OF BOLTS	DIAMETER OF BOLT (mm)
152	165	279	235	4	16
254	273	406	356	8	19
305	324	457	406	8	22
432	457	610	523	12	25
584	610	787	724	12	25
686	711	870	857	16	25
889	914	1 073	1 016	24	25
1 067	1 092	1 251	1 194	28	25
1 524	1 549	1 784	1 702	32	25
1 829	1 861	2 108	2 019	36	35

The above data is a selection of recommended flange sizes for steel pipes from BS 1560 Class 150 to withstand working pressures up to $1\,034\,kN/m^2$ ($10 \cdot 5\,kgf/cm^2$)
Bolt-holes in flanges are marked out in accordance with Standard Tables, and the holes drilled to suit the correct diameter of bolt.

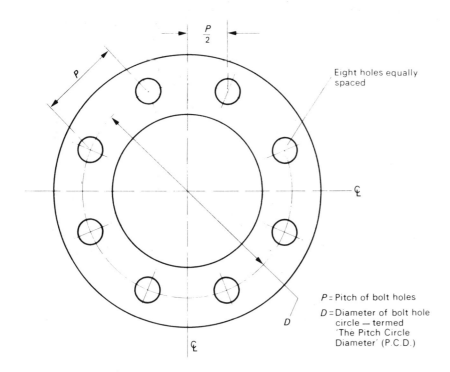

Fig. 5.10 Marking-out bolt holes for flanges

Table 5.2 Constants for bolt hole location (flanges)

NUMBER OF BOLT HOLES	CONSTANT (to be multiplied by Bolt Circle Diameter or P.C.D.)
4	0·407 1
8	0·382 7
12	0·258 8
16	0·195 1
20	0·156 4
24	0·130 5
28	0·112 0
32	0·098 0
36	0·087 2

The centre of the flange is plugged with a suitable piece of wood or piece of flat bar, which is 'tack-welded' in position, to enable the centre of the flange to be located. On flanges up to about 460 mm in diameter horizontal and vertical centre lines (℄) may be marked with the aid of a height gauge in conjunction with an angle plate on a marking out table, and a pair of trammels. On very large diameter flanges, use is made of a large centre-square to locate the centre for the bolt hole circle.

For any specific size of flange for a particular class of work, details such as the BOLT CIRCLE DIAMETER, NUMBER OF BOLT HOLES and DIAMETER OF BOLTS are obtained from the appropriate standard table (see Tables 5.1 and 5.2).

Use of tables: Example
To mark out a 305 mm Class 150 Standard Flange.
From Table 5.1, the **P.C.D.** = 406 mm and the **number of bolts** = 8.
(The diameter of the flange = 457 mm)
The pitch (chord length of bolt holes centres) = the **'P.C.D.'** multiplied by the **'constant'** from Table 5.2.
The 'constant' for 8 holes = 0·382 7
Therefore, 'pitch' = 406 × 0·382 7
  P = 155·76 mm

For any number of holes not shown in the table, multiply the

SINE of HALF THE SUBTENDED ANGLE between a pair of holes by the 'P.C.D.'

The subtended angle = $\dfrac{360}{\text{Number of holes}}$

Example:
Number of holes = 15
Subtended Angle = $\dfrac{360}{15} = 24°$

The constant = Sine 12° = 0·2079 (for 15 holes)

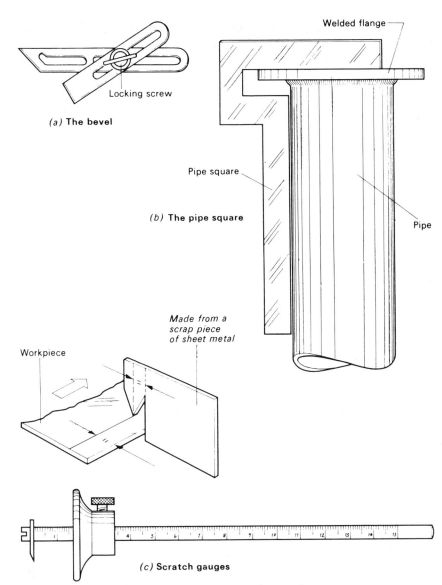

Having located the centre and marked the horizontal and vertical centre lines, the appropriate bolt circle is marked by means of trammels.

The pitch is constant and is usually obtained from tables of 'bolt hole locations' which *provide a constant which has to be multiplied by the diameter of the bolt hole circle to obtain the required pitch.*

To obtain the position of the first bolt hole, divide the pitch by 2, set the dividers to this dimension and mark off from the intersection of the vertical centre line and the bolt circle, and centre punch.

The remainder of the bolt hole centres may now be located with the dividers set at the correct pitch (CHORD LENGTH) and centre-punched in readiness for drilling to required size.

The procedure for drilling bolt holes in very large flanges is outlined in Chapter 9.

5.7 The bevel

A useful marking-off tool is the BEVEL which is frequently used in fabrication work for the marking-off of angles or mitres on steel sections. This simple tool is illustrated in Fig. 5.11(a).

It consists of a blade and base which are set in position by a locking screw.

The bevel is set to the required angle and is used for checking, transferring and marking-out angles.

Fig. 5.11 Useful marking-out and measuring tools

5.8 The pipe square

The PIPE SQUARE is used to check the correct alignment of pipe flanges on assembly, and is illustrated in Fig. 5.11(b).

Care must be taken when using the pipe square for assembly work to ensure that it does not come into contact with the arc-welding electrode or the earthing cable clamp.

5.9 The scratch gauge

This useful tool is similar in principle to the carpenter's tool used for marking parallel lines to a given plane surface.

Such a gauge is usually made of steel with a hardened high carbon steel head. The head has a split bushing through it against which a set screw acts to hold it firm. The steel beam is usually graduated in millimetres. The small scribing point consists of a thin square piece of suitably hardened and tempered steel, which is held firmly against one end of the beam. The scratch gauge is illustrated in Fig. 5.11(c), together with a simple but effective scratch gauge which can readily be produced from a small piece of scrap sheet metal.

5.10 Centre punches

These tools are made from high carbon steel, usually of hexagonal or octagonal section for easy grip, a cylindrical section with a knurled grip is also common.

The centre punch is used for making circular indentations in the surface of metal for location, marking, drilling and general machining purposes. If it is intended to be used on soft metals, the tapered point is generally of smaller dimensions than when the punch is designed for use on tough steels. According to the purpose for which it is to be used, the diameter at the base of the conical point ranges from about 3 to 6.35 mm.

Centre punches for marking centres of holes to be drilled are made from 9.5 mm octagonal or cylindrical high carbon steel from about 127 mm long, and the tapered point is ground to an included angle of 90°. *The indentation the punch makes in the surface of the metal will not only locate the drill point, but will prevent the drill from wandering.*

Centre punches used for marking positions of lines (see *Basic Engineering*, ch. 5, 'Preservation of the scribed line') and centres of circles and arcs to be drawn by dividers or trammels, are the smaller type of punch, made from about 6.35 mm diameter high carbon steel with a sharper point, usually about 60°.

'Dot' or 'nipple' punches are especially useful for repetition work when a TEMPLATE is used. The template has accurately positioned holes, all of which are of the same small diameter. The 'dot' or 'nipple' punch is designed with a parallel point, the end of which has a 'pip' ground on it — such punches are accurately machined, concentric to the parallel point, on a lathe.

The parallel point is made in a number of precise diameters to locate accurately in the template holes which are pre-determined by the diameter of the punch to be used. Thus by locating the punch in the holes in the template, the hole centres on each workpiece can be easily marked out precisely and identically.

In a similar manner, the indentations made by a 'dot' or 'nipple' punch can be used to great advantage to locate any shape of punch used for piercing sheet metal, provided it has a centre pip.

Figure 5.12 gives details of centre punches commonly used in the fabrication industry.

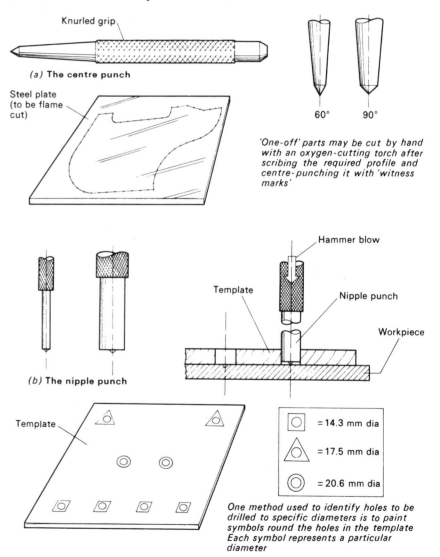

Fig. 5.12 Centre punches (applications)

5.11 The use of spirit-levels and plumb-lines

The theory of the use of spirit-levels and plumb-lines has been clearly explained in *Basic Engineering*, Chapter 5. A typical application of the use of the plumb line in steel fabrication work is shown in Fig. 5.13.

The bottom vessel is securely bolted into the correct position and the top vessel is aligned to it, as shown, with the aid of a plumb-line before being bolted into the required position to receive the connection pipe.

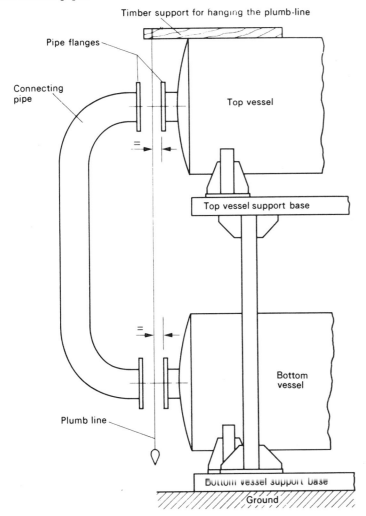

Fig. 5.13 The use of a plumb-line (fabrication)

5.12 The use of the tensioned wire

On large fabricated components, a tensioned wire may be used to check straightness and for checking alignment.

Piano wire or stainless steel wire of about 0·55 mm in diameter is used for this purpose, and when not in use should be kept on a suitable reel.

When in use for measuring or checking, both ends of the wire are hung over supports which are rounded, such as round bar section or pulleys, and weighted sufficiently to keep the wire in TENSION. Alternatively the wire may be secured by means of adjustable clamping devices. Figure 5.14 illustrates the use of a tensioned wire.

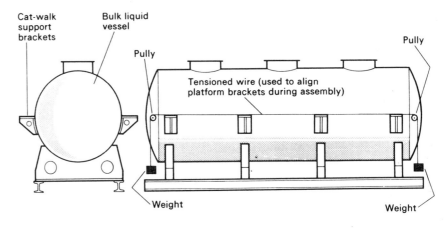

Fig. 5.14 Use of the tensioned wire

5.13 The need for templates

There are several reasons for the use of templates or patterns in the sheet metal and plate fabrication industries. For example:
1. To avoid repetitive measuring and marking-off of the same dimensions, where a number of identical parts or articles are required. *Marking-off large numbers of exactly the same type from a template or pattern is a much quicker method and a great deal more accurate than measuring and marking each part individually.*
2. To avoid unnecessary wastage of material. Very often, when marking a full-size layout directly on to a sheet or plate, from information given on a drawing, *it is almost impossible to anticipate exactly where to begin in order that the complete layout can be economically accommodated. Consequently, large-size layouts tackled in this manner generally result in an extravagant waste of material.*

3. To act as a guide for cutting processes. Profile templates have been discussed in Chapter 7, with regard to 'Oxy-fuel gas cutting'. Guide templates are also invaluable for repetition cutting of contoured shapes and apertures in sheet metal on static 'nibbling machines'.
4. As a simple means of checking bend angles and contours during forming and rolling operations.
5. As a precise method of marking-off hole positions on sheet metal fabrications, platework and structural sections such as angles, channels, columns and beams, gusset plates and angle cleats.

Figure 5.15 illustrates the economical arrangement of patterns on a standard dimensioned sheet of metal.

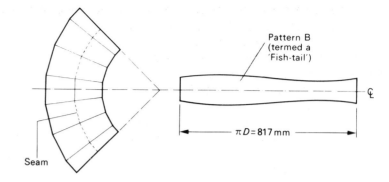

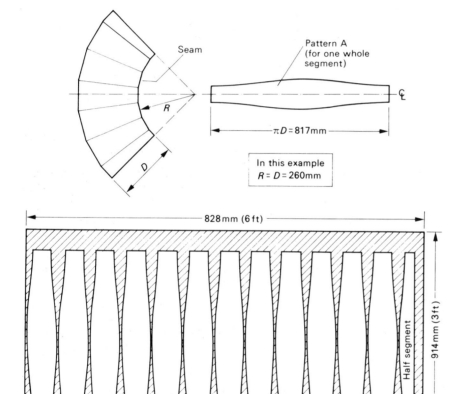

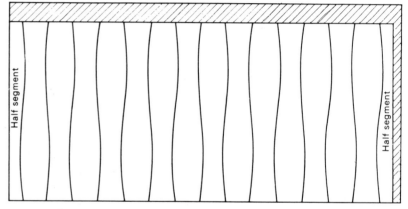

Using Pattern B, the maximum of segments which can be cut from the same standard sheet is 15, as shown in the arrangement above.
When lobster back bends are to be produced in quantity, the adoption of a 'Fish-Tail pattern' will result in a greater economy in the use of material than the pattern A'
A study of the two methods of marking the segments will clearly indicate that the use of a Fish-tail pattern will reduce the shearing operations by approximately 50%

(b) Use of 'fish-tail' segments

Fig. 5.15 The economic arrangement of patterns

The example shown in Fig. 5.15 is a 'LOBSTER BACK BEND' consisting of 5 segments and 2 half-segments.

It is to be made from 1·6 mm-thick mild steel, and all joints are to be butt welded. The seam for the cylindrical segments is to be placed in the throat, as indicated.

It is general practice to make the throat radius R equal to the diameter D of the bend.

NOTE: *In both these examples the metal thickness has been purposely ignored.*

In practice, an allowance must be made for the metal thickness for calculating the correct girth length of the pattern:
(a) For an INSIDE DIAMETER, use $D - T$,
(b) For an OUTSIDE DIAMETER, use $D + T$,
where T = thickness of metal.
D = diameter of the bend.

Figure 5.15(a) shows that when a series of 'whole segments' are cut from a standard sheet there is quite a lot of waste.

Figure 5.15(b) shows the same lobster back bend, but in this case the seam for the cylindrical segments is to be placed along the central axis of the bend, as indicated.

5.14 Template making (large fabrication shops)

Large fabrication workshops are often provided with an area reserved for template making, known as the 'template shop' or 'loft'.

Such shops are usually situated above the normal shop floor level, but those situated at ground level are fitted with an overhead runway and lifting tackle to handle steel plates for the making of steel templates.

A template shop should be well glazed to ensure good lighting during daylight hours, and provided with adequate artificial lighting for use in the darker hours.

Specialist template makers are employed in the template shop to produce accurate templates for use in the various fabrication shops by the croppers, smiths, benders, platers and welders when cutting, marking for drilling, punching, forming and welding the steel parts. *Skilled template makers must possess a sound knowledge of the principles of plane geometry and be able to apply workshop calculations.* They must be able to interpret detailed drawings and also have the ability to use carpenter's tools.

Much of the machinery used in a template shop is of the type normally used for woodworking, such as a circular saw, fret-saw, planing machine and wood-worker's drilling machine. It also includes a cardboard shearing machine to cut the special template paper.

5.15 The setting-out floor

It is essential that the floor used for full-size laying-out consists of floor boarding placed diagonally across the floor joists. *If the floor boards were laid in the conventional manner, lengthways, or square across the shop, the joints between the boards (which tend to shrink) would offer a serious handicap, as most lines are marked on the floor in these directions.* The joints between the boards may easily be mistaken for lines, or some portion of a line may coincide with an open joint. *Such problems are eliminated when the boards are laid diagonally.*

The floor is given a coat of 'lamp black' and 'size' to ensure that the lines (made with a 'chalkline') can be clearly seen. Working from scale drawings the template maker marks out full-size sets of steelwork on this black surface. The laying out of the drawing full size on the template shop floor is called 'lofting'.

5.16 Basic tools used by the template maker

The basic tools used by the template maker are listed in Table 5.3.

Table 5.3 'Tools used by template makers'

TOOLS OR ITEMS	REMARKS
Carpenter's saws, planes, hand-brace and bits.	For making wooden templates.
Joiner's marking gauge.	Used for scribing scrieve lines on batten templates for steelwork.
Steel tape to measure about 15 m.	Steel tapes are available for measuring up to 50 m. Used for marking out large plates and long batten templates.
Various size compasses or dividers.	These are used for marking small-diameter circles, and for dividing lengths on templates for pitch of hole centres.
A pair of trammel heads.	These may be used with any length of beam for marking out large radii.
A protractor.	For measuring and marking angles.
Back gauges.	The adjustable type are more suitable, used for marking the positions of tail holes at standard 'back mark' dimensions from the heel of the section.
Engineer's squares and Flat squares.	For checking the squareness of two planes or marking a line square to another.
A steel straight edge.	For marking straight lines up to 2·5 m in length.
Hammers, centre and nipple punches.	For marking hole centres, and making witness marks.

TOOLS OR ITEMS	REMARKS
Chalk line and soft chalk. French chalk.	For marking long straight lines. French chalk is generally supplied in sticks about 10 mm square and 100 mm long.
Coloured and indelible pencils. Crayons.	Used for marking instructions and information on templates.

5.17 Materials used for templates

Table 5.4 gives details of the materials used for templates together with some of their applications.

Table 5.4 'Materials for templates'

MATERIAL	APPLICATIONS
Template paper	Outlines for small bent shapes, such as brackets, small pipe bends and bevelled cleats, may be set out on template paper. Used for developing patterns for sheet metal work.
Hardboard	Templates for gusset plates to be produced in small quantities.
Timber	Used in considerable quantities for steelwork templates. Easy to drill and cut to shape. Whitewood timber strips (battens) up to 153 mm wide and 12·7 mm thickness are used to represent steel members. Plywood used for making templates for use with oxy-fuel gas profiling machines.
Sheet metal	Used for making patterns for repetition sheet metal components. Templates for checking purposes. Steel, 3·2 mm thick is used for profiling templates on oxy-fuel gas profiling machines fitted with a magnetic spindle head.
Steel plate	Light steel plate fitted with drilling bushes is used as templates for batch drilling of large gusset plates.

In the making of templates considerable use is made of timber: it is easy to drill and cut to shape; relatively light in weight; and fabrication instructions can be pencilled on it.

Suitable wooden battens of various convenient widths and usually 10 mm or 12 mm thickness are cut to represent the steel members outlined on the template floor. These battens are then laid on the appropriate lines on the floor together with the paper or hardboard patterns representing gusset plates and cleat angle connections. All are temporarily nailed to the floor in their exact positions to represent the particular steel structure.

The centres of holes required for making the connections to be bolted or riveted are marked on the assembled templates, which are then removed from the floor to be drilled and have the necessary fabrication instructions marked on them. After being drilled, and the information for the guidance of the fabricators having been marked on them, the whole assembly is replaced in the correct position on the template floor and checked for accuracy. They are then carefully stored until required on the workshop floor.

For economic reasons template battens may be used again after they have served their purpose in the fabrication shop. Long templates in which details are concentrated at several points may be cut and the drilled lengths can be re-used, after planing off the written instructions, for shorter templates. Very often a wooden template may be used again after the holes have been plugged and it has been planed.

5.18 Information given on templates

Information for the guidance of the various craftsmen employed in the fabrication shop and those employed on assembly work on site is marked on the templates and on the finished parts.

On wooden or hardboard templates the necessary information is best marked with an indelible pencil. Coloured pencils are also used for marking information. On sheet metal templates, for example, which are to be used for the marking of various diameter holes, it is common practice to mark rings, triangles or squares around the holes required to be of the same diameter with a distinguishing colour. On steel templates, whitewash or white paint is often used for marking the information.

Typical information 'written up' on templates may be as follows:
1. Job or contract number,
2. Size and thickness of the plate,
3. Steel section and length,
4. Quantity required,
5. Bending or folding instructions,
6. 'This side up', 'left hand' or 'right hand',
7. Drilling requirements,

8. Cutting instructions,
9. Assembly reference mark.

5.19 The use of templates

Many detailed parts of a structure are so simple that they do not require to be set out on the template floor, instead they are marked out direct from drawings at the bench in the fabrication shop. However, templates are made for these simple details where a number of identical parts are required.

A few selected examples of the use of templates will now be considered in more detail.

5.20 Templates as a means of checking

These are usually made of sheet metal or wood, although for some applications template-making paper may be used. Figure 5.16 illustrates examples of the use of templates as a means of checking.

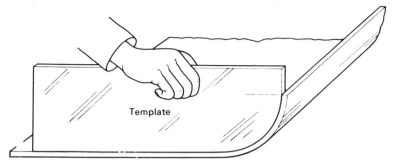

(b) Checking the contour of a radiused corner

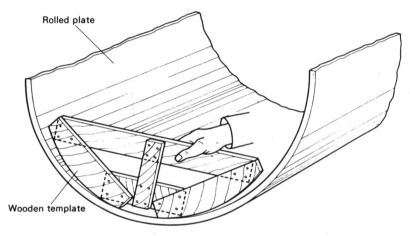

(c) Checking the contour of a rolled plate

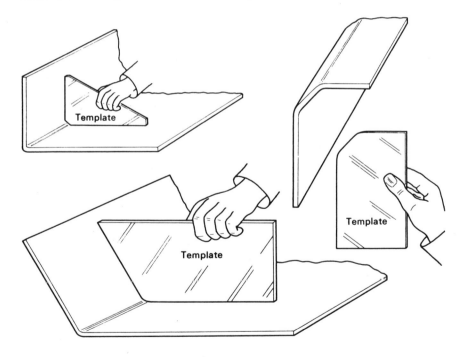

(a) Checking angles with a template
    It is often necessary to make simple bending templates especially if the sheet or plate material *requires bending in several places to definite angles*
    These templates are generally made from sheet metal

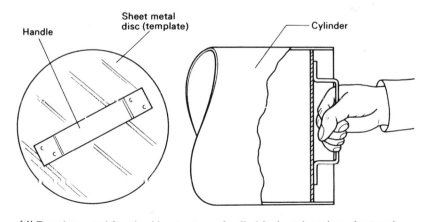

(d) Template used for checking contour of cylindrical work such as ductwork

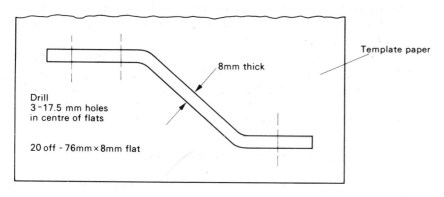

(e) Use of template paper for checking bent flats

Fig. 5.16 The use of templates as a means of checking

5.21 Templates for setting out sheet metal fabrications

Tinplate or light-gauge sheet steel and template-making paper are the materials most commonly used when making templates or PATTERNS for sheet-metal fabrications. For economy reasons, many patterns are developed half-full-size or to scale from the drawing and then the information contained on them is transferred to full-size dimensions when the craftsman marks it off on the job 'in the flat'. Very often, on precision sheet-metal details, the job is marked off from a scale drawing which provides co-ordinates with precise dimensions marked on them. With many sheet-metal developments it is only necessary to use part patterns which are lined up with DATUM LINES.

Figure 5.17 shows examples of the use of 'Patterns' for marking out sheet metal prior to cutting and forming operations.

Figure 5.17(a) shows a smoke-cowl which is to be made out of 1·2 mm-thick mild steel. The edges of the open ends are to be wired with 3·2 mm-diameter wire. The connection flanges are 12 mm for spot welding, and the side seams are to be 6 mm grooved. The completed assembly is to be hot dipped galvanised.

Basically, this component is a combination of 'Tee'-pieces between cylinders of equal diameter, and only requires a part template which may be made from template paper or light-gauge sheet metal. This is then used to mark-out the contours of the intersection joint lines for the parts 'A', 'B' and 'C' whose developed sizes are marked-out in the flat with the appropriate DATUM LINES.

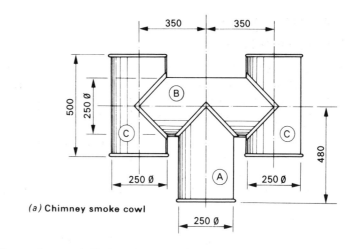

(a) Chimney smoke cowl

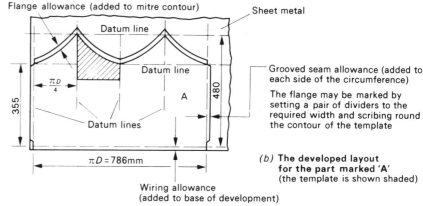

(b) The developed layout for the part marked 'A' (the template is shown shaded)

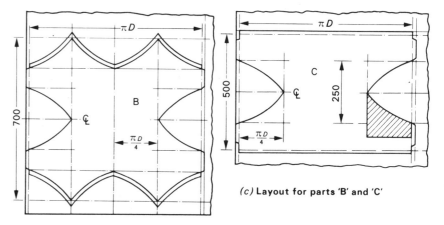

(c) Layout for parts 'B' and 'C'

Fig. 5.17 An example of the use of templates

Procedure for using the template

See figures 5.17(b) and 5.17(c)

1. On the developed part, datum lines are marked to represent quarter circumference lines, and the depth of the mitre. IN THIS EXAMPLE THE CONTOUR OF THE MITRE LINE IS IDENTICAL FOR EACH QUARTER BECAUSE THE INTERSECTION IS BETWEEN EQUAL DIAMETERS.
2. Place the mitre line template in alignment with the datum lines, as illustrated, and mark the contour for the mitre for each quarter in turn, reversing the template where necessary.

An alternative method of marking the flange allowance is to move the template up the width of the flange and mark the contour, using a scriber.

Figure 5.18 shows a square to round transformer. Figure 5.18(a) shows an ISOMETRIC VIEW of the sheet-metal transforming piece to be used for connecting a circular duct to a square duct of equal cross-sectional area.

In this example, the diameter of the circular duct is 860 mm, the length of one side of the square duct is 762 mm, and the distance between the two ducts is 458 mm. The transformer is to be made from galvanised steel 1·2 mm thick.

Figure 5.18(b) represents a scale development pattern on which are marked the full-size ordinate dimensions. Such drawings are supplied by the drawing office for use by the craftsman for marking-out purposes. Any necessary allowances for seams and joints must be added to the layout (two off required).

5.22 Templates for hopper plates

Large steel hoppers are usually of riveted or welded construction made up of four tapered steel plates. The templates for these hopper plates may be made from wooden battens, sheet metal or template paper. The template is laid on the plate and the outline marked with French chalk and 'witness marks' are centre-punched at suitable positions.

Rivet holes (where applicable) are marked through the template with a nipple punch. When paper templates are used, the holes are not provided in the template, as is the case with wooden and metal templates. The centres of the hole positions are marked on a paper template and may be transferred on to the plate by centre-punching through the template. Fig. 5.19 illustrates typical templates for marking off hopper plates.

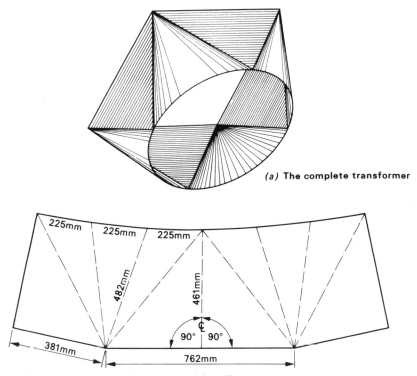

(a) The complete transformer

(b) Pattern or half-template without joint allowances

Fig. 5.18 Square to round transformer

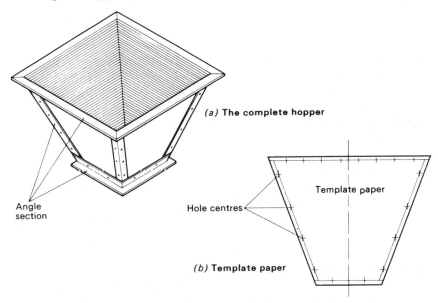

(a) The complete hopper

(b) Template paper

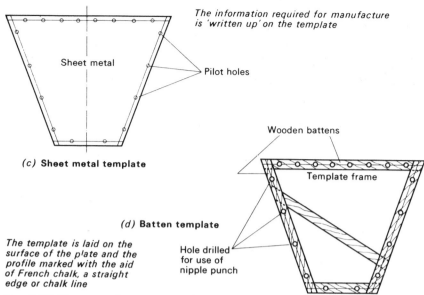

Fig. 5.19 The use of templates for hopper plates

5.23 Box templates

These are made from wood and are simply two flange templates fastened together. They are used for marking up purlin cleats used in constructional steelwork. The hole positions are marked on the box template to standard dimensions, usually supplied by the drawing office, and drilled. When marking off holes from a box template, a nipple punch is used. Figure 5.20 illustrates a box template. These are made from wooden battens, cut to required length and nailed as shown.

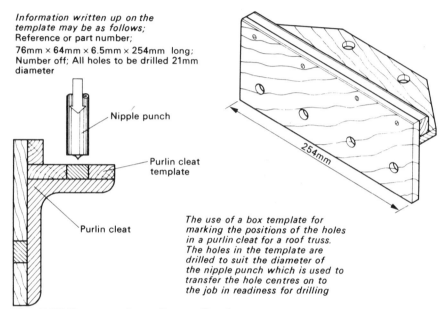

Fig. 5.20 Box templates for purlin cleats

Figure 5.19(a) shows the basic construction of a SQUARE HOPPER between parallel planes symmetrical to the centre line, made from steel plates. Although many such hoppers are of welded construction, the example shown is of riveted construction, using angle sections.

This simple type of hopper is made up of four identical steel plates for which only one template is required. The choice of template material is as follows:

When only one or two off are required, template paper may be used, as shown in Fig. 5.19(b). The profile of one side is developed and marked out on the paper, together with the exact positions for the centres of the holes. The template is then sheared to size on the guillotine.

Light-gauge sheet metal is ideal material for the template where a quantity of identical hoppers are required (see Fig. 5.19(c)).

Small pilot holes are either punched or drilled in the template in the correct positions. The hole positions are then transferred to the hopper plate with the aid of a centre punch.

The batten template shown in Fig. 5.19(d) is relatively light in weight and is used for quantities. It is constructed from suitable wooden battens, in the form of a frame to represent the developed profile of the required hopper plates. All the outside edges of the template are planed straight. The hole positions are drilled to suit the diameter of the nipple-punch.

5.24 Steel templates (ordinary and bushed)

For economy reasons, steel templates are made for part-drilling a batch of plates which may be separated for any additional holes to be drilled. *On the completion of the drilling the template itself becomes part of the fabrication.*

Bushed steel templates are employed for batch drilling over a long period of time, or where a high standard of accuracy is required. They are steel plates with hardened steel bushes tightly press-fitted into them. Hardened steel bushes, termed DRILL BUSHES are available with centre-holes to match a whole range of drill diameters.

When drilling large steel plates, for example GUSSET PLATES for large constructional steelwork, one plate is accurately marked out, carefully aligned as the top plate of a pile of three or four identical plates, the whole pile firmly clamped together and the assembly drilled as one.

Such a pile of plates are usually supported on a simple gantry and positioned under a radial drilling machine for the drilling of the holes, as shown in Fig. 5.21.

On completion of the drilling the top plate (TEMPLATE) is used as a component part of the steel assembly.

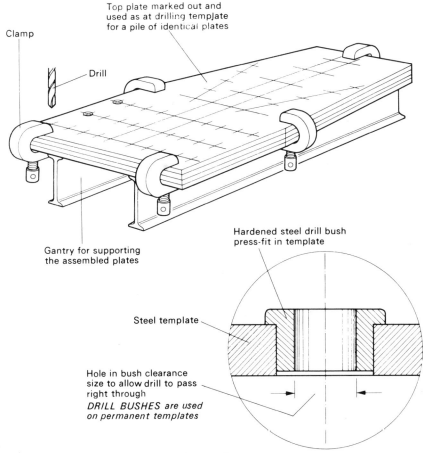

Fig. 5.21 The use of large steel templates

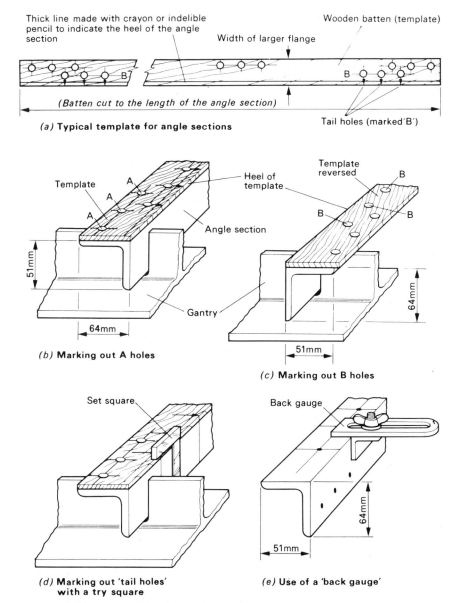

Fig. 5.22 Marking-off hole positions in angle sections

5.25 Marking off holes in angle sections

Angle sections are usually cut to length and mitred, where applicable, before the hole positions are marked. Figure 5.22 illustrates the method employed for marking hole positions on angle sections.

First, a batten template is made as shown in Fig. 5.22(a).
Second, the template is laid on the surface of the larger flange with the heel line of the template on the heel of the angle as shown in Fig. 5.22(b). The holes marked 'A' on the template are marked through on to the surface of the flange with a nipple punch.

105

Third, the angle section is then turned over in the gantry and the template (bottom face up) is laid on the surface of the smaller flange with the heel line of the template in line with the heel of the angle as shown in Fig. 5.22(c), and the 'TAIL HOLES' marked 'B' on the template marked through.

Alternatively when the tail holes are not drilled in the template, their positions are marked off with the aid of a set square, French chalk and a 'BACK GAUGE', as shown at Fig. 5.22(d) and (e), and their centres marked with a centre punch.

The back gauge illustrated is of the adjustable type which may be set to a standard back mark dimension from the heel of the angle. These standard dimensions are usually supplied by the drawing office.

5.26 Marking off holes in channel sections

Channel sections, cut to required lengths, are placed on a simple gantry with the web horizontal. The wooden template is placed in position, with the information uppermost, and clamped into position. Fig. 5.23(a) shows the method of marking off holes in channel section.

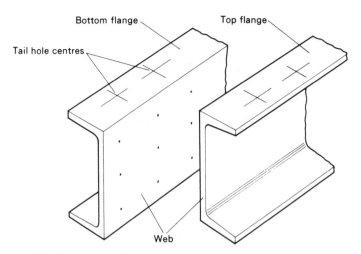

(c) Marking out the flange

Fig. 5.23 Marking-off hole positions in channel sections

The hole positions in the web are marked through the template with a nipple punch, as shown in Fig. 5.23(b). Whilst the template is clamped in position the positions of the tail holes are marked with the aid of a set square and French chalk, i.e. 'square-off' on the faces of both flanges.

When the template has been removed, a back gauge and French chalk are used to mark the position of the tail holes from the heel of the flanges and their centres marked with a centre punch, as shown in Fig. 5.23(c).

5.27 Marking off holes in 'Tee' section

One batten template is generally used to mark off the hole positions on both the flange and the web or stalk, as illustrated in Fig. 5.24(a).

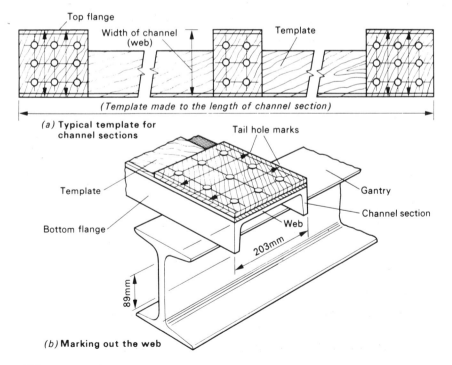

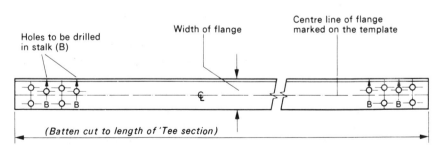

(a) Typical template for tee sections

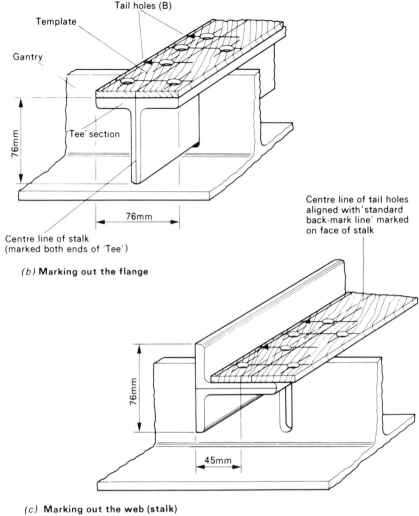

(b) Marking out the flange

(c) Marking out the web (stalk)

Fig. 5.24 Marking-off hole positions in tee sections

Before applying the template, a centre line representing half the thickness of the stalk is marked with French chalk on both ends of the 'Tee' section.

The template (with the instructions uppermost) is laid on the surface of the flange with the centre line aligned with the centre lines marked each end of the 'Tee', and clamped in position, as shown in Fig. 5.24(b). Except for the tail holes (marked 'B'), the holes are marked through to the face of the flange with a nipple punch.

The 'Tee' is then turned on the gantry with the stalk horizontal. A back-mark line is marked with French chalk on the face of the stalk at a standard dimension from the flange, as shown in Fig. 5.24(c). The template is laid on the face of the stalk with the centre-line of the tail holes in line with the back-mark line, and clamped in position. The tail holes are marked through to the face of the stalk with a nipple punch.

5.28 Marking off holes in columns or beams

The procedure for marking off hole positions in columns and beams is illustrated in Fig. 5.25. For reference purposes a beam has a 'top flange' and a 'bottom flange', whilst a column has an 'outside flange' and an 'inside flange'.

Information marked on template: Plain holes and A holes to be drilled on the top flange. Plain holes and B holes to be drilled on the bottom flange C holes to be drilled in web. All flange holes to be drilled 14mm diameter. Web holes to be drilled 17.5mm diameter

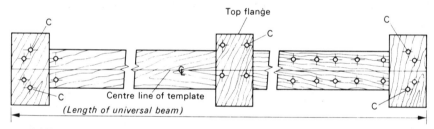

(a) Typical template for columns and universal beams

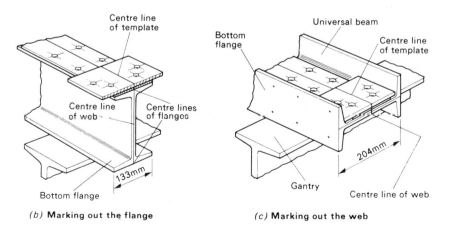

(b) Marking out the flange   (c) Marking out the web

Fig. 5.25 Marking-off hole positions in beams and columns

Figure 5.25(*a*) shows a typical template for marking out columns and universal beams.

First, centre lines are marked on the web at both ends of the section for the purpose of locating with the centre line marked on the template.

Second, the template is laid on the surface of the top flange with the respective centre lines aligned as shown in Fig. 5.25(*b*) clamped in position, and the plain and 'A' holes marked through with a nipple punch.

This procedure is repeated for the bottom flange where the plain and 'B' holes are marked through.

Finally, the beam is laid with the web horizontal on the gantry, and the web holes 'C' marked through in the same manner, as shown in Fig. 5.25(*c*).

5.29 Basic method of laying-out templates for a roof truss

Using the information supplied on the drawings, lines representing the roof truss are marked out on the floor. To ensure proper alignment of holes through templates, the battens (templates representing the angle sections) are drilled and laid on plate templates in the correct position on the lines on the floor. The holes in the plate templates are marked from the hole positions in the battens, and then drilled. The 'tail' or 'back' holes are marked in position for the purlin cleats and shoe connections.

After the templates have been checked for accuracy by replacing on the lay-out lines, they are marked up ready for use by the fabricators. *The edge of each batten template to be set against the heel of the angle section is marked with a line close to that edge.*

Figure 5.26 shows a layout for a simple roof truss, together with information on how the various templates employed are marked-up. Box templates are used for the purlin cleats as described in Fig. 5.20.

Further information on the use of templates can be found in Chapter 9.

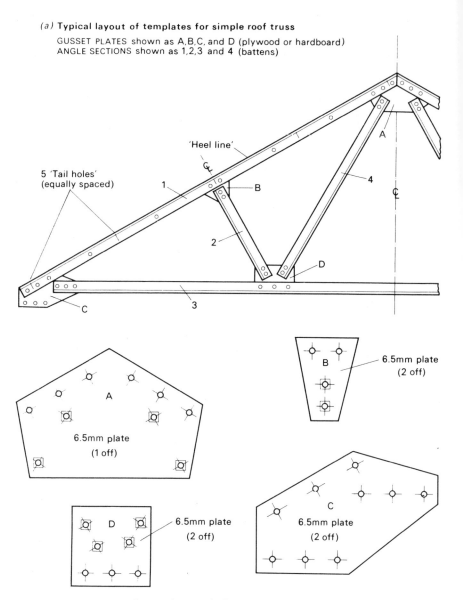

(*a*) Typical layout of templates for simple roof truss
GUSSET PLATES shown as A,B,C, and D (plywood or hardboard)
ANGLE SECTIONS shown as 1,2,3 and 4 (battens)

Instructions marked on gusset templates
All plain holes to be drilled 21 diameter
Holes marked ⊡ to be drilled 17.5mm diameter

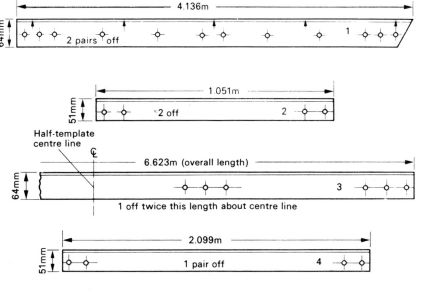

(b) **Batten templates for the angle sections**

Note: For clarity these are not shown to scale

Instructions marked up on templates (generally written in indelible pencil)

Template 1. 64mm × 51mm × 6.35mm × 4.136m long – 2 pairs off.
Tail holes to 29mm Back Gauge mark, drill 17.5mm diameter
Plain holes drill 21mm diameter

Template 2. 51mm × 51mm × 6.35mm × 1.051m long – 2 off.
Drill holes 17.5mm diameter

Template 3. 64mm × 51mm × 6.35mm × 6.623m long – 1 off
Drill holes 21mm diameter
THIS IS A HALF TEMPLATE TO THE CENTRE LINE OF THE SECTION'S OVERALL LENGTH

Template 4. 51mm × 51mm × 6.35mm × 2.099m long – 1 pair off
Drill holes 17.5mm diameter.

Fig. 5.26 Use of templates for structural steelwork

6 Materials

Rapid cooling increases hardness

6.1 Quench hardening plain carbon steels

Figure 6.1 summarises the methods by which the structure, and therefore the properties, of plain carbon steels can be changed by heat treatment processes.

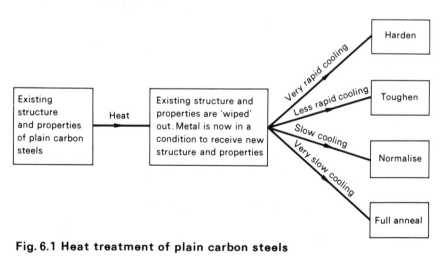

Fig. 6.1 Heat treatment of plain carbon steels

The temperature to which the steel must be heated to wipe out the initial structure and properties will depend upon the carbon content of the steel. Figure 6.2 shows the temperatures from which plain carbon steels should be quench hardened. It will be seen that these are the same temperatures as for annealing. The only difference being in the rate of cooling (Fig. 6.1). The degree of hardness the steel achieves is solely dependent upon:

1. The carbon content.
2. The rate of cooling.

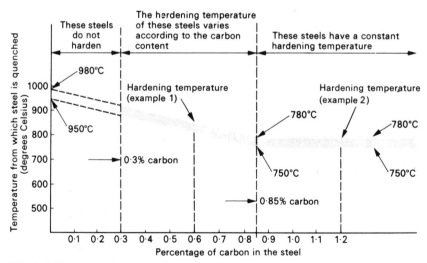

Fig. 6.2 Hardening and annealing of plain carbon steels

6.2 Effect of carbon content

There must be sufficient carbon present to form the hard iron carbides in the steel when it is heated and quenched. The effect of carbon content when a steel is heated and quenched is shown in Table 6.1.

Table 6.1 Effect of carbon content

TYPE OF STEEL	CARBON CONTENT %	EFFECT OF HEATING AND QUENCHING (RAPID COOLING)
Mild	Below 0.25	Negligible
Medium carbon	0.3–0.5	Becomes tougher
	0.5–0.9	Becomes hard
High carbon	0.9–1.3	Becomes very hard

6.3 Rate of cooling

The rapid cooling necessary to harden a steel is known as QUENCHING. The liquid into which the steel is dipped to cause this rapid cooling is called the QUENCHING BATH.

In the workshop, the quenching bath will contain either
(a) Water
(b) Quenching oil (on NO account use lubricating oil)

The more rapidly a plain carbon steel is cooled the harder it becomes. Unfortunately, rapid cooling can lead to CRACKING and DISTORTION.

Therefore, the workpiece should not be cooled more rapidly than is required to give the desired degree of hardness. For plain carbon steels, the cooling rates shown in Table 6.2 are recommended.

Table 6.2 Rate of cooling

CARBON CONTENT %	QUENCHING BATH	REQUIRED TREATMENT
0.30–0.50	Oil	Toughening[1]
0.50–0.90	Oil	Toughening
0.50–0.90	Water	Hardening
0.90–1.30	Oil	Hardening[2]

[1] Below 0.5 per cent carbon content, steels are not hardened as cutting tools, so water hardening has not been included.
[2] Above 0.9 per cent carbon content, any attempt to harden in water could lead to cracking.

6.4 Tempering

Hardened plain carbon steel is very brittle and unsuitable for immediate use. A further process known as TEMPERING must be carried out to greatly increase the toughness of steel at the expense of some hardness.

Tempering consists of reheating the steel to a suitable temperature and quenching again. The temperature to which the steel is heated depends upon the use to which the component is going to be put. Table 6.3 gives suitable temperatures for tempering components made from plain carbon steel.

Table 6.3 Tempering temperatures

COMPONENT	TEMPER COLOUR	TEMPERATURE °C
Edge tools	Pale straw	220
Turning tools	Medium straw	230
Twist drills	Dark straw	240
Taps	Brown	250
Press tools	Brownish-purple	260
Cold chisels	Purple	280
Springs	Blue	300
Toughening (crankshafts)	–	450–600

In the workshop, the tempering temperature is usually judged by the colour of the oxide film that appears on a freshly polished surface of the steel when it is heated.

Some tools, such as chisels, only require the cutting edge hardened; the shank being left tough to withstand the hammer blows.

6.5 Overheating plain carbon steels

It is a common mistake to overheat a steel with the hope that it will become harder. As already stated, the hardness only depends upon the carbon content of the steel and the rate of cooling. Once the correct hardening temperature has been reached, any further increase in temperature only slows up the rate of cooling and tends to reduce the hardness. Further, overheating also causes *crystal growth* resulting in a weak and defective component (*Basic Engineering*, 6.29).

On the other hand, if the hardening temperature is not reached the component will not harden no matter how quickly it is cooled.

6.6 Examples of hardening plain carbon steels

EXAMPLE 1 *To harden and temper a cold chisel made from octagonal chisel steel having a carbon content of 0·6/0·7 per cent.*

From Fig. 6.2 it will be seen that the correct hardening temperature for a 0·6/0·7 per cent plain carbon steel lies between 820°C and 850°C (bright red heat). As the carbon content is fairly low, and a chisel is a simple shape, it can be safely quenched in water to achieve maximum hardness.

To temper the chisel, first polish the cutting end so that the temper colours caused by the oxide film can be observed. The shank is then heated as shown in Fig. 6.3, and the polished end is watched as the temper colours travel towards it.

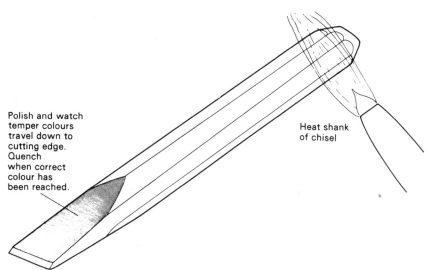

Fig. 6.3 Tempering a cold chisel

If the chisel is to cut metal, the cutting edge should be a brownish purple when it is quenched.

If the chisel is to cut brick or concrete the cutting edge should be purple to give it greater impact resistance.

Letting the heat travel down from the shank ensures that the shank is left in a tough rather than hard condition so that it will not shatter when hit by a hammer.

EXAMPLE 2 *To harden and temper a press tool die made from a 1·2 per cent plain carbon steel.*

From Fig. 6.2 it will be seen that the correct hardening temperature for a 1·2 per cent plain carbon steel lies between 750°C and 780°C (dull red heat). As the carbon content is high, and the die may be intricate in shape, it should be quenched in oil to prevent distortion and cracking.

Unlike Example 1 the press tool die must be of uniform hardness throughout. This can only be achieved by uniform heating. One simple method, as shown in Fig. 6.4, is to use a sand tray to spread the heat source. When the die is the correct tempering colour — dark brown for brass; brownish purple for steel — it is again quenched in oil.

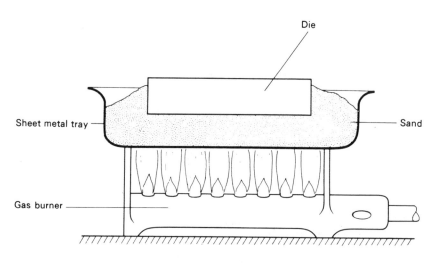

Fig. 6.4 Tempering a die uniformly

6.7 Annealing plain carbon steels

Comparison of Fig. 6.2 in this book with Fig. 6.16, *Basic Engineering*, will show that the temperatures from which plain carbon steels are hardened and annealed are identical. The difference between the processes lies in the rate of cooling (Fig. 6.1). To anneal a plain carbon steel it is cooled very slowly by turning off the furnace; closing the dampers, and allowing both work and furnace to cool down together. Alternatively the component may be buried in lime or ashes. The annealing of plain carbon steel was introduced in Section 6.28, *Basic Engineering*.

EXAMPLE 1 *To anneal a component made from 0·6 per cent carbon steel.*

Since the steel is annealed and hardened from the same temperatures reference can be made to either Fig. 6.2 in this book or Fig. 6.16, *Basic Engineering*. It will be seen that the full annealing temperature for a 0·6 per cent carbon steel is approximately 850°C.

The steel is raised to this temperature in the furnace and soaked until the temperature is uniform throughout. The furnace is turned off; the dampers are closed, and furnace and component cool down slowly together.

EXAMPLE 2 *To anneal a component made from 1·2 per cent plain carbon steel.*

The procedure is the same for Example 1 except for the temperature from which the metal is cooled. Reference to Fig. 6.2 in this book or Fig. 6.16, *Basic Engineering*, shows that the correct temperature for a 1·2 per cent plain carbon steel is 750°C and 780°C. The metal is slowly cooled as before.

6.8 Normalising

The temperature ranges for normalising plain carbon steels are different to those for hardening and annealing. Figure 6.17, *Basic Engineering*, shows suitable normalising temperatures for plain carbon steels, and the principles of normalising are discussed in Section 6.26, *Basic Engineering*.

EXAMPLE *To normalise a 1·2 per cent carbon steel forging after rough machining to relieve any residual stresses.*

Reference to Fig. 6.17, *Basic Engineering*, shows that the normalising temperature for a 1·2 per cent carbon steel is 850°/980°C. It will be noted that this is considerably higher than for annealing or hardening a 1·2 per cent carbon steel. Despite the higher temperature, grain growth is avoided by the faster rate of cooling. After the steel has reached 950°/980°C it is removed from the furnace and allowed to cool down in still air. Care must be taken to avoid:
1. Placing the component in a draught.
2. Placing the component on a surface that will 'chill' it.
3. Restricting the natural circulation of air around the component.

6.9 Hot and cold forging

The process of *forging* is the hammering of metal to shape. That is shaping the metal by *plastic flow*. Metals that are *worked* in this manner become *work hardened*. The temperatures at which metals re-crystallise after being cold worked have already been discussed in Section 6.28, *Basic Engineering*.
1. Forging the metal *below* the temperature of re-crystallisation is called COLD FORGING. Most rivet-heading operations are cold forging. The degree of flow-forming that is possible below the temperature of re-crystallisation is limited since the metal will work harden and eventually crack.
2. Forging the metal *above* the temperature of re-crystallisation is called HOT FORGING. The processes of hot forging are normally performed by a blacksmith. It is these processes that will be considered in this section of the chapter.

Metals whose internal structure has been modified by working (hot or cold) are called WROUGHT METALS.

6.10 The effect of forging on the properties of metals

The rolling and drawing processes to which metal bars are subject are themselves forming processes and effect the 'lay' of the grain in the bar as shown in Fig. 6.5.

(a) **Crystal structure as cast**

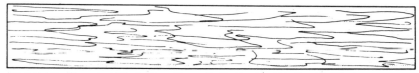

(b) **Crystal structure after rolling** (Crystals elongated in direction of rolling)

Fig. 6.5 Grain orientation

Figure 6.5(*a*) shows a cast bar with its random grain or crystal structure. Figure 6.5(*b*) shows a rolled or drawn bar with its grain directed along the length of the bar.

The component shown in Fig. 6.6(a) may be machined from the bar or 'upset' forged (see Chapter 9). The differences in the lay of the grain between machining and forging the component are shown in Fig. 6.6(b) and 6.6(c).

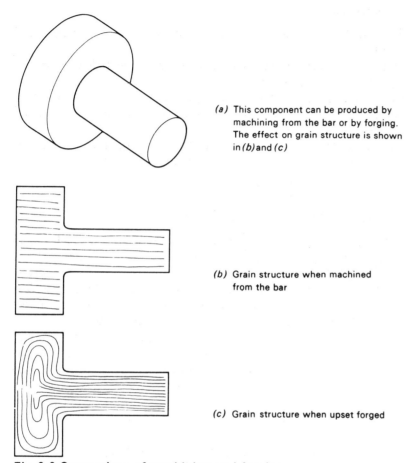

(a) This component can be produced by machining from the bar or by forging. The effect on grain structure is shown in (b) and (c)

(b) Grain structure when machined from the bar

(c) Grain structure when upset forged

Fig. 6.6 Comparison of machining and forging

If the component shown in Fig. 6.6 was to be used for a gear blank, this difference in grain orientation would be very important. Figure 6.7(a) shows that the grain of the machined blank is parallel to the teeth of the gear, causing planes of weakness. The teeth would break off relatively easily. Figure 6.7(b) shows that the grain of the forged blank would be at right-angles to the teeth of the gear. This produces strong teeth. Remember that grain in metal behaves like grain in wood, metal breaks more easily *along* the grain than *across* the grain.

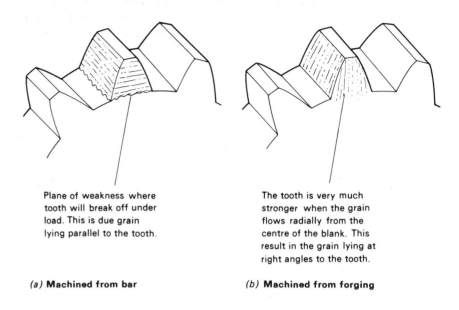

Plane of weakness where tooth will break off under load. This is due grain lying parallel to the tooth.

The tooth is very much stronger when the grain flows radially from the centre of the blank. This result in the grain lying at right angles to the tooth.

(a) Machined from bar

(b) Machined from forging

Fig. 6.7 Effect of grain flow on component strength

The forging process also breaks up and refines the crystal structure of the metal. The finer the crystals become, the tougher and stronger the metal becomes. The temperature at which the metal is forged is important. *Basic Engineering* Fig. 4.27 gives the upper and lower temperatures for forging various metals.

1. If the upper temperature is exceeded burning and grain growth will occur (6.5).
2. If forging is continued below the lower temperature work hardening will occur and the component may crack and become weakened.

It is important to relate the cooling time of the component to the time taken to complete the forging process. Figure 6.8 shows some examples of what can happen.

In Fig. 6.8(a) the forging process and the cooling cycle take the same time. This is ideal as no grain growth occurs after forging has finished, neither has any work hardening and cracking occurred.

In Fig. 6.8(b) the forging time is very much shorter than the cooling time. Grain growth occurs after forging has finished and this reduces the strength of the component. To prevent this, either the component is not heated to the maximum forging temperature, or the forging is subjected to grain refinement heat treatment processes after it has cooled down.

In Fig. 6.8(c) the forging time is longer than the cooling time and reheating is necessary. As in Fig. 6.8(b) the maximum forging temperature of the reheat must be carefully judged, or grain refinement after cooling must be carried out.

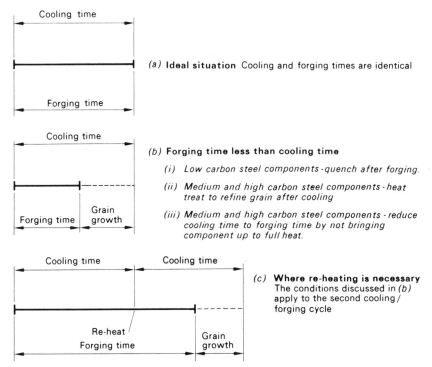

Fig. 6.8 The forging cycle.

The skill of the blacksmith lies in suiting the forging temperature to the process. However, in production forging, or where large components are being forged, grain refinement is carried out after cooling. Although an added expense it is cheaper than scrapping a large and costly component or a large quantity of mass-produced components.

The advantages of the forging process may be summarised as:

(a) *Economy in the use of raw materials.* Less time is used and less swarf is produced when machining from a forging, than when machining from the solid.

(b) *Increased strength* compared with the same component made from a casting or machined from the solid.

Examples of forging operations are to be found at the end of Chapter 9.

6.11 Forging copper alloys and aluminium alloys

Forging has little or no effect on commercially pure copper and aluminium.

1. Neither of these metals is susceptible to quench hardening, so the rate of cooling is unimportant.
2. Since forging is always carried out above the temperature of re-crystallisation both these metals will be in the annealed condition when they have cooled down.

However, most high duty copper and aluminium alloys are heat treatable. Therefore, their properties will be affected by:

(a) the forging temperature
(b) the time they are held at that temperature
(c) the rate of cooling.

Aluminium alloys are normally forged between 340°C and 450°C. Copper alloys are normally forged between 750°C and 810°C. Cooling should be as quick as possible. Alternatively the properties of the alloy can be modified by solution heat treatment after forging. If in doubt always consult the alloy manufacturer's technical literature before forging or heat treating these materials.

Aluminium alloys must be rapidly heated to forging temperature. Slow heating would result in a coarse grain structure. If the alloys are overheated they will be susceptible to cracking during the forging operation. Fairly light hammer blows must be used because heavy blows result in surface cracks which cannot be welded up by subsequent forging (as is the case with ferrous metals).

6.12 The relationship between welding and the properties of metals

When two pieces of metal are joined by *fusion welding* the weld pool and the edges of the parent metal are molten. As the joint cools down, the molten metal solidifies and becomes a minature casting. The fact that a cast metal structure is weaker than a wrought steel structure has already been discussed. Therefore a welded joint is usually weaker than the surrounding metal. Fortunately the chilling effect of the parent metal refines the crystal size of the joint and prevents excessive weakness. Figure 6.9 shows the changes in crystal structure that occur through a welded joint.

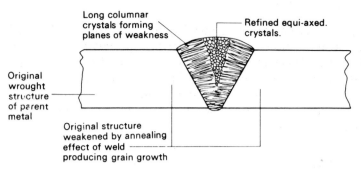

(a) Large single-run weld

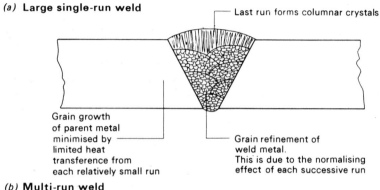

(b) Multi-run weld

Fig. 6.9 Structure of weld metal (mild steel)

All the common physical properties can in their turn affect the ease with which welding can be performed, and the type of process which can be used. Table 6.4 lists the common physical properties of some pure metals.

Table 6.4 Common physical properties of some pure metals

METAL	MELTING POINT °C	BOILING POINT °C	DENSITY g/cm^3	THERMAL CONDUCTIVITY J/cm^2/sec/°C/g
Aluminium	658	2 500	2·7	2·39
Copper	1 083	2 500	8·92	3·86
Iron	1 535	2 900	7·87	0·73
Lead	327	1 750	11·34	0·34
Magnesium	650	1 103	1·74	1·47
Molybdenum	2 620	4 700	10·4	1·45
Nickel	1 455	3 100	8·9	0·59
Silver	961	3 180	10·5	4·45
Zinc	419	907	7·14	1·12

Table 6.4

METAL	LATENT HEAT J/g	SPECIFIC HEAT J/g/°C	COEFFICIENT OF EXPANSION $\times 10^{-6}$
Aluminium	390	0·885	23·6
Copper	210	0·386	16·4
Iron	251	0·458	12·3
Lead	24·4	0·130	29
Magnesium	361	1·030	26
Molybdenum	—	0·272	4·9
Nickel	315	0·458	13
Silver	105	0·235	19
Zinc	109	0·396	31·2

The figures in the table are, of course, modified by the addition of alloying elements to the pure metals. *The important* changes, from a welding point of view, being:
1. A reduction in the thermal and electrical conductivities
2. The appearance of 'melting ranges' in place of 'melting points'.

The first property of interest when welding is the MELTING POINT or MELTING RANGE. Whether the material to be welded has a melting point or a melting range is of very great importance.

The temperature at which a material melts has a considerable influence in determining methods by which it can be welded. As would be expected, ARC METHODS are more suited to the HIGHER MELTING POINT materials and GAS WELDING to those with LOWER MELTING POINTS. For example, the welding of lead by metal-arc processes is not conceivable, but it was one of the earliest metals to be welded by a low-temperature flame. Aluminium is frequently welded by oxy-coal gas or oxy-hydrogen flames, whilst nickel with its higher melting point requires the hotter oxy-acetylene flame or arc welding methods.

The need to maintain a correct relationship between the temperature of melting and that of the arc or flame is associated with the heat required to melt the metal, but not in an obvious way. It is not just a question of the heat content of the molten metal, but is due rather to *the influence of the temperature on the heat losses during welding*. The theoretical amount of heat required to melt the higher melting point alloys is small in relation to the heat extracted from the weld at these temperatures by CONDUCTION, RADIATION and CONVECTION. For this reason, differences in LATENT HEATS and even SPECIFIC HEATS appear to play little part in determining differences in welding behaviour, whilst CONDUCTIVITY is very important in this respect.

COPPER *is outstanding among the more common NON-FERROUS metals in requiring a great deal of heat in welding.* This high rate of heat input and high rate of heat loss by conduction from the welding zone makes it more difficult than usual in maintaining the delicate THERMAL EQUILIBRIUM necessary just to melt the welding edges and not burn through. *The operator sees this as difficulty in controlling penetration*, the molten metal drops right through the joint and leaves a hole or fails to fuse the edges. This applies more to metal-arc welding than to gas-welding, since in the former the heat input cannot be controlled independently of the filler addition. To some extent the effect of HIGH CONDUCTIVITY can be neutralised by PRE-HEATING.

The effect of reduced conductivity is very apparent with the alloys of copper, since some of these are welded relatively easily even with metallic-arc. The same comparison can be made between pure aluminium and aluminium alloys.

In general, the lower the conductivity the less the practical difficulty in welding. As a point of interest, in this respect, when welding PLASTICS, where conductivity is extremely low, it is very easy to char the top surface of the weld while the underside is still cold. This condition, fortunately, does not occur in the fusion welding of metals.

With COLD-WORKED metals, conductivity influences the width of the ANNEALED ZONE on either side of the weld. The higher the conductivity the wider is this zone.

Electrical conductivity appears to be of little importance in fusion welding processes but seems to be a major consideration to resistance welding. For example, where nickel alloys may be satisfactorily 'spot-welded' using a low-power machine similar to that used for mild steel, aluminium requires a machine capable of delivering a very high current over a very short time, and pure copper is virtually unweldable by this process.

The high thermal expansion of copper is a considerable nuisance in all welding processes. It can cause movement of the joint edges which interferes with the proper execution of the weld and gives rise to distortion which must be subsequently removed. Contraction stresses are also responsible for producing cracked welds in those alloys which are 'hot short'. Even in those cases where thermal expansion and contraction do not appear to be an immediate nuisance they may indirectly lead to problems giving rise to RESIDUAL STRESSES. The problems of distortion and residual stresses have been considered in Chapter 4.

Where the components being joined operate at normal atmospheric temperatures they can benefit by being bonded by modern adhesives. The bonding processes are carried out below the temperatures at which changes in the grain structure occur thus preventing weakening of the parent metal. Adhesive bonding is now being used on motor-car body panels and even on such highly-stressed components as wheel studs.

6.13 The effects of welding on the parent metal

Fusion welding operations cause temperature gradients in the plates to be joined. The limits of the gradient are the melting point of the parent metal and room temperature. In the case of steel, a portion of the parent metal will be within the upper and lower critical temperature ranges. The effect that the welding heat exerts on the temperature of the plates or parts to be welded and the amount of heat conduction is illustrated in Fig. 6.10.

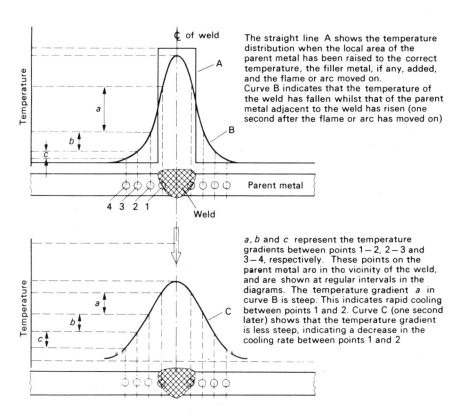

The straight line A shows the temperature distribution when the local area of the parent metal has been raised to the correct temperature, the filler metal, if any, added, and the flame or arc moved on.
Curve B indicates that the temperature of the weld has fallen whilst that of the parent metal adjacent to the weld has risen (one second after the flame or arc has moved on)

a, *b* and *c* represent the temperature gradients between points 1–2, 2–3 and 3–4, respectively. These points on the parent metal are in the vicinity of the weld, and are shown at regular intervals in the diagrams. The temperature gradient *a* in curve B is steep. This indicates rapid cooling between points 1 and 2. Curve C (one second later) shows that the temperature gradient is less steep, indicating a decrease in the cooling rate between points 1 and 2

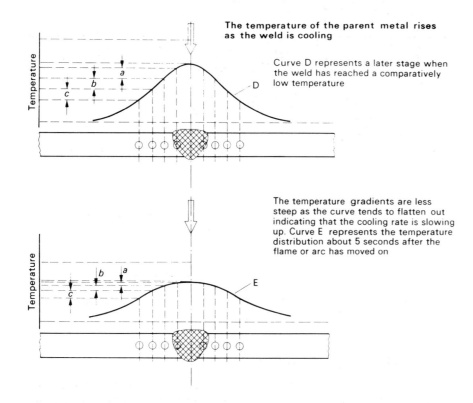

Fig. 6.10 The temperature of all points in the vicinity of a weld at five different times. (instantaneous heat source)

Figure 6.11 shows the approximate temperatures of the parent metal when arc-welding. There is a great variation in the temperature of the parent metal during metallic-arc welding.

Point A (shaded area) indicates the metal immediately adjacent to the molten weld pool heated to a pasty stage.

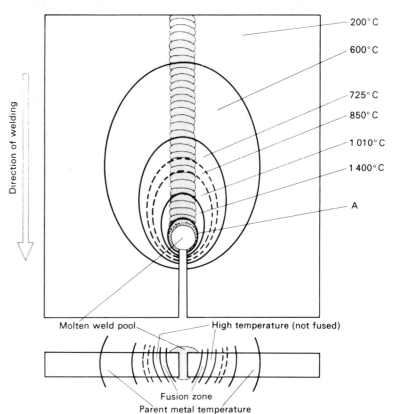

Fig. 6.11 Approximate temperatures of the parent metal (arc-welding)

Figure 6.12 shows the temperature distribution on a fairly thick steel plate. Table 6.5 summarises the effect of the temperature gradient on the properties of the metal in the vicinity of the weld.

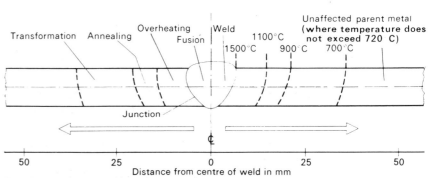

Fig. 6.12 Temperature distribution during oxy-acetylene welding 10 mm thick mild steel

Table 6.5 Effect of welding heat on the temperature of the plate

TEMPERATURE ZONES	REMARKS
Fusion zone	Temperature reaches melting point. The cooling rate is in the order of 350°–400°C/min., which is the maximum quenching range. The weld is less hard than the adjacent area of the parent metal because of loss of useful elements (carbon, silicon and manganese).
Overheated zone	The temperature reaches 1 100°–1 500°C. Cooling is extremely rapid in the order of 200°–300°C/min. Some grain coarsening occurs.
Annealed zone	Here the temperature reaches slightly higher than 900°C. The parent metal has a refined normalised grain structure. The change is not complete because the cooling rate is still high, in the order of 170°–200°C/min.
Transformation zone	The temperature here is between 720°C and 910°C. These are the upper and lower critical temperatures between which the iron in steel transforms from a body-centred-cubic to a face-centred-cubic structure. The parent metal tends to recrystallise.

A comparison of the heat spread associated with the leftward and rightward methods of gas-welding is made in Chapter 11, showing that less heat spread occurs in the plate with the rightward method. With the rightward method, the welding torch is so directed that the large bushy envelope of the flame is directed over the completed weld. *This produces an annealing effect of the weld metal itself.*

When a metal is heated up to its melting point and then allowed to cool suddenly the resulting structure is coarse, the metal is brittle and its mechanical properties are poor. If, however, the metal is allowed to cool down slowly in a special furnace much better properties are obtained. *The welding flame, in the case of rightward welding, has precisely this effect.* With leftward welding, the torch is directed away from the line of the finished weld and so the weld metal must cool abruptly in air.

Figure 6.13 illustrates the effects of welding on the parent metal.

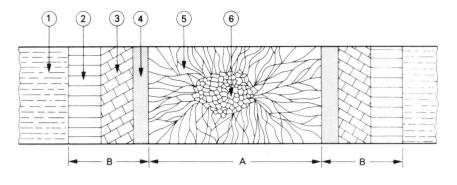

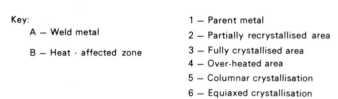

Key:
A – Weld metal
B – Heat-affected zone

1 – Parent metal
2 – Partially recrystallised area
3 – Fully crystallised area
4 – Over-heated area
5 – Columnar crystallisation
6 – Equiaxed crystallisation

Fig. 6.13 The effect of welding on the structure of a welded joint (schematic)

6.14 Welding mild steel

This is one of the easiest metals that can be joined by fusion welding. With care the joint that is formed is almost as strong as the parent metal. The claim by some welders that their joints are stronger than the parent metal is nonsense. Such claims have grown up due to weakening of the parent metal adjacent to the weld by grain growth. This causes any subsequent fracture to occur at the side of the weld rather than through the weld itself. The effects of non-metallic inclusions in the weld and the effects of faulty techniques are considered in Chapter 11.

In arc welding mild steel, the HEAT AFFECTED ZONE will be much narrower than in gas-welding. This is because the heat of the arc is more localised and the temperature is raised more quickly.

When a welding flame is applied to the plate its temperature is raised and so prevents 'chill casting' of the weld deposit occurring. Because the heat from the welding flame is applied for a longer period than the heat from the arc, GRAIN GROWTH will be more pronounced. Figure 6.14 compares macrostructures of a single-run weld in mild-steel plate (*a*) with oxy-acetylene and (*b*) with metallic arc.

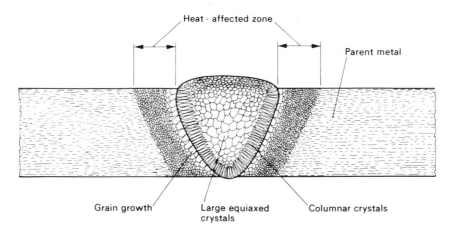

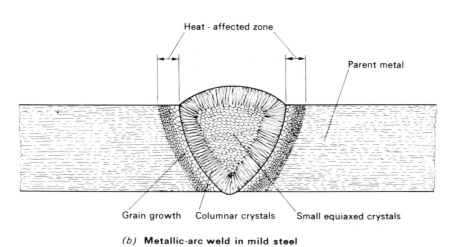

Fig. 6.14 Macrostructures of single run welds

When a weld is made in low carbon steel with the addition of filler material, the following structures will result:
1. Metal which has been molten will be made up of the deposited metal mixed with the parent metal.
2. There will be a FUSION LINE at the junction between the metal which has been melted and the unmelted parent metal.
3. A HEAT-AFFECT ZONE which extends from the 'fusion line' to the parent metal which has not been heated hot enough to change the original structure.
4. Adjacent to the 'fusion line' is a zone of COARSE GRAINS. This metal has been heated nearly to its melting point and 'GRAIN GROWTH' has resulted.
5. Progressing away from the weld through the 'coarse grains' the grains become smaller, and the zone where they are very fine is called the 'REFINED ZONE'. This metal has been heated to the 'TRANSFORMATION TEMPERATURE' just long enough to 'RE-CRYSTALLISE' and then cooled before the grains had time to grow.
6. Progressing away from the 'refined zone' there is a 'TRANSITION ZONE'. In this zone some of the grains were re-crystallised and others were not, resulting in a 'MIXED STRUCTURE'.
7. The last zone is the 'UNAFFECTED ZONE', because this part of the parent metal has not been heated sufficiently to cause any structural changes.

6.15 Welding copper

Reference to *Basic Engineering*, Table 6.19, shows that there are several grades of copper available. Ordinary tough pitch copper contains oxygen in the form of copper oxides. These give the metal its increased strength. Unfortunately these oxides react with the welding flame to produce steam. This causes *'gas porosity'* in the weld leading to weakness. This can be reduced by using a slightly oxidising flame and a filler rod containing phosphorus.

Where it is known that copper components are going to be welded they should be made from one of the 'de-oxidised' grades and a neutral flame used.

The high thermal conductivity of copper (seven times that of steel) can also be a disadvantage when welding.
1. A larger jet is required to build up the temperature of the joint to fusion point. There is a tendency to melt the filler rod but not the edges of the parent metal. This results in planes of weakness at the edges of the joint.
2. Like most non-ferrous metals, copper depends upon cold-working to give it strength. The heat conducted back from the joint anneals the parent metal resulting in general weakness in the vicinity of the component as shown in Fig. 6.15. The structure of the weld metal can be improved by lightly hammering the metal whilst it is still red hot, and continuing with slightly heavier blows as the metal cools. This refines the crystal structure as when forging.

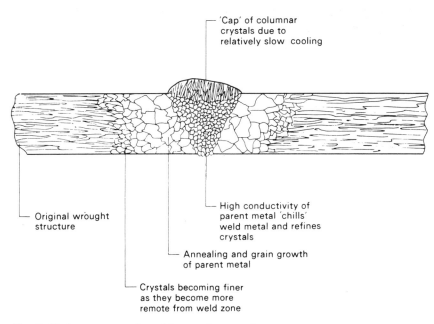

Fig. 6.15 Structure of weld zone (copper)

6.16 Welding aluminium

Like copper, aluminium also has a high conductivity and depends upon cold working to improve its strength. Therefore the conditions shown in Fig. 6.15 also apply to aluminium. Further, aluminium oxidises very easily and has to be protected from atmospheric oxygen by the use of fluxes and a reducing flame setting. Unfortunately, aluminium absorbs hydrogen more readily than any other metal when in the molten state. The hydrogen comes from various sources such as the welding flame, fluxes and atmospheric moisture. As the weld cools, dissolved hydrogen is expelled resulting in 'gas porosity'. The fact that the conditions for preventing oxidation and gas porosity in aluminium conflict with each other is the reason why aluminium and its alloys are such difficult metals to weld.

The HEAT AFFECTED ZONE in joints made by metallic arc welding rarely extends beyond 50 mm from the centre line of the weld. Gas-welding usually affects a wider area. With regard to TENSILE properties, the weakest point of the joint would be immediately adjacent to the weld.

In welded joints made in commercially pure aluminium and in NON-HEAT-TREATABLE aluminium alloys there are usually three zones:

1. **Weld metal** The weld bead with its 'as-cast' structure where parent metal is alloyed with deposited weld metal.
2. **Annealed zone** Region where the heat from welding has caused RE-CRYSTALLISATION or annealing.
3. **Unaffected zone** Region where heating has not affected the structures.

Figure 6.16 shows the various zones in a cross-section of a weld in aluminium.

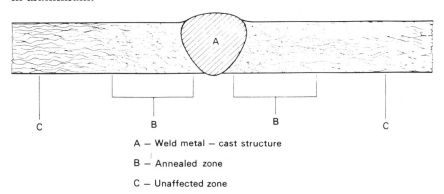

A – Weld metal – cast structure
B – Annealed zone
C – Unaffected zone

Fig. 6.16 The various zones in a cross-section of a weld in a non-heat-treatable aluminium alloy

The HEAT-TREATABLE aluminium alloys contain alloying elements which exhibit a marked change in solubility with temperature change. These elements are very soluble in aluminium at elevated temperatures but have low solubility at room temperature. They tend to separate out as various microconstituents in the parent metal structure. The high strengths of the heat-treatable alloys are due to the controlled solution and precipitation of some of these microconstituents. Most of the difficulties in welding these alloys are due to uncontrolled melting, solution and precipitation of the alloying elements.

The alloying elements in heat-treatable aluminium alloys are dissolved in the aluminium at high temperature by a process commonly known as 'SOLUTION HEAT TREATMENT'. They are maintained in solid solution by rapidly quenching from this temperature. The solution of specific elements or compounds in aluminium governs the strength of these alloys in the 'as-quenched' condition. Additional increases in strength are affected by 'precipitation' of a portion of the soluble elements in finely divided form.

'Solution heat treatment', 'natural age hardening' and 'precipitation hardening' have been discussed in *Basic Engineering*, Section 6.29.

Welds in heat-treatable alloys generally exhibit five microstructural zones:
1. **Weld metal** Weld bead with 'as-cast' structure where parent metal is alloyed with deposited weld metal.
2. **Fusion zone** Region where partial melting of the parent metal occurs, primarily at the grain boundaries.
3. **Solid solution zone** Region where the heat from welding is high enough to dissolve soluble constituents which are partially retained in solid solution if cooling is sufficiently rapid.
4. **Partially annealed or overaged zone** Region where the heat from welding has caused precipitation and/or coalescence of particles of soluble constituents.
5. **Unaffected zone** Region where heating has not affected the structure.

Figure 6.17 shows the various zones in a cross-section of a weld in a heat-treatable aluminium alloy.

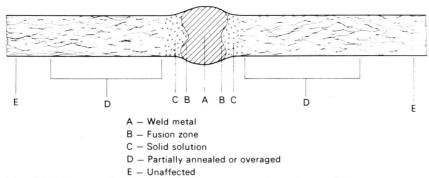

A — Weld metal
B — Fusion zone
C — Solid solution
D — Partially annealed or overaged
E — Unaffected

Fig. 6.17 The various zones in a cross-section of a weld in a heat-treatable aluminium alloy

The five zones are generally quite evident in welds made in heat-treatable alloys in which copper and zinc are the major alloying elements.

Alloys of the MAGNESIUM-SILICON type exhibit microstructural changes in the HEAT-AFFECTED ZONE that are somewhat different, the principle heat effect is generally OVERAGING. The partially annealed and overaged zones being of much greater widths.

The speed of welding has a marked effect upon the properties of welds in heat-treatable alloys. High welding rates not only decrease the width of the heat-affected zone, but they minimise effects such as grain boundary precipitation, overaging and grain growth.

The heat-treatable aluminium alloys may be reheat-treated after welding to bring the heat-affected zone back to approximately its original strength.

6.17 Weld cracking

There are several reasons for the intercrystalline cracking of welds. These are:
1. Cracking due to contractional strains when a weld cools down whilst it is too rigidly restrained.
2. A coarse-grained deposit with large columnar crystals. These have a relatively small grain boundary area and are less able to resist the cooling strains even when the weld is not rigidly constrained. A multi-run weld helps to prevent this as shown in Fig. 6.9.
3. The presence of grain boundary films with a low melting point. Iron sulphide in mild-steel welds gives rise to hot cracking. Steels for welding should have a high manganese and a low sulphur content to prevent iron sulphide forming.
4. Pure aluminium is not so susceptible to cracking as some aluminium alloys and it can be strengthened by reheating to approximately 500°C and quenching in water. This is solution treatment as described in *Basic Engineering*, Section 6.20. It refines and strengthens the grain structure.

6.18 The effects of flame cutting on the properties of metals

The effect on the structure of the metal is similar to that in the weld zone. In addition, hard oxide films are also formed along the *flame cut* edge of the *metal* and this can make machining difficult. Figure 6.18 shows a section through the edge of a flame cut component, whilst Fig. 6.19 shows how it should be machined.

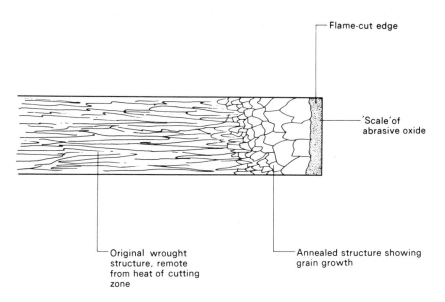

Fig. 6.18 Effect of flame-cutting on crystal structure

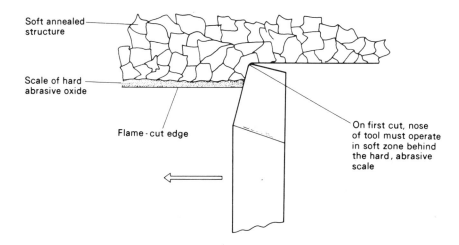

Fig. 6.19 Machining a flame-cut blank

7 Material removal

Better ask for a sheet metal punch next time Fred

The fundamental principles of material removal have been simply explained in *Basic Engineering*. In this chapter some of the more common methods employed in the fabrication industry for material removal will be considered in greater detail.

7.1 The use of the chisel for cutting sheet metal

When used for cutting sheet metal, the flat chisel must be held at a slight angle to the line of cut, as shown in Fig. 7.1(a). The reasons for inclining the chisel are:
1. To provide a shearing angle.
2. To make the chisel move along the line of cut smoothly and continuously.

If the chisel is held vertically, a separate cut is made each time a hammer blow is delivered, and the 'line' becomes a series of irregular cuts. A block of soft iron is generally used to support the sheet metal whilst it is being cut.

The chisel is also inclined when cutting slots or apertures of various shapes or sizes. In this case the removal of the material is simplified by punching or drilling a series of holes, as near together as possible, before the chisel is used. The advantages of pre-drilling or pre-punching the sheet metal are shown in Fig. 7.1(b).

If the slot was not relieved by pre-drilling, chiselling would distort the plate where the amount of metal left outside the slot is small.

Figure 7.1(c) shows how sheet metal may be held in the vice for cutting with a chisel. Care must be taken to ensure that the line along which the cut is to be made is as near to the top of the vice jaws as possible, otherwise the metal will be bent and the cut edge badly burred over.

(a) Cutting sheet metal with a flat chisel

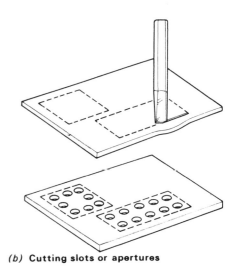

(b) Cutting slots or apertures

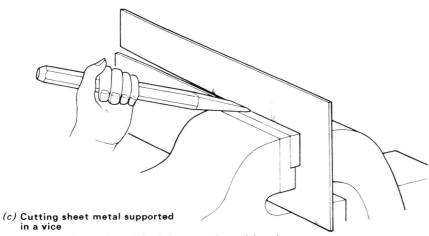

(c) Cutting sheet metal supported in a vice

Fig. 7.1 The use of a chisel for cutting thin plate

7.2 Blacksmith's chisels

Blacksmith's chisels are commonly referred to as 'sets', and these are employed for cutting off bars of iron or steel. The blacksmith's chisel is usually fitted with a handle. Some have a wooden handle or shaft similar to that of a hammer, but most 'sets' are provided with a sturdy twisted wire handle.

There are two types of blacksmith's chisel, the 'COLD SET' and the 'HOT SET'.

The 'hot set' is employed when cutting hot metal, and does not require to be hardened and tempered. Its cutting edge is keener than that of a 'cold set'. Hot sets are manufactured from a tough variety of steel in order that it may cut through relatively soft red-hot metal with ease. A hot set cannot be used on cold metal because the cutting edge would soon become blunted.

The 'cold set', as the name implies, is employed when cutting cold metal, therefore it has to be correctly hardened and sharpened.

Figure 7.2 illustrates the two types of blacksmith's chisel.

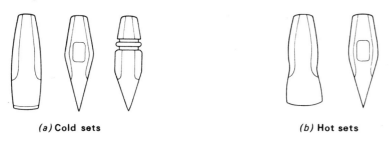

(a) Cold sets   *(b)* Hot sets

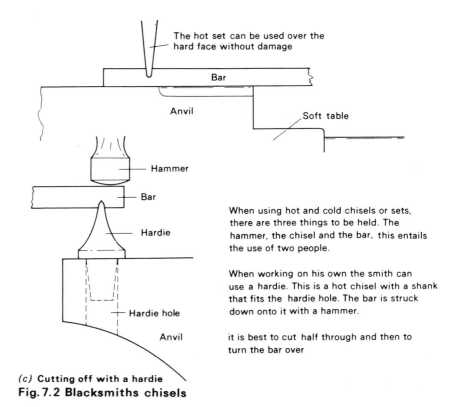

(c) Cutting off with a hardie

Fig. 7.2 Blacksmiths chisels

When using cold sets, the work is supported on the soft table of the anvil *to avoid damaging the cutting edge as the chisel breaks through.*

Figure 7.2(*c*) illustrates cutting with a hot set.

HARDIES

The use of sets requires an extra pair of hands, therefore special chisels are available with a tapered square shank which fits into a square hole in the anvil. These are called 'hardies' and allow the craftsman to work on his own.

The work is struck down on to the hardie as shown in Fig. 7.2(*c*).

7.3 Rivet removal

Cold chisels are useful for removing defective rivets in sheet-metal work. Figure 7.3 illustrates the basic steps in removing a rivet. The removal of rivets in platework will be discussed in 7.33.

125

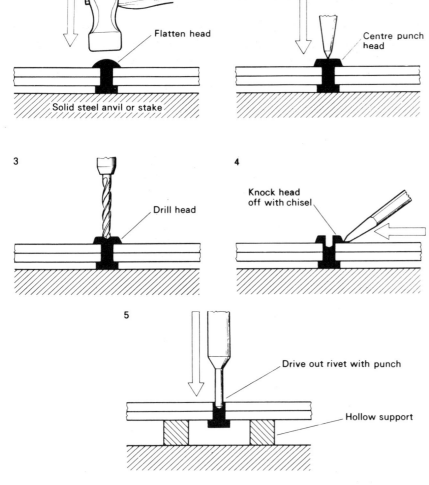

Fig. 7.3 Removing a rivet

7.4 Snips and hand shears

Various types of snips and hand shears are available, the most common are:

Snips

These can be obtained in very many sizes and types up to about 400 mm in length. Figure 7.4 illustrates the basic types of snips together with their application.

1. STRAIGHT SNIPS Sometimes called 'tinman's shears'. Used for making straight cuts and large external curves.
2. UNIVERSAL COMBINATION SNIPS Often referred to as 'gilbows'. The blades are designed for universal cutting — straight line or internal and external cutting of contours. May be 'right-hand' or 'left-hand', *easily identifiable as the top blade is either on the right or the left.*
3. PIPE SNIPS As the name implies used for trimming cylindrical or conical work in sheet metal. Have smaller thinner blades than universal snips.

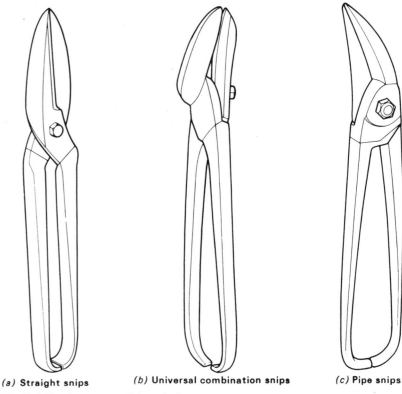

(a) Straight snips   (b) Universal combination snips   (c) Pipe snips

Fig. 7.4 Basic types of hand shears

Figure 7.5(a) shows the correct grip when using 'snips'. The scissors action is provided by the finger movement indicated by the arrows. To produce the maximum cutting force, the hand must be as far as possible from the pivot and the metal being cut must be kept close up to the pivot. Figure 7.5(b) shows how the *principle of moments* (*Basic Engineering*, 7.6) is exploited to magnify the effort of the craftsman's hand.

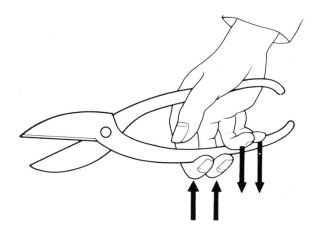

(a) **Correct grip**
The scissors action is provided by the finger movement indicated by the arrows

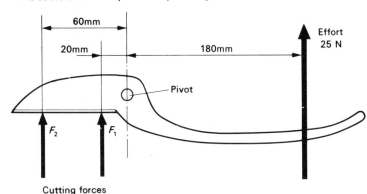

(b) **Force magnification**

1. *Shearing Force (F_1) near to pivot*

$$\text{Anti-clockwise forces} = \text{Clockwise forces}$$
$$F_1 \times 20 = 25 \times 180$$
$$F_1 = \frac{25 \times 180}{20}$$
$$F_1 = \underline{225\ N}$$

2. *Shearing force (F_2) remote from pivot*

$$\text{Anti-clockwise forces} = \text{Clockwise forces}$$
$$F_2 \times 60 = 25 \times 180$$
$$F_2 = \frac{25 \times 180}{60}$$
$$F_2 = \underline{75\ N}$$

Fig. 7.5 Principle of the hand shear

Hand-operated shears

These are operated by means of a lever (handle) and are generally for bench mounting. Various types of hand-operated shears are illustrated in Fig. 7.6.

1. *Hand-lever bench shears* Used for straight line cutting of sheet metal up to 3 mm thick.

Note: The movable top blade is curved to provide a constant SHEAR ANGLE at cutting position (may be used for CONVEX cutting).

2. *Lever-operated throatless bench shears* May be used for either CONVEX or CONCAVE cutting by suitable manipulation of the sheet. The heavier type of this machine is capable of cutting 4·7 mm thick mild steel.

3. *Hand-lever corrugated bench shears* Used for straight shearing or corrugated material — capacity 1·6 mm in mild steel.

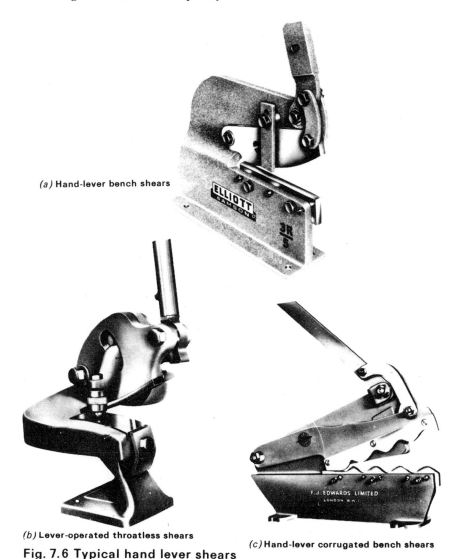

(a) **Hand-lever bench shears**

(b) **Lever-operated throatless shears**

(c) **Hand-lever corrugated bench shears**

Fig. 7.6 Typical hand lever shears

7.5 Basic principles (shearing)

There is a wide range of machines for shearing sheet metal from the basic elementary snips to static and portable power machines. In each case the basic principle of metal cutting, whether the machine is operated by hand or by power, is the shearing action of a moving blade in relation to a fixed blade.

The standard type of bench shear and all guillotines are used for *straight-line cutting*. The basic principle of these machines is that one blade is fixed (bottom blade) and the moving blade (inclined to the fixed blade) is brought down to meet the fixed blade, as shown in the line diagram, Fig. 7.7(a).

The moving cutting member of a shearing machine may be actuated by:
1. HAND-LEVER — bench shearing machines
2. FOOT TREADLE — treadle guillotines
3. ELECTRIC MOTOR or HYDRAULIC — power guillotines

If the cutting members of a guillotine or shearing machine were arranged parallel to each other, the **area under shear** would be the CROSS-SECTION of the material to be cut, i.e. 'length × thickness', as shown in Fig. 7.7(b).

The top cutting member of a shearing machine is always inclined to the bottom member to give a 'SHEARING ANGLE' of approximately 5 degrees. Figure 7.7 shows that with this arrangement of the blades, *the area under shear is greatly reduced and, consequently the force required to shear the material is considerably reduced.*

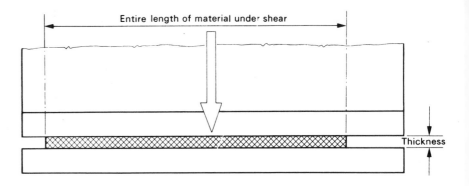

(b) **Parallel cutting blades**

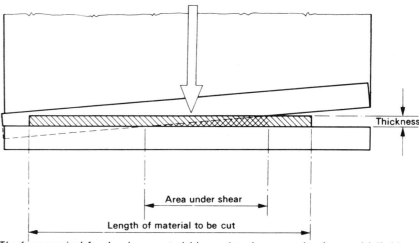

The force required for shearing a material is equal to the area under shear multiplied by the 'shear strength' of the material

(c) **Top cutting blade inclined**

Fig. 7.7 The effect of shear angle (shearing machines)

Figure 7.8 explains how the shearing action is used to cut metal. The shear blades are provided with a RAKE ANGLE of approximately 87 degrees and there must be CLEARANCE between the cutting edges of the blades to assist in the cutting action. The importance of clearance will be discussed later in this section.

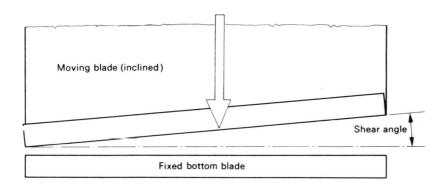

(a) **Shear blade movement**

STAGE 4

Fractures begin to run into the work-hardened metal from the points of contact of the cutting members. *When these fractures meet, the cutting members penetrate the whole of the metal thickness.*

With all shearing machines a sufficient force must be applied to the moving blade to overcome the **shear** **strength** *of the material and cause it to shear along the line of action.* In respect of snips and bench shears, these are designed in such a manner that the fulcrums of the moving blades are carefully positioned to provide an optimum MECHANICAL ADVANTAGE to enable a relative small EFFORT (hand power) to be applied in order to cut the metal with comparative ease. Figure 7.9 shows a simplified diagram of the lever system of a hand-operated shearing machine.

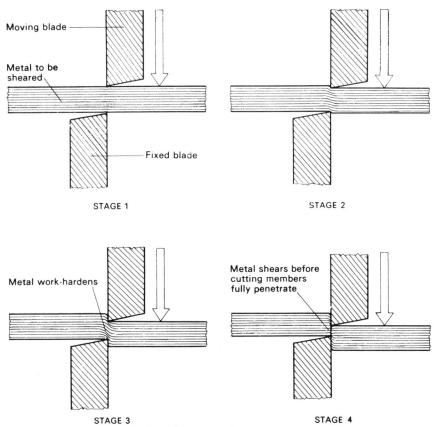

Fig. 7.8 The action of shearing metal

STAGE 1

As the top cutting member is moved downwards and brought to bear on the metal with continuing pressure, the top and bottom surfaces of the metal are deformed, as shown in Fig. 7.8.

STAGE 2

As the pressure increases, the internal fibres of the metal are subjected to deformation. This is 'PLASTIC DEFORMATION' prior to 'SHEARING'.

STAGE 3

After a certain amount of plastic deformation the cutting members begin to penetrate, as shown in Fig. 7.8. The uncut metal 'WORK-HARDENS' at the edges.

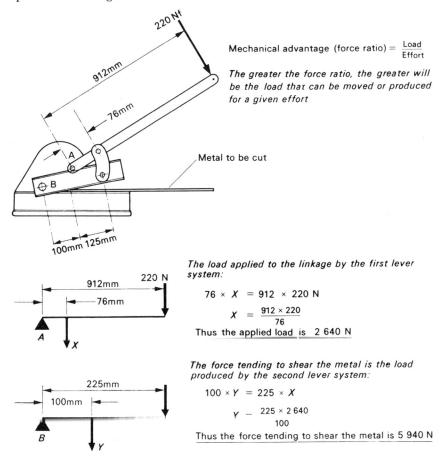

Mechanical advantage (force ratio) = $\frac{\text{Load}}{\text{Effort}}$

The greater the force ratio, the greater will be the load that can be moved or produced for a given effort

The load applied to the linkage by the first lever system:

$76 \times X = 912 \times 220$ N

$X = \frac{912 \times 220}{76}$

Thus the applied load is 2 640 N

The force tending to shear the metal is the load produced by the second lever system:

$100 \times Y = 225 \times X$

$Y = \frac{225 \times 2640}{100}$

Thus the force tending to shear the metal is 5 940 N

Fig. 7.9 The lever system of hand-operated bench shears

A lever is a simple machine which applies the principle of moments to give mechanical advantage.

By means of a lever, a comparatively small force, *usually termed 'effort' applied at a relatively large* distance *from the* fulcrum *(pivot), will either overcome or balance a* greater force *or* load *at a* small distance *from the* fulcrum.

The simple shearing machine shown in Fig. 7.9 makes use of a double application of 'THE SECOND ORDER OF LEVERS' in which THE LOAD IS BETWEEN THE FULCRUM AND THE EFFORT.

Never extend the arms of a pair of snips or a bench shear in order to cut a greater thickness of metal than that for which they are designed. *This careless action will only result in deflecting the cutting blades and bending or shearing the fulcrum bolt.*

Figure 7.10 shows details of the cutting blades of hand shears. It will be noticed that the design principles are the same as those employed in respect of guillotine shear blades.

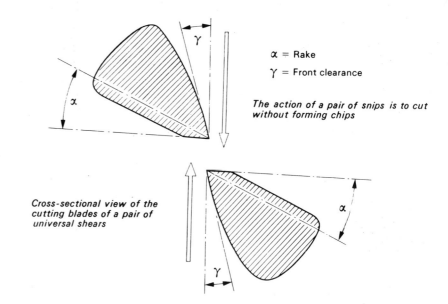

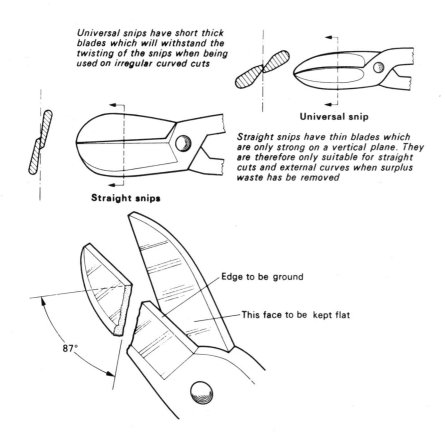

Fig. 7.10 Details of the cutting blades of hand shears

Blade clearances are very important and should be set to suit the material being cut. An approximate rule is that the clearance should not exceed 10 per cent of the thickness to be cut and must be varied to suit the particular material.

Correct clearance is essential in order to obtain optimum shearing results.

Figure 7.11 shows the results of incorrect and correct setting of shear blades, and Table 7.1 gives details of clearance in relation to material thickness.

Table 7.1 Blade clearances for optimum cutting

METAL THICKNESS		BLADE CLEARANCE (Tested by Feeler gauges)			
		Low Tensile Strength (e.g. BRASS)		High Tensile Strength (e.g. STEEL)	
in.	mm	in.	mm	in.	mm
0·015	0·381	0·0003	0·0075	0·0005	0·013
0·032	0·813	0·0015	0·038	0·0018	0·046
0·065	1·651	0·0020	0·051	0·0025	0·064
0·100	2·540	0·0022	0·056	0·0030	0·076
0·125	3·175	0·0030	0·076	0·0040	0·10
0·250	6·350	0·0055	0·14	0·0070	0·18

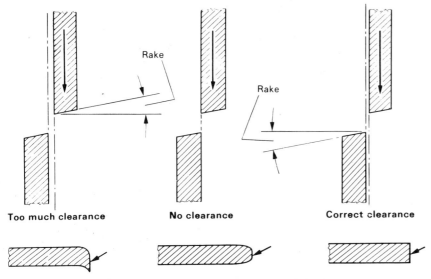

1. Excessive clearance causes a burr to form on the underside of the sheet
2. With no clearance, overstrain is caused, the edge of the sheet becomes flattened on the underside
3. With the correct clearance optimum shearing results are obtained

Fig. 7.11 Results of incorrect and correct setting of shear blades

7.6 Summary of shearing machine requirements

1. All sheet metal shearing machines have two cutting members.
2. The operational clearance between the cutting members is always important (5–10 per cent of material thickness). *The maximum clearance should be given consistent with the quality of the shearing required.*
3. The sharpness of the cutting members in contact with the material. The cutting members should always be maintained with reasonably sharp edges. *Sharp blades tend to produce a more intense local strain, and better cutting results.*
4. Most cutting members have a face shear or rake-way from the cutting edge. Together with blade sharpness this assists in the cutting action by increasing the local strain. A small amount of face shear (about 2 degrees) on the blades is useful where there is a risk of the blades or cutters being brought together almost to touch when shearing very light gauge material.

With regard to shearing machines in general, two principal basic requirements for a good machine (irrespective of type) are as follows:

1. The rigidity or robustness of the cutting member mountings, so that *deflection of the cutting members is kept to a minimum during the cutting action.* That is, once the cutting members are set in their correct position they will stay there.
2. A satisfactory means of adjustment should be provided between the cutting members. *This enables the cutting clearance to be adjusted and the position of the cutting members to be maintained with regard to the feed of the material.*

7.7 Rotary shears

With these types of machines a pair of ROTARY CUTTERS replaces the conventional flat blades used in those shearing machines which are essentially for straight-line cutting. *One advantage of rotary shears is that there is no restriction on the length of cut.*

The cutters rotate producing a continuous cutting action with very little distortion of the material. These machines may be hand or power driven and consist of two basic types, and these are illustrated in Fig. 7.12.

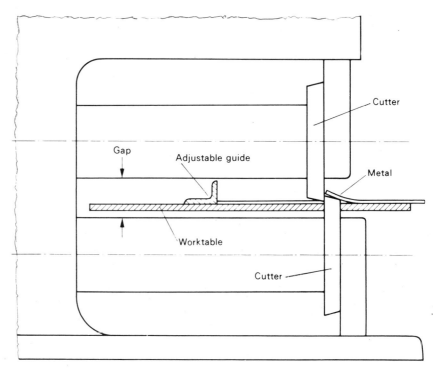

(a) **Bench rotary machine for straight-line cutting (horizontal spindles)**

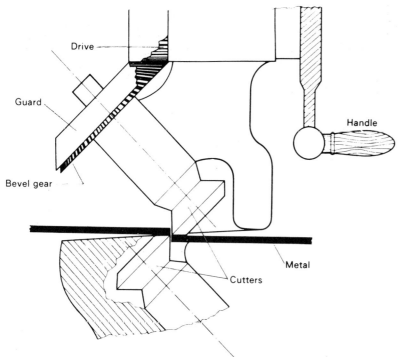

(b) Hand-operated rotary throatless shear (inclined spindles)

Fig. 7.12 Rotary shears

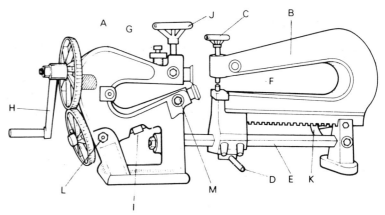

Circle cutting machine (hand operated)

- A — Cutting head
- B — Bow
- C — Pallet adjustment wheel
- D — Clamping handle (bow)
- E — Bar
- F — Clamping pallets
- G — Stop screw
- H — Operating handle
- I — Lower cutter adjusting screw
- J — Handwheel
- K — Handwheel (circle diameter control)
- L, M — Lower cutter bearing adjustment nuts

A good practical method of adjusting the cutters is to aim at cutting a true circle in paper. If a machine will do this in a satisfactory manner, then it will shear sheet metal without burring the edge of the disc

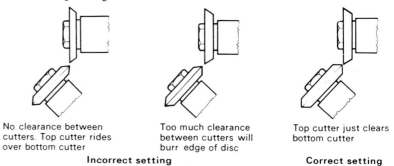

No clearance between cutters. Top cutter rides over bottom cutter | Too much clearance between cutters will burr edge of disc | Top cutter just clears bottom cutter
Incorrect setting | | **Correct setting**

Fig. 7.13 Circle cutting machine

Shearing essentials (rotary cutters)

1. The edges of the cutters must overlap by the smallest amount consistent with clean cutting.
   Excessive overlap tends to distort the material.
   Insufficient overlap does not shear the material.
2. There must be clearance between the working edges of the cutters (see 7.5, 'Blade clearance') and a means to adjust this.
3. Both cutters must run dead true, both on face and diameter.
4. The shafts on which the cutters are mounted must have no lateral movement. Neither the shafts nor their mountings should deflect under load.
5. The cutters must be kept reasonably sharp. In the case of machines which have no adjustment between the spindles, the cutters should be sharpened by grinding on the face only.

Figure 7.13 shows details of a circle-cutting machine which operates on the rotary cutter principle.

7.8 Portable shears and nibblers

Portable power cutting tools are either electrically or pneumatically operated and may be used for straight-line cutting and for cutting irregular curves. *The pneumatic machines are much safer for work 'on site', the power source being a compressor.*

There are two basic types of portable shearing machines — the 'shear type nibbler' and the 'punch type nibbler'.

The shear type nibbler

This portable power tool is used for rapid and accurate straight-line or curved cutting of material up to 4·5 mm thickness. It is basically a short-stroke power shear fitted with a rapidly reciprocating cutting blade, so that each stroke makes a cut approximately 3 mm in length. The speed of the cutting blade is between 1 200 and 1 400 strokes per minute, and the linear cutting speed for material up to 1·62 mm thickness is approximately 10 metres per minute, and for 4·5 mm thickness the cutting speed is reduced to a maximum of approximately 4·6 metres per minute, and produces a cleanly cut edge.

The shear type nibbler is fitted with a pair of very narrow flat blades, one of which is usually fixed, and the other moving to and from the fixed blade at fairly high speeds. Generally these blades have a very pronounced RAKE to permit piercing of the material for internal cutting, and since the blades are so narrow, the sheet material can be easily manoeuvered during cutting.

The top blade is fixed to the moving member or ram, and the bottom blade on a spiral extension or 'U-frame'. This extension is shaped like the body of a *'throatless shear'*, to part the material after cutting.

The lighter machines have a minimum cutting radius of 16 mm, and the heavier ones about 50 mm.

There is usually provision for vertical adjustment to allow for re-sharpening of the blade by grinding, and an adjustment behind the bottom blade to allow for setting the cutting clearance.

Figure 7.14 shows details of the 'shear type nibbler'.

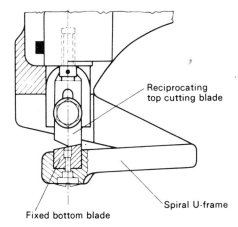

The line diagram opposite shows a cross-section of the cutting head of a typical portable vibrating shears driven electrically or pneumatically. The spiral U-frame is designed to assist in parting the metal after it has been sheared

Basic details of cutting blades are given in the diagram below

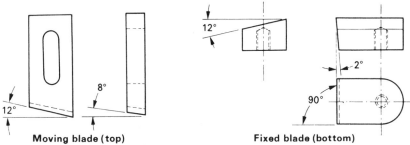

Fig. 7.14 Portable shearing machine

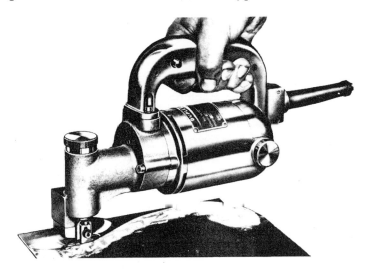

The punch type nibbler

This portable nibbling machine does not operate on the same principle as the shear type. *A punch and die is employed instead of shearing blades, and the nibbling principle is a special application of punching.*

The advantage of these machines is that they will effect certain operations that cannot be accomplished on other shearing machines. For example, they may be used to cut out apertures which could only otherwise be produced by means of specially designed punches and dies set up in a powerful press.

These portable power tools are used for rapid and accurate straight-line or curved cutting of material from approximately 1·02 mm to 3·2 mm thickness.

Like the shear type machine the top cutting tool (a punch) reciprocates at fast short strokes. Punch type nibblers are available in various sizes and the punch reciprocates at a rate of 350 to 1 400 strokes per minute over a die nibbling out the material by the simple principle of overlapping punching, and only a slight finishing is necessary to produce a smooth clean edge.

Although these machines are generally used for cutting material up to 3·2 mm thickness, there are heavy-duty machines available capable of cutting steel up to 6·35 mm thickness.

Standard punches of 4·8, 6·35 and 9·5 mm diameter are employed with different sizes of machines, and the maximum linear cutting speed is approximately 1·8 metres per minute.

One main advantage of nibbling over shearing is that there is less distortion of the work.

Figure 7.15 shows details of the Punch type nibbler.

Some models are fitted with a controlled drip feed to LUBRICATE the punch.

Portable nibbling machines are also available with narrow rectangular punches. Rectangular punches produce a cut without any ragged edges such as would be experienced with a circular punch.

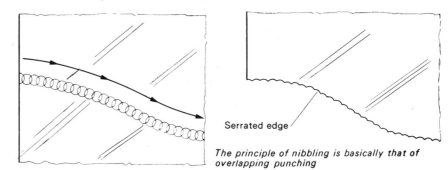

The principle of nibbling is basically that of overlapping punching

The width of the cut produced by nibbling machines is determined by the diameter of the punch in relationship with the thickness of the material to be cut.   For example:

Capacity of machine – 2mm Width of cut – 8mm Approximate cutting speed – 1·8mm/min
Capacity of machine – 3·2mm Width of cut 9·5mm Approximate cutting speed – 1·5mm/min

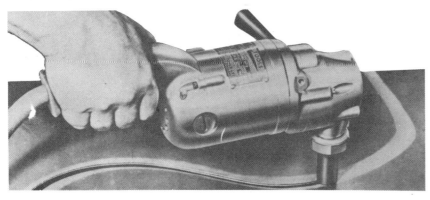

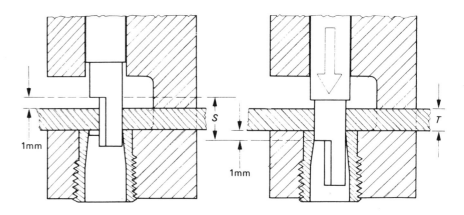

Setting of nibbling punch S = Length of stroke T = Metal thickness

The stroke is adjusted to give movement of approximately 1mm above the material and 1mm through the material, as shown in the diagram above

(b)

Fig. 7.15 Portable nibbling machine

7.9 Principles of piercing and blanking

Hand and power presses similar to those shown in Fig. 7.16 can also be used for cutting sheet metal and thin plate when fitted with suitable punches and dies (press tools).

The operation of cutting out the metal blank is called 'BLANKING'.

The operation of punching holes in the metal is called 'PIERCING'.

If a piece of sheet metal is placed on the lower cutting member (the 'DIE') and the top cutting member (the 'PUNCH'), correctly located with regard to the bottom member, is brought to bear on the metal with continuing pressure, after a certain amount of deformation, the ELASTIC LIMIT of the metal is exceeded and the top cutting member penetrates the metal. **The metal is sheared before the punch fully penetrates the whole of the metal thickness.** The degree of penetration can vary widely with the cutting conditions and the material. Usually the thicker the material, the smaller the penetration required before fracture. The average degree of penetration is between 30 and 60 per cent.

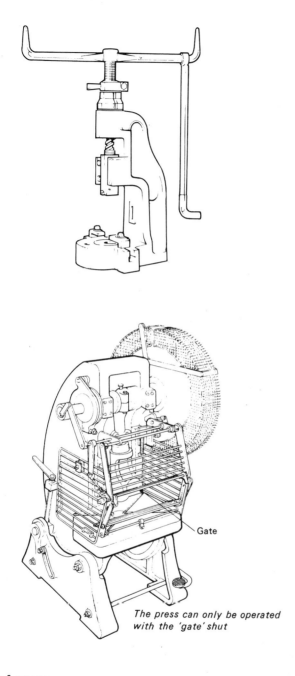

Fig. 7.16 Types of press

Figure 7.17 shows a typical piercing tool and the principle upon which it works.

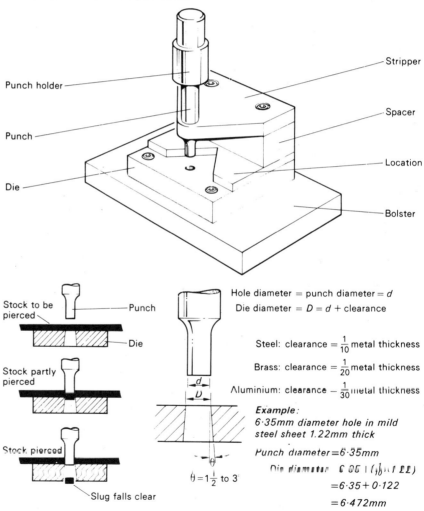

Fig. 7.17 Simple piercing tool for use in a fly press

Hole diameter = punch diameter = d
Die diameter = $D = d$ + clearance

Steel: clearance = $\frac{1}{10}$ metal thickness

Brass: clearance = $\frac{1}{20}$ metal thickness

Aluminium: clearance = $\frac{1}{30}$ metal thickness

Example:
6·35mm diameter hole in mild steel sheet 1.22mm thick

Punch diameter = 6·35mm

Die diameter = 6·35 + (1/10 × 1.22)
= 6·35 + 0·122
= 6·472mm

The principles of punches and dies can be applied to the punching of any shape, and the general arrangement is basic whether a hole is required to be round, square, rectangular or some irregular figure.

The chief difference between piercing and blanking is determined by the clearance between the punch and die, as follows:
1. In the case of piercing, the size of the hole is governed by the size of the punch, and the correct cutting clearance must be provided in the die.
2. In the case of blanking, the size of the blank is governed by the die, and the correct cutting clearance must be provided on the punch.

Clearance of punches and dies

'Clearance' (see Fig. 7.17), *is the distance that determines the quality of the cut and the load on the machine being used.*

When piercing or blanking hard materials, the clearance between the punch and die will be greater than that required for soft materials, since a harder material has greater resistance and stiffness and shows less tendency to form burrs.

If the clearance is too small then the top cutting member has to penetrate much deeper into the material thickness before severance is obtained, and *the load on the machine is increased.*

If the clearance is too great, for the particular material being cut, then the fractures from the two cutting members will not meet, and a certain amount of tearing-apart of the material results. The cut edges will be very rough but, *providing the clearance is not too great, the load on the machine is reduced.*

There is only one correct clearance allowance for any particular material, and this is dependent upon its thickness and its properties, particularly its DUCTILITY.

In general clearance varies between 5 to 10 per cent of the material thickness — this is best tested for quality of cut. *The greater the clearance the less the load required, and the thicker the material to be pierced or blanked the greater will be the clearance.*

Sometimes the faces of either the punch or die are ground at a slight angle in order to obtain greater punching efficiency. This SHEAR ANGLE (approximately 10 degrees) is illustrated in Fig. 7.18, and is essential when blanking thick material.

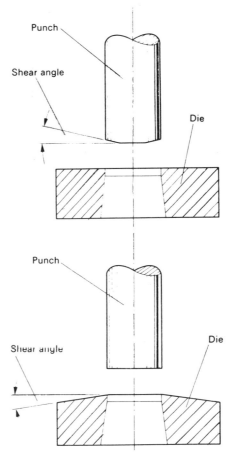

When it is required to leave the metal sheet or strip flat, the DIE FACE IS GROUND FLAT and the PUNCH IS GIVEN THE NECCESSARY 'SHEAR' as shown opposite

If the punched metal or blank is required to be left flat, the PUNCH IS GROUND FLAT and the NECESSARY 'SHEAR' IS PUT ON THE DIE as shown opposite

Fig. 7.18 Shear on punches and dies

7.10 Metal removal on the spinning lathe

The process of metal spinning inevitably results in a COMPRESSION or STRETCHING (or both) of the metal disc from which the spun article is produced. Therefore, it is not possible, by mathematical or other means (except when spinning the simplest of contours), to accurately determine the correct diameter of the disc required to produce a particular article. In practice the craftsman relies, to a great extent, on his experience to enable him to estimate an approximate diameter which will produce an equivalent area to that of the finished article PLUS AN ALLOWANCE FOR EDGE TRIMMING.

The process of trimming consists of the removal of the outer edge of the disc, either during spinning operations or on completion, with a suitable trimming or cutting tool, as illustrated in Fig. 7.19.

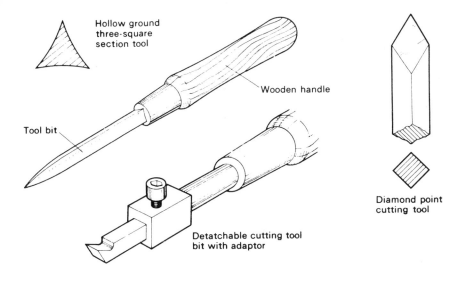

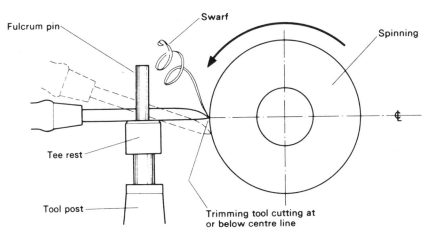

Fig. 7.19 Trimming and cutting tools (metal spinning)

Precautions

In the interest of SAFETY, when using a trimming tool:
wear a suitable glove on the hand holding the tool as a protection against trimming swarf;
never attempt to use a trimming tool above the work centre line;
always wear a safety visor for face and eye protection.

7.11 The milling machine

The milling machine produces plane (flat) surfaces that may be parallel, perpendicular, or at an angle to the work table. Unlike any of the machines previously described in *Basic Engineering*, the milling machine uses a MULTI-TOOTH cutter. Since the cutter has a number of cutting edges, it can remove metal rapidly.

Figure 7.20 shows the two basic types of milling machine. There are other types available for special applications.

It will be seen that the machine gets its name from the plane that the spindle axis lies in.

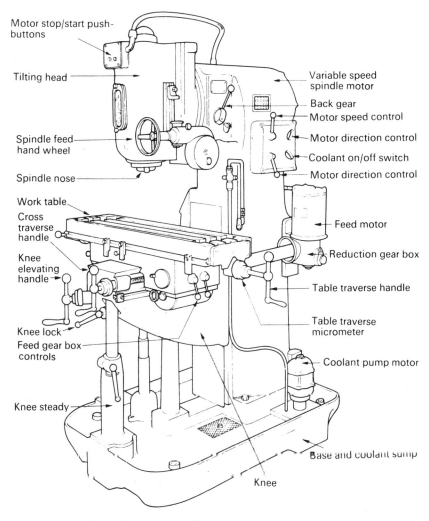

(a) Typical vertical machine

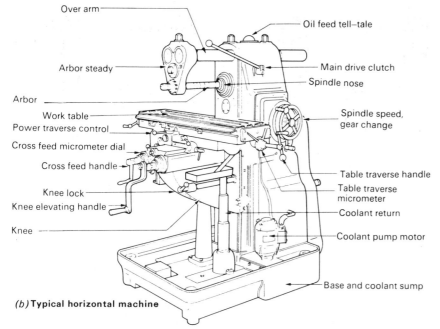

Fig. 7.20 The milling machine

The geometry of the spindle nose and the various types of cutter mounting have been outlined in *Basic Engineering*, Section 7.18. The action of the milling cutter tooth in generating the milled surface is shown in Fig. 7.21. By using a large number of teeth and a fine feed, an acceptably flat surface is produced.

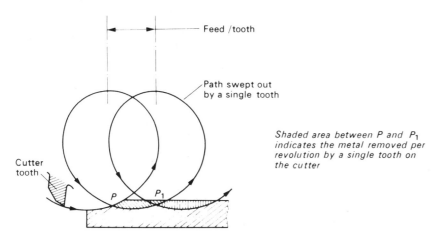

Fig. 7.21 Action of the milling cutter

7.12 The milling cutter

Figure 7.22(a) shows the geometry of a typical milling cutter tooth. The tooth form is a compromise between strength and adequate swarf clearance. The need for a secondary clearance angle to prevent interference between the heel of the tooth and the workpiece is shown in Fig. 7.22(b).

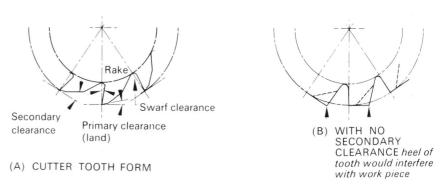

Fig. 7.22 Milling cutter tooth form

To reduce chatter, milling cutters often have helical teeth. The difference between a straight-tooth cutter and a helical-tooth cutter is shown in Fig. 7.23.

As in the single-point tool, the straight-tooth cutter is easier to design, manufacture and maintain. However, the helical tooth form not only reduces the cutting forces, it smooths out the load on the machine since as one tooth ceases to cut the next tooth commences without any break.

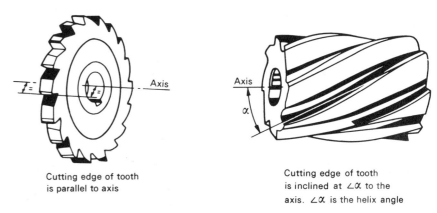

Fig. 7.23 The helical tooth cutter

7.13 Types of milling cutters

Unlike the tool bit used on the lathe the milling cutter cannot be readily adapted to new shapes to suit a particular job. Therefore milling cutters are made in a very large range of standard shapes and sizes. When a component is designed, care should be taken to ensure that it can easily be made with one of these standard cutters. Figure 7.24 shows a small range of some typical milling cutters.

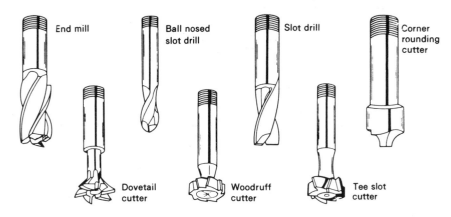

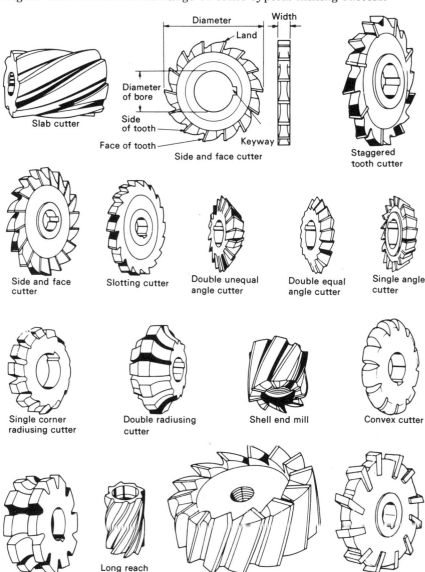

Fig. 7.24 Some typical cutter types

7.14 Mounting milling cutters

The forces acting on a milling cutter, removing metal rapidly, are very great. Therefore, the cutter arbor must be adequately supported and the cutter correctly mounted to avoid inaccuracies and chatter. The correct method of cutter mounting is shown in Fig. 7.25.

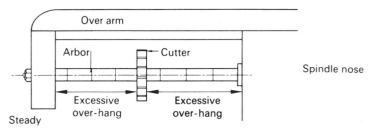

Excessive over-hang results in deflection of the arbor. This causes inaccuracy, chatter, and poor surface finish

Bad mounting

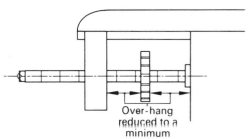

By reducing the over-hang between the cutter and the spindle nose and the cutter and the steady to a minimum, the arbor is rigidly supported and deflection and chatter are reduced to a minimum

Good mounting

139

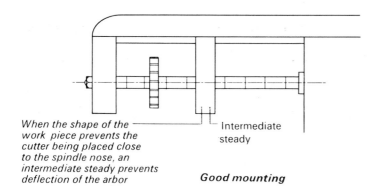

When the shape of the work piece prevents the cutter being placed close to the spindle nose, an intermediate steady prevents deflection of the arbor

Good mounting

Fig. 7.25 Cutter mounting (milling)

The restraints and locations for milling cutter mountings have already been discussed in *Basic Engineering*, Section 7.18.

7.15 Milling techniques

Figure 7.26 shows the difference between *up-cut milling* (conventional milling) and *down-cut milling* (climb milling).

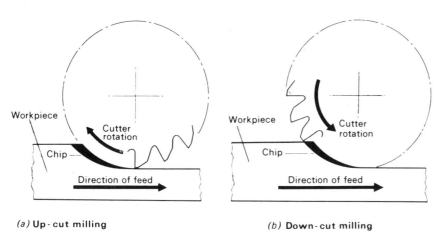

Fig. 7.26 Milling techniques

Up-cut milling

This is the traditional method and has the great advantage that there is no tendency to drag the work into the cutter and, therefore, it can be used on worn and cheap machines lacking in rigidity. However, it has a number of disadvantages.

1. The cutter tends to rub before it bites into the metal resulting in cutter wear and poor finish.
2. The cutting forces are at a maximum as the chip parts from the workpiece, resulting in transmission bounce and the traditional milling machine 'rumble'. This leads to heavy machine and cutter wear and poor surface finish.
3. The cutting forces tend to lift the component off the work table.
4. The feed mechanism must drive the workpiece against the full thrust of the cutter.

Down-cut milling

This has a number of advantages:
1. The cutter bites into the workpiece immediately and the load is eased off the tooth gradually. This leads to smooth operating conditions resulting in longer cutter and machine life and good surface finish.
2. The cutting forces press the workpiece down onto the table, giving maximum rigidity.
3. The feed mechanism only has to control the rate of feed as the cutter tends to draw the work through.

There is one major disadvantage to this technique. It must only be used on a modern machine in good condition and fitted with a BACK-LASH ELIMINATOR.

Never use down-cut milling unless you know for sure that the machine is designed for this technique and is in good condition.

7.16 Rate of material removal

To keep the cost of production as low as possible during a machining process, the waste material must be removed as rapidly as possible. Unfortunately high rates of material removal usually result in a poor finish and relatively low dimensional accuracy. For example a hole is quickly roughed out with a twist drill, but if a high degree of accuracy and finish is required the hole must be finished with a reamer which only removes a small volume of material.

The factors controlling the rate of material are:
1. The finish required.
2. The depth of cut.
3. The tool geometry.
4. The properties and rigidity of the cutting tool and its mounting.

5. The properties of the workpiece material.
6. The rigidity of the workpiece.
7. The power and rigidity of the machine tool.

In a machining operation the same rate of material removal may be achieved by using:

1. A high rate of feed and a shallow cut as shown in Fig. 7.27(*a*). Unfortunately this not only leads to a rough finish, but imposes a greater load on the cutting tool as shown in Fig. 7.27(*b*).
2. A low rate of feed and a deep cut as shown in Fig. 7.28(*a*). This gives a better finish and, if the size is correct, avoids having to take a finishing cut. It also reduces the load on the cutting tool as shown in Fig. 7.28(*b*). Unfortunately a deep cut is liable to cause excessive chatter.

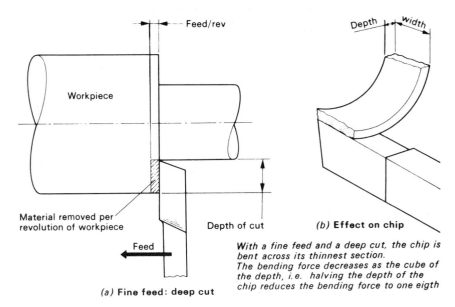

(a) **Fine feed: deep cut**

(b) **Effect on chip**

With a fine feed and a deep cut, the chip is bent across its thinnest section.
The bending force decreases as the cube of the depth, i.e. halving the depth of the chip reduces the bending force to one eigth

Fig. 7.28 Effect of deep cuts

Suitable rates of cutting speeds and feeds for high-speed steel twist drills were given in *Basic Engineering*, Section 8.14. Table 7.2 gives typical feeds and speeds for use with high-speed steel milling cutters. It will be seen that the feed is given as a rate per tooth. This is because milling cutters are relatively difficult to regrind and great care must be taken to avoid overheating and overloading the teeth.

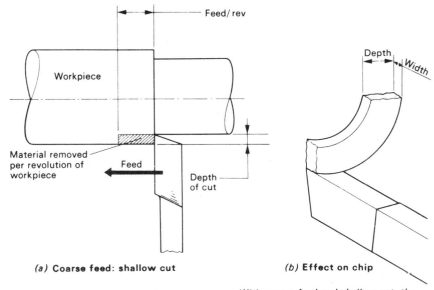

(a) **Coarse feed: shallow cut**

(b) **Effect on chip**

With coarse feed and shallow cut, the chip is bent across its deepest section. The bending force increases as the cube of the depth of the chip, i.e. doubling the depth of the chip increases the bending force eight times

Fig. 7.27 Effect of high feed rates

Table 7.2 Cutting speeds and feeds for H.S.S. milling cutters

MATERIAL BEING MILLED	CUTTING SPEED m/min	FEED PER TOOTH (CHIP THICKNESS) (Millimetres)					
		Face mill	Slab mill	Side & face	Slotting cutter	Slitting saw	End mill
Aluminium	70—100	0·2—0·8	0·2—0·6	0·15—0·4	0·1—0·2	0·05—0·1	0·1—0·4
Brass (alpha) (ductile)	35—50	0·15—0·6	0·15—0·5	0·1—0·3	0·07—0·15	0·035—0·075	0·07—0·3
Brass (free-cutting)	50—70	0·2—0·8	0·2—0·6	0·15—0·4	0·1—0·2	0·05—0·1	0·1—0·4
Bronze (phosphor)	20—35	0·07—0·3	0·07—0·25	0·05—0·15	0·04—0·07	0·02—0·04	0·04—0·15
Cast Iron (grey)	25—40	0·1—0·4	0·1—0·3	0·07—0·2	0·05—0·1	0·025—0·05	0·05—0·2
Copper	35—45	0·1—0·4	0·1—0·3	0·07—0·2	0·05—0·1	0·025—0·05	0·05—0·2
Steel (mild)	30—40	0·1—0·4	0·1—0·3	0·07—0·2	0·05—0·1	0·025—0·05	0·05—0·2
Steel (medium carbon)	20—30	0·07—0·3	0·07—0·25	0·05—0·15	0·04—0·07	0·02—0·04	0·04—0·15
Steel (alloy-high tensile)	5—8	0·05—0·2	0·05—0·15	0·035—0·1	0·025—0·05	0·015—0·025	0·025—0·1
Thermo-setting plastic (low speed due to abrasive properties)	20—30	0·15—0·6	0·15—0·5	0·1—0·3	0·07—0·15	0·035—0·075	0·07—0·3

Notes:
1. The above feeds and speeds are for ordinary H.S.S. cutters. For *super* H.S.S. cutters the feeds would remain the same, but the cutting speeds could be increased by 10% to 15%.
2. The *lower* speed range is suitable for heavy, roughing cuts.
   The *higher* speed range is suitable for light, finishing cuts.
3. The feed is selected to give the required surface finish and rate of metal removal.

EXAMPLE 1   *Calculate the spindle speed in rev/min for a milling cutter 125 mm diameter, operating at a cutting speed of 30 m/min. (Take π as 3.)*

$$N = \frac{1\,000\,S}{\pi D}$$

$$= \frac{1\,000 \times 30}{3 \times 125}$$

$$= \underline{80 \text{ rev/min}}$$

where: N = spindle speed
S = 30 m/min
π = 3
D = 124

EXAMPLE 2   *Calculate the table feed in mm/min for a 12-tooth cutter revolving at 80 rev/min when the feed/tooth is 0·1 mm.*

Feed/rev   = Feed/tooth × number of teeth
= 0·1 × 12
= 1·2 mm/rev

Table feed/min   = Feed/rev × rev/min
= 1·2 × 80
= $\underline{96 \text{ mm/min}}$

EXAMPLE 3 *Calculate the time taken to complete a cut with a slab mill given the following data.*

Diameter of cutter	125 mm
Number of teeth	6
Feed/tooth	0·05 mm
Cutting speed	45 m/min
Length of cut	270 mm

(Take π as 3)

$$N = \frac{1\,000\,S}{\pi D}$$

where: N = spindle speed
S = 45 m/min
π = 3
D = 125 mm

$$= \frac{1\,000 \times 45}{3 \times 125}$$

$$= \underline{120 \text{ rev/min}}$$

Feed/rev = Feed/tooth × Number of teeth.

= 0·05 × 6

= $\underline{0·3 \text{ mm/rev}}$

Table feed/min = Feed/rev × rev/min

= 0·3 × 120

= $\underline{36 \text{ mm/min}}$

Time to complete 270 mm cut

$$= \frac{\text{length of cut}}{\text{table feed/min}}$$

$$= \frac{270}{36}$$

$$= \underline{7·5 \text{ min}}$$

7.17 The abrasive wheel

The abrasive wheel or grinding wheel, as a means of metal removal, was introduced in *Basic Engineering*, Sections 8.16 to 8.18 inclusive.

A grinding wheel consists of two constituents:
1. The *abrasive* that does the cutting.
2. The *bond* that holds the abrasive particles together.

The specification of a grinding wheel gives a clue as to its construction and suitability for a particular operation. For example a wheel carrying the marking:

$$38A60 - J5VBE$$

would indicate that the wheel has an aluminium oxide abrasive; that the abrasive grit is medium to fine in grain size; that the grade is soft; that the structure shows medium spacing; that a vitrified bond is used. How the code marked on the wheel can indicate all this information required in selecting the correct wheel for a particular job will now be examined in some detail.

7.18 Abrasive

This must be chosen depending upon the material being cut.
1. 'Brown' aluminium oxide for general purpose grinding of tough materials.
2. 'White' aluminium oxide for grinding hard die steels.
3. 'Green' silicon carbide for very hard materials with low tensile strength such as cemented carbides.

MANUFACTURER'S TYPE CODE	BS CODE	ABRASIVE	APPLICATION
—	A	Aluminium oxide	A high strength abrasive for hard, tough materials
32	A	Aluminium oxide	Cool; fast cutting, for rapid stock removal
38	A	Aluminium oxide	Light grinding of very hard steels
19	A	Aluminium oxide	A milder abrasive than 38A used for cylindrical grinding
37	C	Silicon carbide	For hard, brittle materials of high density such as cast iron
39	C	Silicon carbide (green)	For very hard, brittle materials such as tungsten carbide

Note: In the above examples, the manufacturer's type code is based upon Norton abrasives.

7.19 Grit size

The number indicating the size of the grit represents the number of openings per linear 25 mm in the sieve used to size the grains. The larger the grit size number, the finer the grit.

COARSE	MEDIUM	FINE	VERY FINE
10	30	70	220
12	36	80	240
14	40	90	280
16	46	100	320
20	54	120	400
24	60	150	500
		180	600

7.20 Grade

This indicates the strength of the bond and therefore the 'hardness' of the wheel. In a hard wheel the bond is strong and securely anchors the grit in place and therefore reduces the rate of wear. In a soft wheel the bond is weak and the grit is easily detached resulting in a high rate of wear.

The bond must be carefully related to the use of the wheel. If it is too hard the wheel will glaze and become blunt, if it is too soft it will wear away too quickly. This would be uneconomical and also cause loss of accuracy.

VERY SOFT	SOFT	MEDIUM	HARD	VERY HARD
EFG	HIJK	LMNO	PQRS	TUWZ

7.21 Structure

This indicates the amount of bond present between the individual abrasive grains and the closeness of the individual grains to each other. An open structure wheel will cut more freely. That is, it will remove more metal in a given time and produce less heat. It will not produce such a good finish as a closer structured wheel.

CLOSE SPACING	MEDIUM SPACING	WIDE SPACING
0 1 2 3	4 5 6	7 8 9 10 11 12

7.22 Bond

1. *Vitrified bond* This is the most widely used bond and is similar to glass in composition. It has a high porosity and strength giving a wheel suitable for high rates of stock removal. It is not adversely affected by water, acid, oils or ordinary temperature conditions.
2. *Rubber bond* This is used where a small degree of flexibility is required in the wheel, as in cutting off wheels and centreless grinding control wheels.
3. *Resinoid bond* This is used for high-speed wheels. Such wheels are used in foundries for dressing castings. Resinoid bond wheels are also used for cutting off. They are strong enough to withstand considerable abuse. They are cool cutting even when metal is being removed rapidly.
4. *Shellac bond* This is used for heavy duty, large diameter wheels where a fine finish is required. For example, the grinding of mill rolls.

TYPE OF BOND	B S CODE
Vitrified bond	V
Resinoid bond	B
Rubber bond	R
Shellac bond	E

7.23 Wheel selection

The correct selection of a grinding wheel depends upon many factors.

1. Material to be ground

(a) An aluminium oxide abrasive should be used on materials with a high tensile strength.
(b) A silicon carbide abrasive should be used on materials with a low tensile strength.
(c) A fine grain wheel can be used on hard brittle materials.
(d) A coarser grain wheel should be used on soft ductile materials.
(e) *Grade.* Use a hard wheel for soft materials and a soft wheel for hard materials.

(f) *Structure.* A close structured wheel can be used on hard, brittle materials. An open structured wheel should be used on soft, ductile materials.
(g) The bond is seldom influenced by the material being ground.

2. Arc of contact

(a) Figure 7.29 shows what is meant by 'arc of contact'.
(b) *Grain size.* For a small arc of contact a fine grain can be used. For a large arc of contact a coarse grain should be used.
(c) *Grade.* For a small arc of contact a 'hard' wheel can be used. For a large arc of contact a 'soft' wheel should be used.
(d) *Structure.* For a small arc of contact a close grained wheel can be used. For a large arc of contact an open structured wheel should be used.

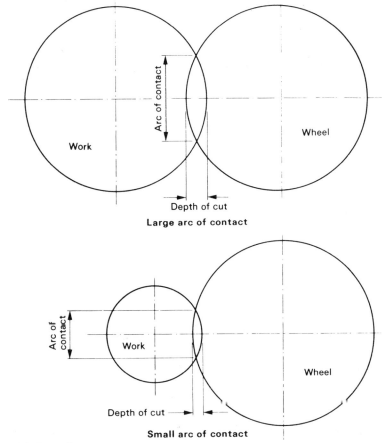

Fig. 7.29 Arc of contact

3. Type of grinding machine

A heavy, rigidly constructed machine can produce accurate work using softer grade wheels. This reduces the possibility of 'burning' the workpiece. Also broader wheels can be used increasing the rate of metal removal.

Lightly constructed machines or machines with worn wheel spindle bearings require a harder wheel.

4. Wheel speed

(a) *Grade.* Increasing the surface speed of the wheel has the same effect as increasing the hardness of the wheel.
(b) *Bond.* The strength of the bond should be sufficient to resist the bursting effect caused by rotating the wheel at high speed. Do not exceed the safe working speed marked on the wheel.

5. Process

As well as being used for tool sharpening and precision grinding, the abrasive wheel is also used for the following processes:
(a) Dressing the weld and grinding the weld bead flush.
(b) Edge preparation.
(c) Cutting off awkward sections.

Abrasive wheels used for these purposes are subject to far more abuse than those used for precision grinding. They are also required to remove metal more quickly, but they do not have to leave such a good finish or work so accurately as precision grinding wheels.

7.24 Weld grinding

1. A suitable straight wheel for heavy-duty dressing using a portable hand grinder (7.27) could be specified as:

$$A \quad 16 \quad R \quad 4 \quad B7$$

Reference to the earlier sections of this chapter will show that such a wheel would be free cutting and resistant to impact.

A	Heavy duty aluminium oxide.
16	Coarse grain for rapid metal removal.
R	Hard structure for long life.
4	Medium spacing for free cutting.
B7	Tough, shock resistant, cool-cutting resinoid bond.

2. A suitable reinforced, dished wheel for fine finishing on light gauge sheet metal-work similar to automobile body building could be specified as:

A   241   S   10   BD

Reference to the earlier sections of this chapter will show that such a wheel would give a good finish and be cool cutting.

A	General purpose aluminium oxide.
241	Very fine grain for fine finish and close control.
S	Hard structure for long life.
10	Wide spacing for cool cutting.
BD	Reinforced resinoid bond for strength and cool cutting.

7.25 Cutting off

A suitable cutting-off wheel for mild steel would be:

A   24   Q   8   B   or   R

Reference to the earlier sections of this chapter will show that such a wheel would have cool, free cutting characteristics and the ability to withstand the deflections inherent in cutting off processes.

A	Heavy duty, general purpose aluminium oxide.
24	Fairly coarse grain for rapid metal removal.
Q	Hard structure to resist excessive wear to which such a thin wheel is subjected.
8	Fairly wide spacing for cool cutting.
B	Resinoid bond for strength and cool cutting. (wheels over 1 mm wide)
R	Rubber bond to be unaffected by deflections of the wheel when less than 1 mm wide. It is also used to give less burr and a better finish in wider wheels.

7.26 Abrasive wheel cutting-off machines

A typical abrasive wheel cutting-off machine is illustrated in Fig. 7.30.

Fig. 7.30 Abrasive wheel cutting-off machine

These machines are driven by high power electric motors, the abrasive wheel revolving at speeds in excess of 3 000 rev/min.

An adjustable stop is fitted to the counterbalanced head to compensate for wheel wear. A 'vee' belt drives the cutting head from the motor. The diameter of the abrasive wheel is generally about 400 mm and its thickness approximately 3 mm. The section to be cut is clamped in a screw vice, which has a swivelling back-plate for mitre cutting. The head is brought down by hand to sever the material. The high-speed abrasive wheel cuts through steel sections like a knife cutting through butter. Table 7.3 shows typical cutting speeds.

Table 7.3 Typical cutting times (abrasive cutting-off machine)

MATERIAL SECTION			CUTTING TIME (in seconds)
	in.	mm.	
Tube	1½ O.D.	38 O.D.	2·5
Angle	3 x 3 x ¼	76 x 76 x 6·35	5
R.S.J.	3 x 1½ x ³⁄₁₆	76 x 38 x 4·76	5
Flat bar	2¼ x ⅜	57·2 x 9·53	2
Round bar	1 Diameter	25·4 Diameter	2·5
Channel	4 x 2 x ¼	101·6 x 50·8 x 6·35	9

7.27 Portable grinding machines

Portable grinding machines are often used for smoothing down welded joints and seams and generally do much of the fabrication workshop jobs which would otherwise be done by the laborious methods of chiselling and filing.

These portable tools are basically of two types, ELECTRIC and PNEUMATIC, and the three most commonly used variations of portable grinders are:

1. THE STRAIGHT GRINDER
2. THE ANGLE GRINDER
3. THE SANDER GRINDER

Figure 7.31 illustrates typical electrically-powered portable grinders.

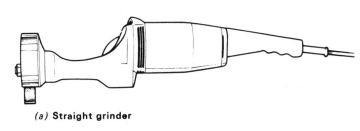

(a) Straight grinder

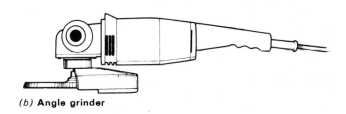

(b) Angle grinder

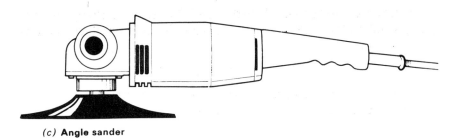

(c) Angle sander

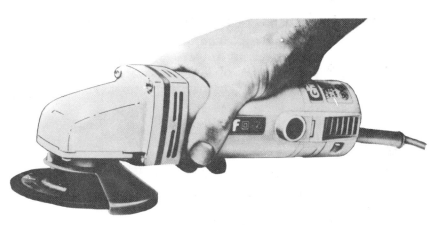

(d) Lightweight general duty high speed 'grinderette'

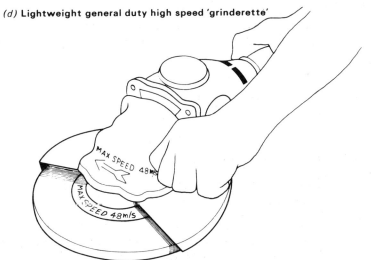

Portable grinders and all abrasive wheels larger than 55mm diameter must be marked with the maximum speed specified by the manufacturer

Fig. 7.31 Portable electric grinding machines

Figure 7.31(a) shows a 'straight' portable grinder. This uses an ordinary grinding wheel, cutting on its periphery as in tool grinding.

Usually two sizes are available:
1. Grinding wheel diameter 102 mm
   Grinding wheel thickness 19 mm
   Spindle speed (running light) 8 400 rev/min
2. Grinding wheel diameter 127 mm
   Grinding wheel thickness 19 mm
   Spindle speed (running light) 6 600 rev/min

Both these machines have the same nett mass (less wheel and guard) of 5·6 kg.

When grinding materials of high tensile strength, such as alloy, carbon and high grade steels an ALUMINOUS OXIDE grinding wheel is used. For grinding materials of low tensile strength, such as cast and chilled irons, brass, soft bronze and cemented carbides, a SILICON CARBIDE wheel must be used.

Do not allow the wheels to wear more than 25 mm below their normal diameter. To exceed this amount of wear lowers the peripheral speed of the wheel below its efficient cutting action.

Figure 7.31(b) shows an 'angle' portable grinder using a depressed centre, reinforced wheel. It will be seen that such wheels are pierced with a honeycomb of small holes to ventilate both wheels and work for cool cutting.

These machines have been designed for use with depressed centre reinforced high-speed abrasive discs. They are available in three sizes.
1. Grinding disc capacity 127 mm
   Spindle speed (running light) 11 500 rev/min
2. Grinding disc capacity 178 mm
   Spindle speed (running light) 8 500 rev/min
3. Grinding disc capacity 228 mm
   Spindle speed (running light) 6 500 rev/min

The grinding disc thickness may be 5, 6 or 10 mm, and the nett mass of these machines (less disc and guard) is 5·2 kg.

Only slight pressure, sufficient to keep the disc cutting, should be applied to the machine. *When working on a flat surface tilt the machine slightly — just sufficient to present a cutting angle to the edge of the disc.*

Figure 7.31(c) shows the 'angle sander'.

These are designed for use with resin bonded fibre-backed SANDING DISCS on a flexible backing pad made of virtually indestructible POLYURETHANE.

The sanding disc capacity is 178 mm.

Spindle speed (running light) 4 200 rev/min.

The abrasive sanding discs are available with various coarseness or fineness of abrasive grits, as follows:

0/80, ½/60, 1/50, 2/36, 3/24 and 5/16.

As a rough guide for selecting grit sizes, 0/80 is a fine abrasive disc suitable for finishing operations on metal surfaces, and will remove scratches left by a medium abrasive disc. ½/60 is a medium abrasive disc, generally used for removing weld reinforcements on light fabrications. For rough grinding of metal surfaces, abrasive discs 2/36, 3/24 and 5/16 are most suitable. Abrasive discs may be 'CLOSE COATED' or 'OPEN COATED', the latter with medium to coarse grits are used for fast stripping down of old paint and cutting down solder, where the NON-CLOG characteristics of the coating are advantageous.

Figure 7.32 shows typical applications of portable electric grinding machines.

(a) Grinding welds with angle grinder

(b) **Angle grinder used for bevel grinding**
Preparation of internal surfaces of large diameter pipe ends for welding

(c) **Grinding edges of flame-cut apertures with a straight grinder**

(d) **Use of portable sander for metal surface finishing**

ALWAYS USE THE CORRECT SIZE AND TYPE OF WHEEL FOR THE JOB — if it is too hard or too fine it becomes glazed. The operator must then use excessive pressure resulting in more breakages and reduced productivity.

Pneumatic grinders must have a mechanical governor to prevent the spindle exceeding its maximum speed.

7.28 Thermal-cutting processes

In the fabrication industry the main types of thermal-cutting processes employed are:

1. Flame cutting   : (*a*) Oxygen   (*b*) Powder
2. Electric arc   : (*a*) Oxygen   (*b*) Air/carbon   (*c*) Plasma
3. (*a*) Electron beam   (*b*) Lazer

In this chapter only FLAME CUTTING by oxygen/fuel-gas processes will be considered.

7.29 Flame cutting

Of all the methods used for material removal in the fabrication industry, the flame-cutting process plays a prominent part in the preparation of mild-steel plate material for welded fabrications. It is readily applicable to very large thicknesses of material and allows a multiplicity of shapes or contours to be cut — two points which restrict the use of guillotines.

It is faster than machining operations, which is an important advantage in connection with plate edge preparations for welded joints.

As explained in *Basic Engineering*, Chapter 4, most metals oxidise. The rate of oxidation in air depends upon the type of material and the temperature. The properties of the oxides formed are different from that of the parent metal.

Oxygen combining with a metal at a *slow rate*, as in the case of the *rusting of iron*, is referred to as 'OXIDATION', whereas if the formation of the oxide is very rapid, it is referred to as 'COMBUSTION' or 'BURNING'.

Generally, *a rise in temperature of the metal has the effect of accelerating the rate of oxidation.* In the case of mild steel when heated at a temperature of 890°C (bright cherry red), complete combustion takes place if it is in an atmosphere of pure oxygen, and a magnetic oxide of iron is formed.

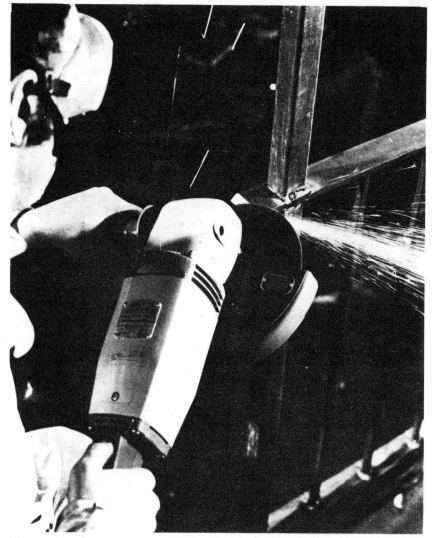

(*e*) Use of angle grinder - preparing steel balustrade section for welding on site

Fig. 7.32 Application of portable grinding machines

Safety

Hand powered tools with exposed rotating heads must be switched off and have stopped revolving before being laid down. Otherwise, they can spin themselves off scaffolding, for instance, causing damage and injury.

Power grinders and cutting off wheels must have guards for protection and to prevent oversize wheels being used.

The process

Flame cutting is made possible by two basic facts:
1. That IRON BURNS IN PURE OXYGEN WHEN HEATED TO ITS 'IGNITION' TEMPERATURE (890°C).
2. HEAT IS GIVEN OUT BY THE EXOTHERMIC REACTION BETWEEN THE IRON AND OXYGEN.

$$\text{Iron} + \text{Oxygen} \rightarrow \text{Iron Oxide} + \text{Heat}$$

The process consists of creating a local hot spot on the surface of the steel with a flame to ignition temperature and directing a high pressure jet of pure oxygen on to this PRE-HEATED spot. *A vigorous and rapid chemical reaction takes place*, the steel burns and oxides are formed. *The exothermic reaction produces a great deal more* heat. This heat is sufficient to melt the oxides formed on the metal, and also to heat and melt the metal itself over a very localised area. The melted oxides are fluid and are blown away by the force of the oxygen stream, so that more metal is exposed.

The action of 'oxygen cutting' is shown in Fig. 7.33.

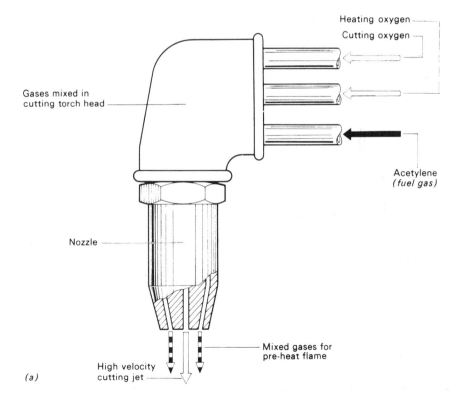

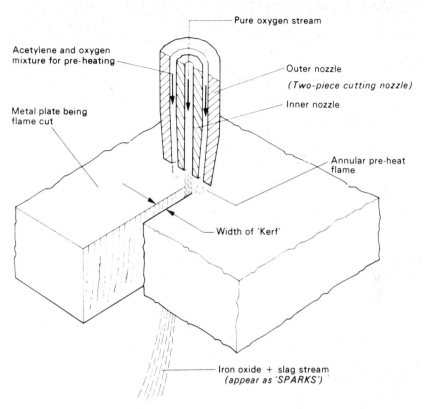

Fig. 7.33 The action of oxygen cutting

7.30 Flame-cutting equipment

Since the oxygen-cutting process involves directing a high-pressure jet of oxygen continuously on to an area of steel that has been previously heated to ignition temperature, the basic equipment combines a pre-heating flame and a pure oxygen source. Both these requirements are provided by a specially designed cutting torch and nozzle combination which is connected to a FUEL GAS and an OXYGEN supply, in the same manner as a gas-welding torch. The cutting torch and nozzle combinations are of varying design, depending on the specific type of cutting to be employed.

The equipment required for gas welding has been described in Part A of Chapter 11, and oxy-gas cutting equipment is basically similar except for the following items:

Pressure regulators

OXYGEN

An oxygen pressure regulator complete with gauges giving a higher outlet pressure (up to 6·5 bar) than for welding is required. *The standard* acetylene *pressure regulator and gauges, giving outlet pressures up to 0·28 bar, used for welding is also used for oxy-acetylene cutting.*

PROPANE

If propane is the fuel gas used for cutting, a special pressure regulator must be used giving outlet pressures up to 0·6 bar.

NATURAL GAS

When natural gas is used for cutting torches, no pressure regulator is required (mains pressure) *but a non-return valve must be fitted in the supply line to prevent flash back.*

Both OXY-PROPANE and OXY-NATURAL GAS cutting processes use the same type of OXYGEN pressure regulator as the OXY ACETYLENE cutting process.

Cutting torch

The cutting torch may be either a HIGH-PRESSURE or LOW-PRESSURE type. It is similar in construction to that employed for welding.

The *'high-pressure cutting torch'* uses acetylene or propane as the fuel gas, and these are supplied from cylinders. It has a 'Gas Mixer' which is usually incorporated in the torch head.

The *'low-pressure cutting torch'* embodies an *'injection mixer'*, and may be used for either acetylene or natural gas supplied at low pressure.

Injection mixers work on the principle of using the flow energy of one gas (usually OXYGEN) to draw in the second gas (usually the fuel gas).

Figure 7.34(*a*) the basic details of any oxy-acetylene cutting torch together with some of the points to be observed in hand cutting.

Cutting nozzles

The cutting torch nozzles are so designed that they have a central

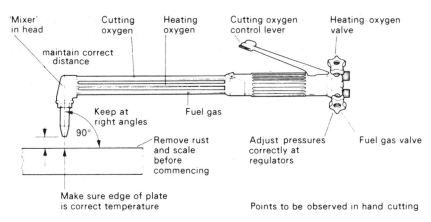

(a) **The cutting torch**
The size of the torch used depends upon whether it is for light duty or heavy continous cutting and the volume of oxygen used is much greater than of fuel gas (measured in LITRES PER HOUR)

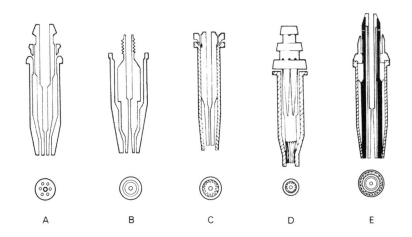

(b)   **nozzle design features**
A   One-piece ACETYLENE cutting nozzle - parallel bore, 3 - 9 pre-heat holes, no skirt.
B   Two-piece ACETYLENE cutting nozzle - venturi bore, pre-heat annulus, no skirt.
C   Two-piece NATURAL GAS moggle - venturi bore, pre-heat flutes, long skirt
D   Two-piece PROPANE nozzle - parallel bore, pre-heat slots, long skirt
E   Two-piece PROPANE nozzle - parallel bore, pre-heat flutes, long skirt, oxygen curtain

Fig. 7.34 Oxygen cutting-torch details

port around which is either an annulus or several smaller ports. *The smaller ports are circular holes in the case of acetylene cutting nozzles, and take the form of annular slots for propane.* Through the smaller ports is fed the fuel gas mixed with the correct proportion of oxygen for the purpose of pre-heating the metal. The central port of orifice through which the main jet of oxygen is released will vary in diameter according to the size of the cutting nozzle required. *The orifice diameter is increased as the thickness of the plate to be cut is increased* — in general, a particular size of nozzle is used to cut a small range of thicknesses.

Figure 7.34(*b*) gives details of typical cutting nozzles.

7.31 Flame adjustment

The procedure used for lighting a welding torch is adopted when lighting a cutting torch, but with one important difference. The fuel gas regulator is set to the correct working pressure in the normal way and *the oxygen regulator is set to the correct working pressure with the cutting oxygen valve on the torch in the open position.*

The fuel gas is lit and the flame adjusted, in the same manner as for a welding torch, until it ceases to smoke. The heating oxygen valve is then opened and adjusted (similar to a neutral flame setting) until there is a series of nicely defined white inner cones in the flame (in the case of the multi-port type nozzle) or a short white conical ring, if the nozzle is of the annular port type.

At this stage the cutting oxygen valve is opened and the flame readjusted to a neutral condition. *The oxygen cutting valve is then closed and the torch is ready for use.*

Figure 7.35 illustrates flame adjustment for cutting.

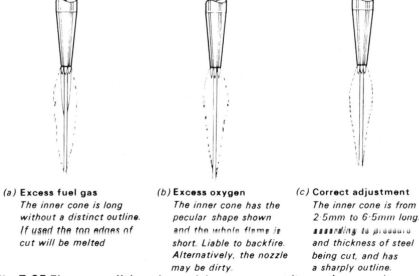

(a) **Excess fuel gas**
The inner cone is long without a distinct outline. If used the top edges of cut will be melted

(b) **Excess oxygen**
The inner cone has the peculiar shape shown and the whole flame is short. Liable to backfire. Alternatively, the nozzle may be dirty.

(c) **Correct adjustment**
The inner cone is from 2·5mm to 6·5mm long, according to pressure and thickness of steel being cut, and has a sharply outline.

Fig. 7.35 Flame conditions (set with oxygen cutting valve open)

When OXY-PROPANE is used for cutting, the correctly adjusted pre-heating flame will be indicated by a small non-luminous central cone with a pale blue envelope.

In the case of OXY-NATURAL GAS the flame is adjusted until the luminous inner cone has a clear definite shape, usually up to 8—10 mm in length.

The correct procedure to extinguish the flame is as follows:
1. Turn off the cutting oxygen.
2. Close the fuel gas control valve.
3. Close the heating oxygen control valve.

GAS PRESSURES

Table 7.4 lists the approximate gas pressures for cutting mild-steel plate.

Table 7.4 Approximate pressures for hand cutting steel plate

PLATE THICKNESS		NOZZLE SIZE		GAS PRESSURES			
				Acetylene lbf/in²	bar	Oxygen lbf/in²	bar
in.	mm.	in.	mm.				
Acetylene							
1/8	3·2	1/32	0·75	2	0·14	15	1·05
1/4	6·4	1/32	0·75	2	0·14	25	1·8
1/2	12·5	3/64	1·0	2	0·14	30	2·1
1	25·4	1/16	1·5	2	0·14	35	2·5
2	51	1/16	1·5	2	0·14	45	3·2
3	76	1/16	1·5	2	0·14	50	3·5
4	100	5/64	2·0	2	0·14	60	4·2
6	150	5/64	2·0	2	0·14	75	5·3
Propane							
1/8	3·2	1/32	0·75	3	0·21	25	1·8
1/4	6·4	1/32	0·75	3	0·21	25	1·8
1/2	12·5	3/64	1·0	3	0·21	40	2·8
1	25·4	1/16	1·5	3	0·21	45	3·2
2	51	1/16	1·5	3	0·21	50	3·5
3	76	1/16	1·5	3	0·21	60	4·2
4	100	5/64	2·0	4	0·28	70	4·9
6	150	5/64	2·0	4	0·28	80	5·6

PLATE THICKNESS		NOZZLE SIZE		GAS PRESSURES			
				Acetylene		Oxygen	
in.	mm.	in.	mm.	lbf/in²	bar	lbf/in²	bar
Natural gas							
1/8	3·2	1/32	0·75	Mains	—	25	1·8
1/4	6·4	1/32	0·75	Mains	—	25	1·8
1/2	12·5	3/64	1·0	Mains	—	30	2·1
1	25·4	1/16	1·5	Mains	—	45	3·2
2	51	1/16	1·5	Mains	—	55	3·9
3	76	5/64	2·0	Mains	—	60	4·2
4	100	5/64	2·0	Mains	—	65	4·6
6	150	3/32	2·5	Mains	—	70	4·9

Note:
The above figures are given only as a guide since the actual requirements may vary according to the nature of the work.
Some pressure regulators are fitted with gauges which are calibrated in kg/cm². *1 kg/cm² is approximately equal to 1 bar.*

7.32 Factors influencing the quality of the cut

The success of the flame-cutting operation depends upon:
1. Selecting the correct size of cutter nozzle for the thickness of the material being cut.
2. Operating the cutting torch at the correct oxygen pressure.
3. Moving the cutting torch at the correct cutting speed.
4. Maintaining the nozzle at the correct distance from the plate surface.

If the torch is adjusted and manipulated correctly, a smooth narrow cut, termed the 'KERF' (see 7.35), is produced.

The PRE-HEATING FLAME has two specific functions:
1. *To transmit sufficient heat to the surface of the steel to compensate for the heat loss due to* **thermal conductivity**.

In this respect ACETYLENE, because of its high flame temperature, has the following advantages over PROPANE and NATURAL GAS:
(a) Faster pre-heat starts.
(b) More able to penetrate priming, scale and rust.
(c) Provides for faster cutting speeds on thinner plate where the HEAT SPREAD across the plate is the controlling factor.

The disadvantage of the high-temperature acetylene preheat flame is the tendency for excessive melting of the top edges of the cut.

This is generally associated with the use of too large a pre-heating flame once the cut has started, and this problem may be eliminated by correct flame control and distance from the workpiece.

2. *To provide sufficient heat to raise a small local area of the surface to 'ignition temperature'.*

The pre-heating flame must be of sufficient intensity in order to break up surface scale and to maintain the steel at ignition temperature irrespective of surface irregularities.

Only those materials whose combustion or ignition temperature is below their melting point can be flame cut. Otherwise the material would melt away before OXIDATION could take place, making it impossible to obtain a cleanly cut edge.

NON-FERROUS METALS CANNOT NORMALLY BE FLAME CUT. Cast iron and stainless steel require special procedures and even then a 'flame-cut edge' of the same quality as with plain carbon steel is difficult to obtain.

Figure 7.36 illustrates the effects of variation in flame cutting procedure.

VISUAL CUTTING EFFECT	GENERAL INDICATIONS
Good quality cut	The edges are sharp and the kerf is smooth. Only light and easily removable scale. Drag lines are almost vertical with a very smooth appearance
Cutting nozzle too high	Excessive melting of top edge. Undercutting at top of cut face. Sharp bottom edge to kerf
Cutting nozzle too low	Small bead along the top edge, which is slightly rounded. Very slight undercutting effect with sharp bottom corner

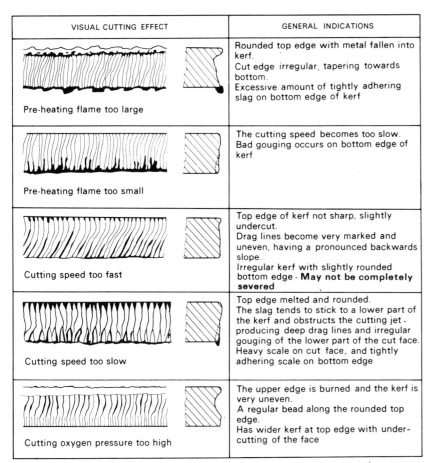

Fig. 7.36 Effects of variations in flame-cutting procedure

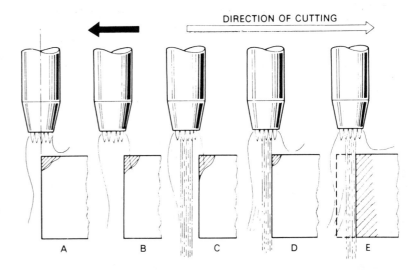

A   Pre-heat to ignition temperature
B   Move the cutting torch backwards, just clear of edge
C   Open the cutting oxygen valve
D   Commence cutting
E   Continue cutting

(a) Starting a cut on the edge of heavy plate

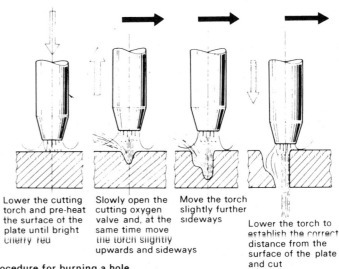

(b) Procedure for burning a hole

Fig. 7.37 Cutting techniques

7.33 Applications of flame-cutting (oxy-fuel gas)

Hand cutting

Before commencing to cut thick plate, it is necessary for the edge to be well heated. The procedure for starting the cut is illustrated in Fig. 7.37(*a*).

When starting a cut away from an edge, a hole should either be drilled, or produced with the cutting torch. Figure 7.37(*b*) shows the procedure for piercing a hole with the cutting torch.

When round bar is to be flame cut, it is advisable to make a nick with a cold chisel at the point where the cut is to start. This enables the cut to be started much more easily.

Once the cut is commenced, the cutting torch should be moved steadily and at a uniform speed, with the small cone of the pre-heating flame just clear of the work surface. *There must be no vibration of the cutting head as such movement will result in a ragged cut and, in some cases, the cut being halted.*

Various attachments are available which, when fitted to the hand-cutting torch, ensure a steady rate of travel and enable the operator to execute straight lines, bevels or circles with much greater ease. These very useful devices (Fig. 7.38) are generally attached to cutting nozzle. The following are some of the devices in common use:

SINGLE CUTTING SUPPORT

This simple device may either be a 'spade support' or a single 'roller guide' which can be adjusted vertically for 'stand off'. Figure 7.38(a) shows a single cutting support used in conjunction with a 'straight edge'.

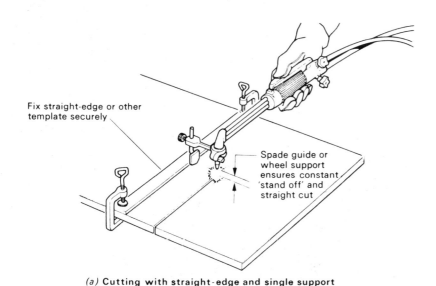

Fig. 7.38 Useful cutting torch attachments

CIRCLE-CUTTING DEVICE

Basically there are three types of devices which may be used for radial cuts and circle cutting.

1. *Small circle-cutting attachment* This is simply a pivot which is attached to the shank of the torch at a particular distance from the

nozzle according to the radius required. Such a device is illustrated in Fig. 7.38(b).

2. *Large circle-cutting attachment* This extremely useful device is illustrated in Fig. 7.38(c).

3. *Special cutting device for small holes* Some cutting torches are supplied in kit form and include a specially designed interchangeable shank in which the cutting head is 'in line' instead of being at the normal 90 degrees. This special head is used in conjunction with a pivot for cutting small holes, as illustrated in Fig. 7.39.

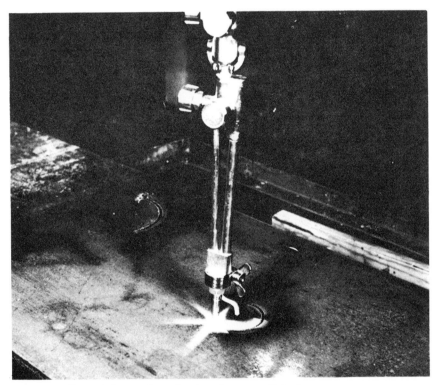

Fig. 7.39 Special cutting device for small holes

DOUBLE CUTTING SUPPORT

This device consists of a U-frame and two rollers which makes it easier to move the cutting torch along the workpiece at the correct distance from the plate. Like the single support, it has provision for vertical adjustment. Certain types of double cutting supports are designed for square and bevel cutting. These are commonly referred to as 'bevel-cutting attachments'. Figure 7.40 illustrates the versatility of a bevel-cutting attachment.

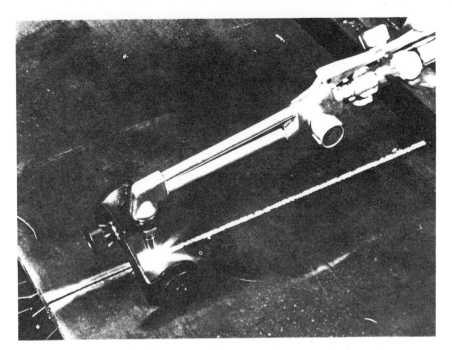

(a) Cutting a square edge

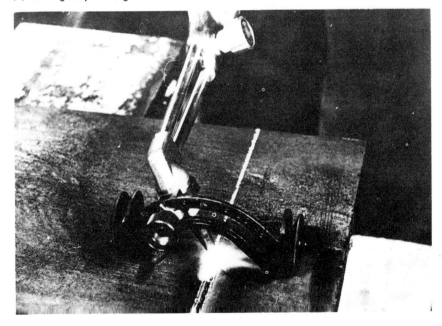

(b) Cutting a bevel edge

Fig. 7.40 Use of a double cutting support

Rivet removal

When carrying out repair work on riveted structures it is often necessary to remove rivets. They may be quickly removed with the aid of a cutting torch by following the procedure illustrated in Fig. 7.41.

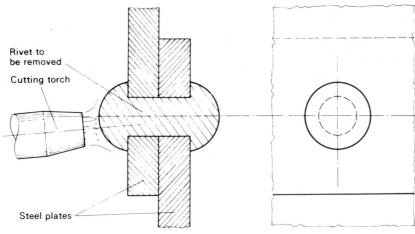

1. The centre of the rivet head is heated until bright cherry red, and the edge of the hole becomes clearly visible (A)
2. The cutting oxygen jet is turned on, and with the torch slightly inclined towards the centre, a cut is carefully made around the edge of the hole, without damaging it (B)
3. On reaching the halfway point the cut is made towards the centre (C)
4. With the cutting torch slightly inclined inwards complete the cut around the edge of the hole (D)
5. The rivet is easily removed by knocking out with a sharp hammer blow on a solid steel punch, as indicated in (E)

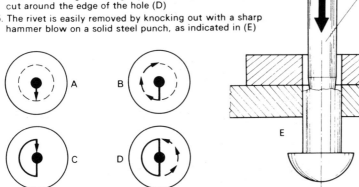

Fig. 7.41 Rivet removal by flame cutting

Flame gouging

This process is very similar to flame cutting except that instead of severing the metal, a groove is gouged out of the surface of the plate. The principle of operation is the same as that used in oxy-fuel gas-cutting processes, except that a special type of nozzle is used in a standard cutting torch. A pre-heating flame is used to bring the metal to ignition temperature, the cutting oxygen is switched on, OXIDATION occurs and cutting is commenced. It is a very useful process for the removal of defective welds and defects in the parent metal. Various details of the flame gouging process are illustrated in Fig. 7.42.

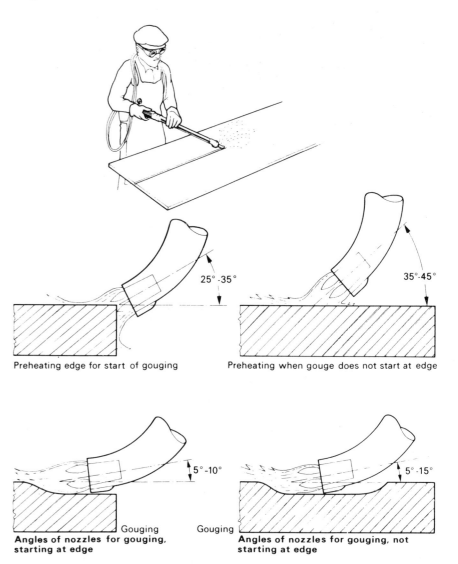

Preheating edge for start of gouging

Preheating when gouge does not start at edge

Angles of nozzles for gouging, starting at edge

Angles of nozzles for gouging, not starting at edge

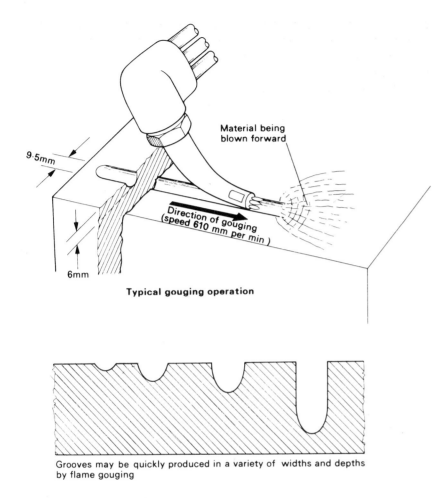

Fig. 7.42 Flame gouging

7.34 Cutting machines (oxy-fuel gas)

Cutting machines basically consist of one or more cutting torches and a means for supporting and propelling them in the required directions with high precision.

There are many machines available, ranging from simple lightweight portable types to static types which are large and versatile with several cutting heads.

A typical portable cutting machine is illustrated in Fig. 7.42(a). It is basically a 'SELF PROPELLED LIGHTWEIGHT TRACTOR' which can be adapted to carry a range of cutting or welding equipment. The machine details are shown in Fig. 7.43.

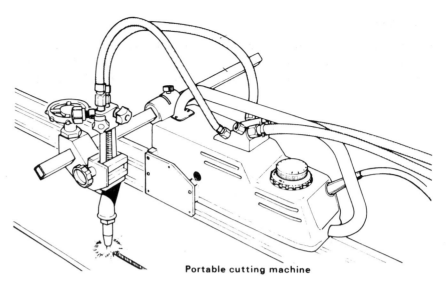

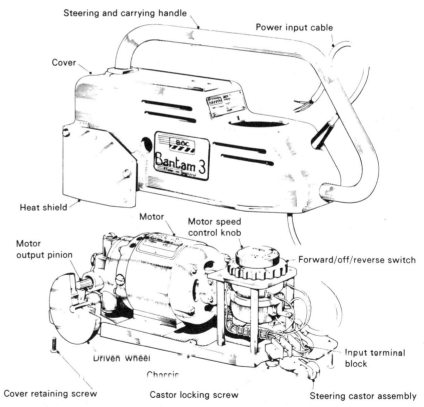

Fig. 7.43 Portable type oxygen-cutting machine

The machine will make runs of any length and can be controlled for straight-line work by the use of, either an extruded aluminium track, or by guiding the tractor along a straight line. One or two cutting heads may be used. Contours may be followed by hand steering the tractor along a drawn line and radial or circle-cutting by using a radius bar attachment. *When using the machine for contour or radial cutting, the castor wheel must be free to swivel and all three wheels should be in contact with the work.*

This machine is very useful for plate edge preparation for welding, as bevel cutting may be carried out by setting the swivel cutting head to the required angle. Two cutting heads may be fitted, one slightly in advance of the other, to enable 'nose and bevel cutting' to be done, and it is important that:

(a) *When making an 'under bevel' cut the leading cutting head should cut the bevel and the second head the nose.*

(b) *When making an 'over bevel' cut the leading cutting head should cut the nose and the second head the bevel.*

Figure 7.44 illustrates typical cutting applications of a portable cutting machine.

Table 7.5 gives cutting data for use with a portable cutting machine.

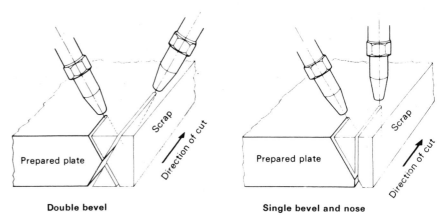

Diagrammatic arrangement of cutting nozzles and sequence of cut for plate edge preparation

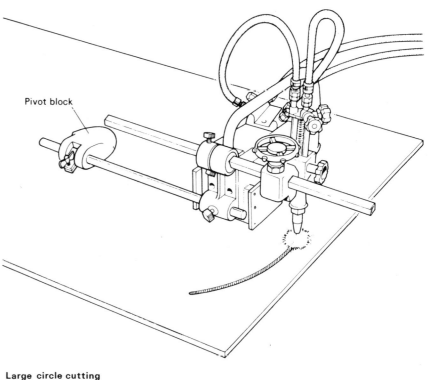

Large circle cutting

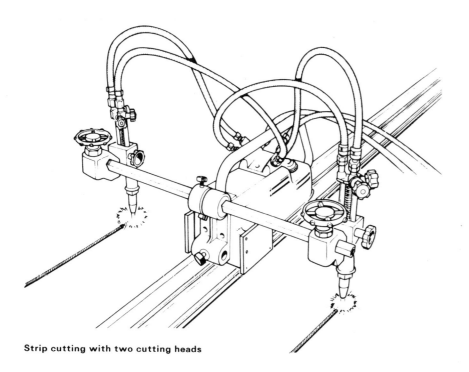

Strip cutting with two cutting heads

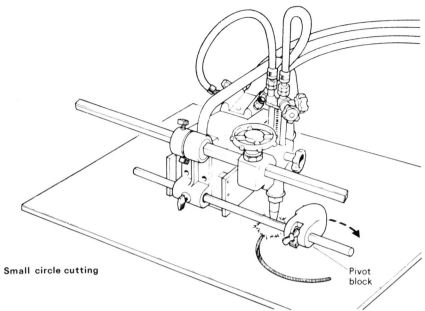

Small circle cutting

Fig. 7.44 Typical cutting applications (portable machines)

The larger static cutting machines are often referred to as 'FLAME PROFILE CUTTING MACHINES', as their greatest use is the profile cutting of shapes within a relatively large area in steel plate up to considerable thickness.

The machine illustrated in Fig. 7.45 is a cross-carriage profile cutting machine capable of flame-cutting shapes from steel plate up to a maximum thickness of 150 mm and, if desired, two cutting heads may be mounted and operated simultaneously. This machine can profile cut shapes with a plate area of 1·2 m x 1·2 m and the cutter heads are so mounted as to permit cutting up to 45 degrees on either side of the vertical.

Table 7.5   Operating data for portable machines

English		Acetylene						Propane					
Plate thickness (in)		1/4	1/2	1	2	4	6	1/4	1/2	1	2	4	6
Nozzle size (in)		1/32	3/64	1/16	1/16	5/64	5/64	1/32	3/64	1/16	1/16	5/64	3/32
Speed in ins/min		21	14	11	8	7	6	18	13	10	7	6	5
Regulator	Fuel Gas	2	2	2	2	2	2	2	2	4	4	4	4
Pressures	Heating Oxygen	5	5	6	6	8	8	9	9	10	10	12	12
lbf/in²	Cutting Oxygen	20	30	35	40	55	65	20	30	35	40	55	65

		Speed Control Settings									
		1	2	3	4	5	6	7	8	9	10
Roller Diameter in inches		Cutting Speed (in/min) These columns correspond with the above settings in relation to the appropriate size roller									
Spindle	1/8in	3	6	9	13	16	19	22	25	28	31
Roller	1/4in	6	13	19	25	31	38	44	50	57	63
Magnet	3/8in	2	5	7	9	12	14	16	19	21	24
Roller	1/2in	3	6	10	13	16	19	22	25	29	32
Wheel tracing		4	9	13	18	22	26	31	36	40	44

Metric		Acetylene Nozzle						Propane Nozzle					
Plate Thickness (mm)		6	12	25	50	100	150	6	12	25	50	100	150
Nozzle		1/32	3/64	1/16	1/16	5/64	5/64	1/32	3/64	1/16	1/16	5/64	3/32
Cutting Speed (cm/min)		53	36	28	20	18	15	46	33	25	18	15	13
Regulator	Fuel Gas	0·14	0·14	0·14	0·14	0·14	0·14	0·14	0·14	0·28	0·28	0·28	0·28
Pressures	Heating O_2	0·35	0·35	0·42	0·42	0·56	0·56	0·63	0·63	0·7	0·8	0·8	0·8
Bar (b)	Cutting O_2	1·4	2·1	2·5	2·8	3·9	4·6	1·4	2·1	2·5	2·8	3·9	4·6

		Speed Control Setting									
		1	2	3	4	5	6	7	8	9	10
Roller Dia (mm)		Cutting Speed (m/min) (Corresponding with the related Speed Control Setting)									
Spindle	3·175	0·076	0·15	0·23	0·33	0·41	0·48	0·56	0·64	0·71	0·79
Roller	6·35	0·15	0·33	0·48	0·64	0·78	0·97	1·12	1·27	1·42	1·60
Magnet	9·525	0·05	0·13	0·18	0·23	0·30	0·36	0·41	0·48	0·53	0·61
Roller	12·7	0·076	0·15	0·23	0·33	0·41	0·48	0·56	0·64	0·74	0·81
Wheel Tracing		0·10	0·23	0·33	0·46	0·56	0·66	0·78	0·91	1·02	1·12

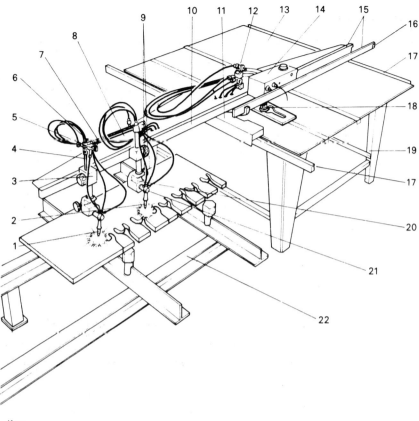

Key:
1. Cutter (optional)
2. Cutter
3. Cutter mounting block
4. Cutter mounting tube
5. Trimming valve (heating oxygen)
6. Trimming valve (fuel gas)
7. Trimming valve (cutting oxygen)
8. Web plates
9. Grip knob
10. Cutter mounting bar
11. Cross carriage unit
12. ON/OFF cock (cutting oxygen)
13. Tracing table
14. Drive unit
15. Traversing carriage rails
16. Rail connecting fitting
17. Cross carriage rails
18. Following head
19. Mounting frame
20. Cutter head mounting block
21. Cutter head
22. Work rests (optional)

Fig.7.45 Typical light-duty oxygen cutting machine

Three interchangeable tracing heads may be utilised with the machine illustrated in Fig. 7.45 depending upon the type of TEMPLATE used for profile cutting, as follows:

1. *A wheel tracing head*   Used for tracing direct from an outline drawing. This method of tracing is most economical for 'one-off' production. It consists of a fine-toothed tracking wheel, steered by hand along the line to be cut. It is driven through a vertical spindle assembly and bevel gear linkage contained within the main body. The body screwed connection remains static to the drive unit while the main body, complete with tracing wheel, is free to turn through 360 degrees. A pointer to assist the operator in guiding the tracing wheel along its path is also incorporated.

Provision is made on the head for attaching a circle-cutting radius bar, capable of cutting circles and flanges up to 350 mm diameter.

2. *A spindle tracing head*   Used with either wooden, hardboard or non-ferrous templates. This is a revolving vertical spindle attached to the drive unit. It has a detachable knurled roller that rotates round either the inner or outer profile of a template. Generally there are two diameters of roller available, either 3 mm or 6 mm, and the larger roller should be used where possible except when the template corners require the smaller one. The spindle rotates in a stationary outer sleeve whose lower end is knurled to provide a non-slip hand grip to enable the roller to be held against the edge of the template in order to follow the profile.

3. *A magnetic head*   Used for steel templates only and *has the advantage over the other tracing heads in that it automatically follows the template profile.* The vertical driving spindle is magnetised by a solenoid encased in the outer shell of the main body of the driving head. The head is generally supplied with two interchangeable knurled steel rollers, usually 9 mm and 12 mm in diameter.

Both the 'spindle tracing head' and the 'magnetic head' find their greatest use in repetition work, and these are illustrated in Fig. 7.46.

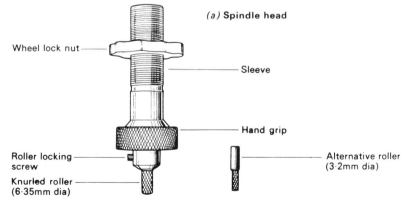

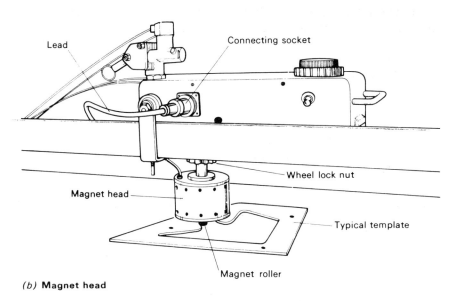

(b) **Magnet head**

Fig. 7.46 Profile cutting machine tracing heads

Static flame profiling machines require three pressure regulators:
1. FUEL GAS PRESSURE REGULATOR (acetylene or propane)
2. HEATING OXYGEN PRESSURE REGULATOR
3. CUTTING OXYGEN PRESSURE REGULATOR

Table 7.6 gives useful cutting data when using 'mixer type' nozzles for flame cutting with the machine illustrated in Fig. 7.45.

Table 7.6   Cutting data for profile cutting machines

FUEL GAS – ACETYLENE

VERTICAL CUTTING

PLATE THICKNESS		NOZZLE		GAS PRESSURES				NOZZLE HEIGHT ABOVE PLATE		CUTTING SPEED APPROX.	
				ACETYLENE		OXYGEN					
in	mm	mm	in	lbf/in^2	Bar (6)	lbf/in^2	Bar (b)	in	mm	in/min	m/min
1/8	3.2	0.75	1/32	2	0.14	20	1.41	1/4	6	28	0.71
1/4	6.4	0.75	1/32	2	0.14	30	2.11	1/4	6	21	0.53
1/2	12.5	1.0	3/64	2	0.14	30	2.11	1/4	6	16	0.41
1	25.4	1.5	1/16	2	0.14	40	2.80	5/16	8	15	0.38
2	51	1.5	1/16	2	0.14	55	3.80	5/16	8	11	0.28
4	100	2.0	5/64	3	0.21	70	4.9	3/8	9.5	6.7	0.15-0.18

30° BEVEL CUTTING

1/8	3.2	0.75	1/32	2	0.14	25	1.75	1/4	6	20	0.51
1/4	6.4	0.75	1/32	2	0.14	30	2.11	1/4	6	18	0.46
1/2	12.5	1.0	3/64	2	0.14	50	3.50	1/4	6	16	0.41
1	25.4	1.5	1/16	2	0.14	50	3.50	5/16	8	13	0.33
2	51	1.5	1/16	2	0.14	70	4.92	5/16	8	9	0.23

FUEL GAS – PROPANE

VERTICAL CUTTING

				PROPANE		OXYGEN					
1/8	3.2	9.75	1/32	3	0.21	25	1.76	1/4	6	20	0.51
1/4	6.4	0.75	1/32	3	0.21	25	1.76	1/4	6	19	0.48
1/2	12.5	1.0	3/64	3	0.21	40	2.81	1/4	6	16	0.41
1	25.4	1.5	1/16	3	0.21	45	3.17	5/16	8	14	0.36
2	51	1.5	1/16	3	0.21	50	3.52	5/16	8	10	0.25
4	100	2.0	5/64	4	0.28	75	5.3	3/8	9.5	6.7	0.15-0.18

30° BEVEL CUTTING

1/8	3.2	1.0	3/64	3	0.21	25	1.76	1/4	6	16	0.41
1/4	6.4	1.0	3/64	3	0.21	30	2.11	1/4	6	11	0.28
1/2	12.5	1.5	1/16	3	0.21	50	3.52	1/4	6	11	0.28
1	25.4	1.5	1/16	3	0.21	55	3.87	5/16	8	10	0.25
2	51	1.5	1/16	3	0.21	70	4.92	5/16	8	7	0.18

Note: The above figures are given only as a guide since the actual requirements may vary according to the nature of the work.

7.35 Template and allowance for profile cutting

Suitable templates may be manufactured from steel or light alloy sheet, plywood or hardboard, depending on the type of spindle head to be used.

In general, an EXTERNAL TEMPLATE is used when the piece to be cut from the plate is the component and an INTERNAL TEMPLATE when the piece cut from the plate is not required for the component.

Composite templates may be used where the component to be cut has both an external and an internal profile, and also includes a hole or holes.

To obtain the best results for accurate cutting, templates should conform to the following basic requirements:
1. Minimum thickness 3 mm.
2. Edges must be square. Plywood or hardwood templates should have their edges prepared with a coarse sandpaper finish, and metal template edges should have a good file finish but not too smooth, *in order to provide sufficient frictional grip for the knurled steel rollers.*
3. When the inside corner of the component to be cut is radiused, *the corner radius on the template must be greater than that of the roller.*
4. Correct allowances must be made in respect of the width of the KERF and the diameter of the tracing roller.

Allowances for flame cutting

These will vary according to the width of the kerf, the diameter of the tracing roller and whether an internal or external template is used.

The kerf This is a term used to define the width of the metal consumed in the cutting process. It may vary between 1½ and 2 times the diameter of the cutting oxygen orifice of the cutting nozzle used; for example, *a 1·6 mm diameter nozzle will produce a kerf of between 2·4 mm and 3·2 mm in width.*

Allowances (to compensate for the kerf and the diameter of tracing roller) must be made on the size of the template and these will differ PLUS or MINUS depending whether an internal or external template is used.

Figure 7.47 illustrates template allowance, the component profile is shown in heavy outline with its applicable template shown hatched.

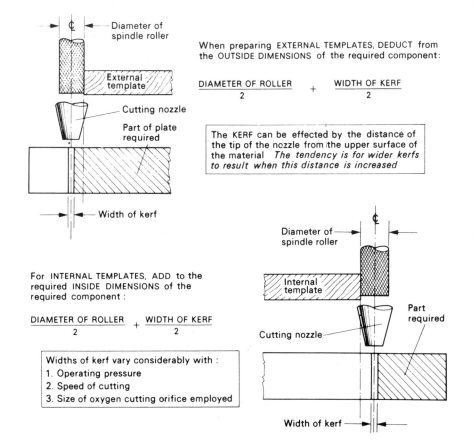

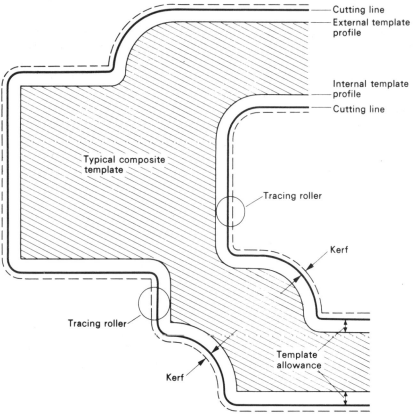

Fig. 7.47 Template allowances

For future use it is advisable to mark templates with the following information:
1. Nozzle type and size.
2. Fuel gas used.
3. Tracing roller diameter.
4. Thickness of plate cut.
5. Speed of cut.
6. Part number (if applicable) of component.

When using a WHEEL TRACING HEAD, allowances must be made on the drawing dimensions for the kerf width. As a general guide, allow the diameter of the cutting nozzle orifice per side, plus or minus for external and internal cuts respectively.

When CIRCLE-CUTTING with a RADIUS BAR, allowances are the same as those for wheel tracing.

7.36 Comparison of flame cutting with shearing

Since the introduction of the first cutting torch at the beginning of the twentieth century, oxygen cutting has brought about far-reaching changes in the industrial practices of cutting and shaping of steel.

The flame-cutting process is more versatile than any other known cutting or shearing process. A multiplicity of components in a wide range of sizes and thicknesses can easily be shaped by oxygen cutting which could otherwise only be performed with great difficulty by other material removal processes. *Because of its versatility, one oxygen-cutting machine can replace several mechanical types of cutting machine.*

The main disadvantage of shearing machines is that they are of very limited capacity with regard to the thickness of the material being cut. *Guillotines can be used for straight-line cutting only, and the length of cut is governed by the capacity of the machines.*

However, shearing machines have the advantage over oxygen cutting in respect of material up to 4 mm thick because it is difficult to produce a clean flame cut edge below this thickness. The pre-heat flame tends to melt the top edges of the cut causing them to fuse together.

Oxygen cutting is faster than sawing and it *cuts greater thicknesses than the shear or the guillotine.* For example, 25 mm thick plate can be cut at the rate of 15·6 m/hour, 100 mm thick at 7·8 m/hour. The cutting of steel 600 mm thick is frequently undertaken with a high degree of accuracy.

Bevel-edge cutting is no problem with the oxygen-cutting process, whereas, with the exception of certain specially designed portable shears, shearing machines only produce square cut edges.

Portable and static nibbling machines may be used in conjunction with a profile template but are only capable of producing one component at a time. Flame profile cutting machines are capable of producing a number of components from a template simultaneously.

Some large profile machines are capable of performing multi-torch operations in which as many identical shapes can be cut simultaneously as there are torches in use. Ten or more cutting torches may be used in normal operations.

In the ship-building industry and in some large plate fabrication shops, large automatic machines are used for cutting several large plates simultaneously. This is usually done from redwood 1:10 or 1:100 scale TEMPLATE DRAWINGS.

The thinnest steel which can be cut satisfactorily by oxygen cutting is 3·2 mm, but thinner sheets can be successfully cut by using the 'STACK CUTTING METHOD'. *In this way as many as 100 sheets may be cut simultaneously.* The advantage and economy of this method for the production of regular or irregular shaped thin plates in repetition is obvious. *The stack cutting method involves piling a number of thin plates one on top of the other, clamping them tight together, and making the cut as if the clamped plates were one piece of solid metal.* The quality and accuracy of the cut will depend upon the stack thickness. With a stack not exceeding 50 mm a cut edge tolerance of 0·8 mm is obtained, while stacks from 50 to 100 mm can be cut with an edge tolerance of 1·6 mm. THE EDGES OF THE STACK CUT SHEETS ARE SQUARE AND FULL WITH NO BURRS SUCH AS ARE OFTEN PRESENT ON SHEETS PRE-FABRICATED BY SHEARING METHODS. *This advantage is of particular importance on sheet metal discs that are subsequently to be flanged or dished.*

The versatility of oxygen cutting cannot be over emphasised, and by adopting machine-cutting processes, small workshops in particular, can extend their activities to a much greater variety of work on material considerably heavier and more intricate, thus increasing their capacity with regard to material removal.

The author is indebted to The British Oxygen Company for much of the information offered on oxygen cutting in this chapter.

8 Restraint and location

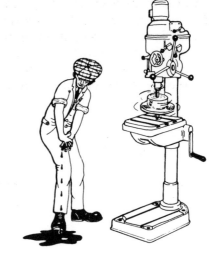

All components should be securely restrained by clamps and abutments before machining

The basic concepts of restraint and location have already been considered in Chapter 7 of *Basic Engineering*. This chapter sets out to consider restraint and location as it applies to components being drilled, milled and fabricated in greater detail.

8.1 Drilling process

To successfully drill a hole in a component so that it is correctly positioned, four basic conditions must be satisfied.
1. The drill must be *located* so that its axis is coincident with the axis of the drilling machine spindle.
2. The drill and the spindle must rotate together without slip occurring. That is the drill must be *restrained* by the spindle.
3. The work piece must be *located* so that the centre lines of the hole are in alignment with the spindle axis as shown in Fig. 8.1.
4. The work piece must be *restrained* so that it is not dragged round by the drill.

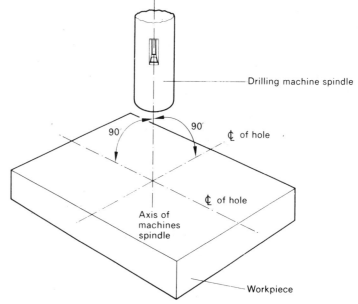

Fig. 8.1 Basic drilling alignments

8.2 Restraint and location of the drill

The use of the Morse taper for the location of drills, reamers and other tools in the drilling machine spindle was introduced in *Basic*

Engineering (Fig. 7.34). It will be seen from Fig. 8.2(a) that a taper location can compensate for variations in size due to manufacturing tolerances and reasonable wear. However, even a small amount of dirt in the taper can cause considerable misalignment, as shown in Fig. 8.2(b). Therefore the shank and spindle bore should be wiped clean before the drill is inserted. This also prevents undue wear and damage to the spindle.

it rotates. If the drill 'digs in' to the work piece so that slip occurs the spindle bore will be quickly scored up and become useless for locating and restraining the drill.

Straight-shank drills and other small tools are often held in self-centring chucks. The principle of the drill chuck as a device for locating and restraining small drills is explained in Fig. 8.3.

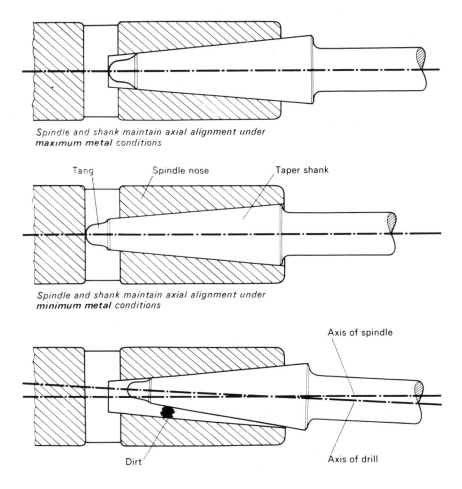

Fig. 8.2 Taper location

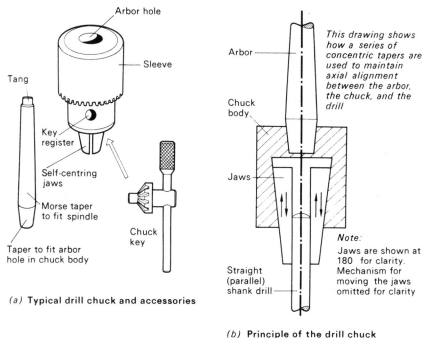

(a) **Typical drill chuck and accessories**

(b) **Principle of the drill chuck**

Fig. 8.3 The drill chuck

The narrow angle of taper of the drill shank causes it to wedge in the spindle of the drilling machine. This provides sufficient restraint to prevent the drill dropping out of the spindle and to prevent relative movement (slip) between the drill and the spindle as

8.3 Restraint and location of a drilled component

Various methods of work-holding on the drilling machine were introduced in *Basic Engineering*, Fig. 7.29. These methods will now be examined in greater detail.

1. Figure 8.4(a) shows the *restraints* acting on a simple component held in a machine vice. The geometric alignments necessary to *locate* the component correctly relative to the axis of the drilling machine spindle are shown in Fig. 8.4(b).

167

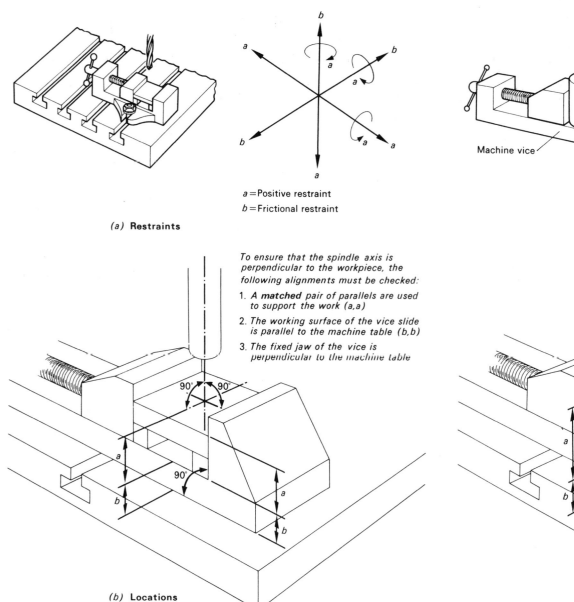

Fig. 8.4 Workholding in the vice — restraints and location

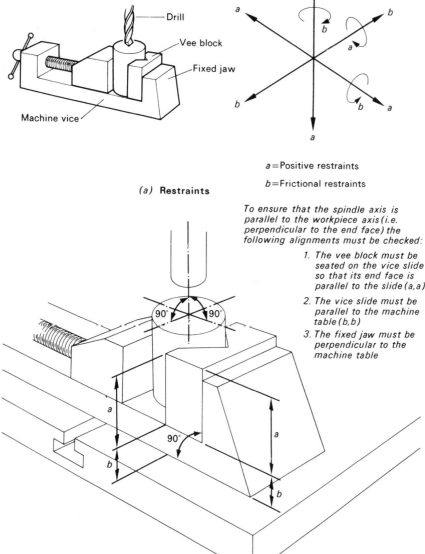

Fig. 8.5 Workholding cylindrical work in the vice

2. Figure 8.5(a) shows the *restraints* acting on a cylindrical component held in a machine vice. The geometric alignments necessary to *locate* the component correctly, relative to the axis of the drilling machine spindle, are shown in Fig. 8.5(b).

3. Figure 8.6(a) shows the *restraints* acting on a simple component clamped directly to the drilling machine table. The geometric alignments necessary to *locate* the component correctly, relative to the axis of the drilling machine spindle, are shown in Fig. 8.6(b).

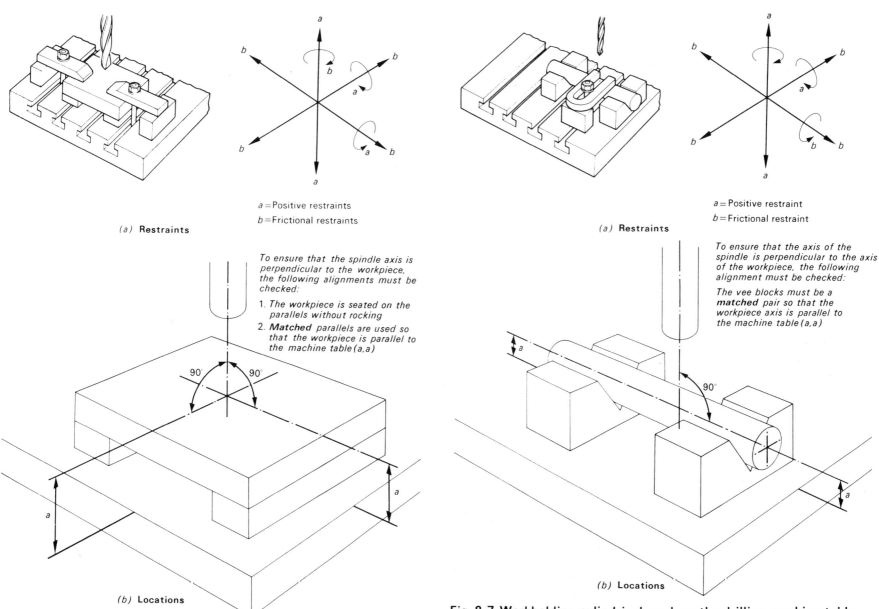

Fig. 8.6 Workholding on the drilling machine table

Fig. 8.7 Workholding cylindrical work on the drilling machine table

4. Figure 8.7(a) shows the *restraints* acting on a cylindrical component supported on vee blocks. The geometric alignments necessary to *locate* the component correctly, relative to the axis of the drilling machine spindle, are shown in Fig. 8.7(b).

5. The relatively chunky precision engineering components shown in Figs 8.1 to 8.7 inclusive are mounted on parallels, or other datum surfaces to ensure that the hole axis is perpendicular to the drilling machine table. However, sheet metal components are not sufficiently rigid to locate in this manner and are better

169

mounted on *flat* wood to ensure support right up to the edge of the hole as shown in Fig. 8.8.

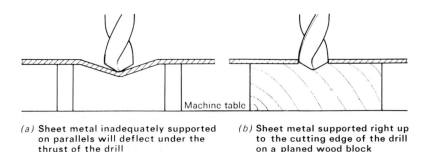

(a) Sheet metal inadequately supported on parallels will deflect under the thrust of the drill

(b) Sheet metal supported right up to the cutting edge of the drill on a planed wood block

Fig. 8.8 Drilling sheet metal

8.4 Milling

Figure 8.9 shows a variety of milling machine set-ups and the corresponding *restraints* acting on the components. Further examples of work setting on the milling machine are included in Chapter 9.

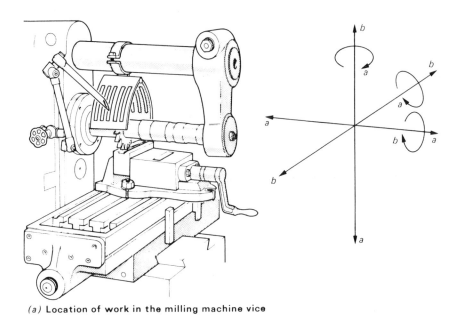

(a) Location of work in the milling machine vice

(b) Location of work on the milling machine table

(c) Location of work on the dividing head

a = positive restraint
b = frictional restraint

Fig. 8.9 Restraint and location of work on the horizontal milling machine

The forces acting on the workpiece when using a vertical milling machine are very different to those previously described. As well as tending to push the work along the table, the cutter also tends to spin the job round. Figure 8.10 shows the forces acting on a component set on a vertical milling machine, and the positioning of suitable abutments to resist the cutting forces.

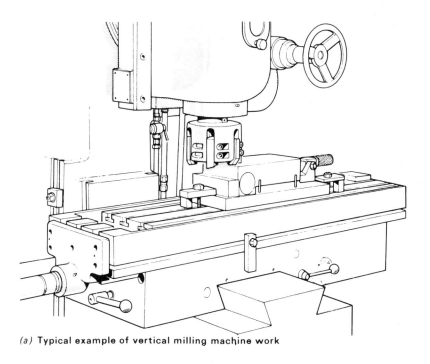

(a) Typical example of vertical milling machine work

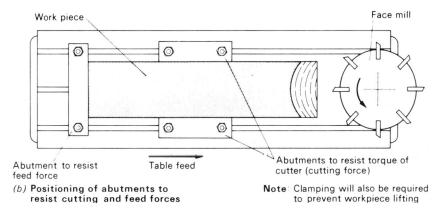

(b) Positioning of abutments to resist cutting and feed forces

Note: Clamping will also be required to prevent workpiece lifting

Fig. 8.10 Workholding on the vertical milling machine

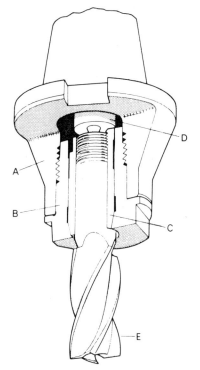

A **The main body of the Clarkson AUTOLOCK Chuck**, which houses the precision-made, self-locking parts of the chuck.
A wide range of tapers is offered from stock, and special fittings can be made to order

B **The locking sleeve.**
A precision fit, positions the collet and mates with the taper nose of the collet

C **The collet.**
Is of split construction and internally threaded at the rear end

D **The male centre.**
Hardened and ground, it serves to centre the cutter and anchors the extreme end to ensure rigidity and true running

E **The AUTOLOCK cutter.**
Any tendency of the cutter to turn in the chuck during operation increases the grip of the collet on the shank of the cutter, thus ensuring maximum feeds and speeds.
The cutter cannot push up or pull down during operation

Fig. 8.11 Mounting of screwed shank, solid end mills

The mounting of milling cutters has already been considered in Section 7.18, *Basic Engineering*. Figures 8.11 and 8.12 summarises some different methods of mounting cutters on milling machines.

171

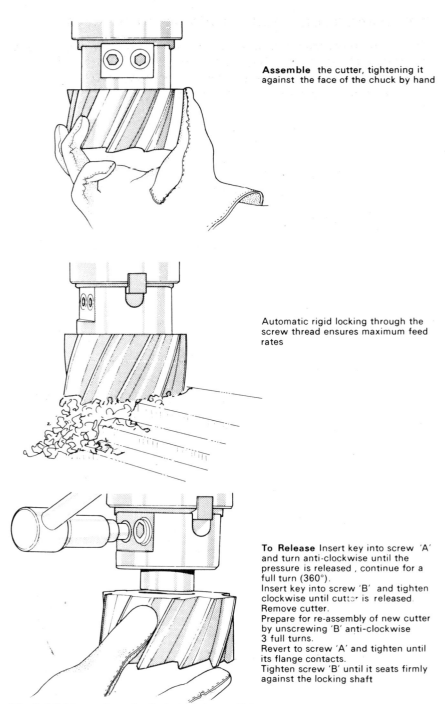

Assemble the cutter, tightening it against the face of the chuck by hand

Automatic rigid locking through the screw thread ensures maximum feed rates

To Release Insert key into screw 'A' and turn anti-clockwise until the pressure is released, continue for a full turn (360°).
Insert key into screw 'B' and tighten clockwise until cutter is released. Remove cutter.
Prepare for re-assembly of new cutter by unscrewing 'B' anti-clockwise 3 full turns.
Revert to screw 'A' and tighten until its flange contacts.
Tighten screw 'B' until it seats firmly against the locking shaft

Fig. 8.12 Mounting shell end and small face mills

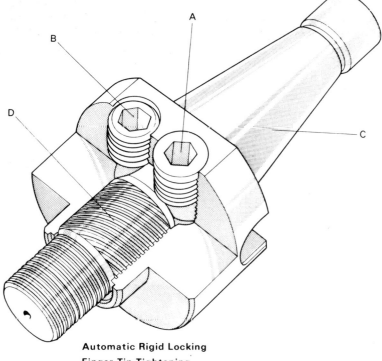

Automatic Rigid Locking

Finger-Tip Tightening

Finger-Tip Release

Virtually Indestructable

Wide Range of Cutters, all held on one Chuck

A **STOP SCREW** Positions the locking shaft exactly in line

B **RELEASE SCREW** Sets the Clarkson release system in motion, turning the Locking Shaft on its fast helix thread, thus moving the cutter away from the face of the chuck

C **CHUCK BODY** Super alloy steel specially hardened and ground to precision limits

D **LOCKINGSHAFT** Specially designed and manufactured by Clarkson from unbreakable 90 tons tensile alloy steel

8.5 Clamping work for folding

For simple bending or folding of thin sheet metal the work may be held in *'bending bars'* or in a pair of *'angle bars'* which, in turn, are usually held in an engineer's vice, as illustrated in Fig. 8.13. When bending in a *folding machine* the work is held by the clamping beam as explained in Chapter 10.

8.6 Hand tools used for clamping (general)

One of the most common tools used for clamping or holding work is the *engineer's hand vice*. Hand vices are an invaluable aid in fabrication work, and no craftsman should be without them. Figure 8.14 shows a hand vice together with typical applications.

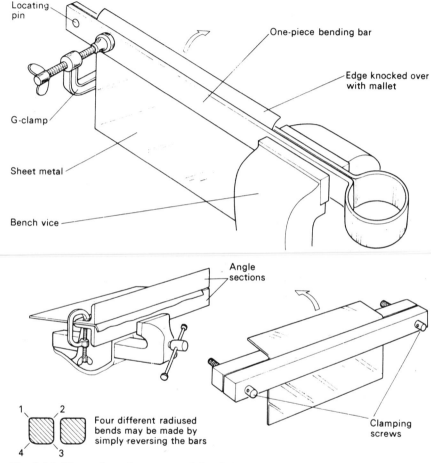

Fig. 8.13 Clamping sheet metal for folding by hand

A pair of angle sections are useful for holding sheet metal when making short bends by hand.

The metal is positioned and clamped between the angle sections by means of a G clamp each end. The whole assembly is then held in the vice and the edge or flange knocked over with a mallet, as shown in Fig. 8.13.

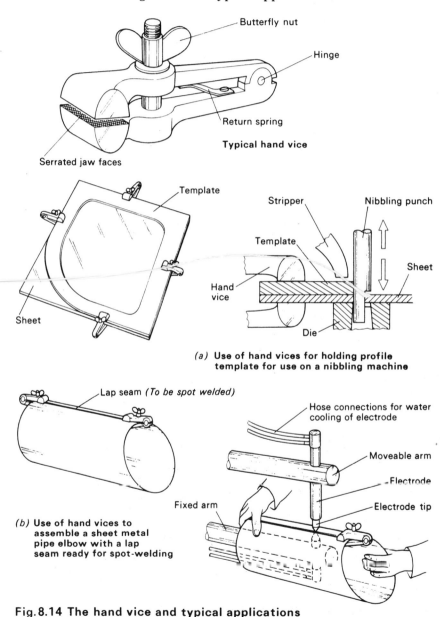

Fig. 8.14 The hand vice and typical applications

Another type of hand vice depends for its gripping power upon its 'toggle action' (Fig. 8.15). With an ordinary type of hand vice, the friction force available to grip the component is somewhat limited. Hand vices with a toggle action tend to grip more securely, and unlike the ordinary hand vice, provide a means of quick release. Figure 8.15 illustrates various types of toggle-action hand vices.

Considerable use is made of G clamps, and these are available in a range of sizes. *When using G clamps care should be taken to select the correct size of clamp for the job.* For example, when large clamps are used where a much smaller size is called for, there is a tendency to put too great a strain on the screw. This results in a permanent misalignment of the two gripping surfaces, as indicated in Fig. 8.16(*a*).

Applications of the use of G-clamps are illustrated in Fig. 8.16(*b*).

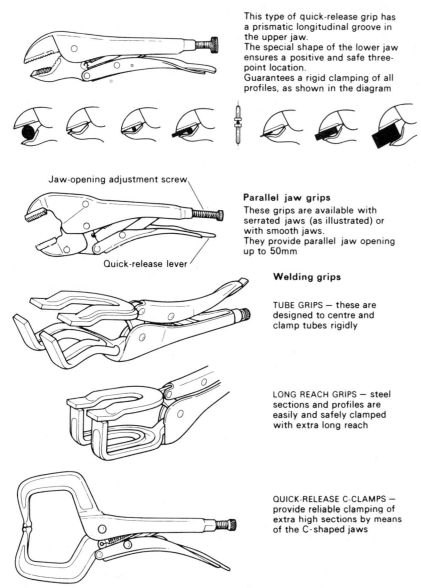

Fig. 8.15 Toggle-action hand vices or grips

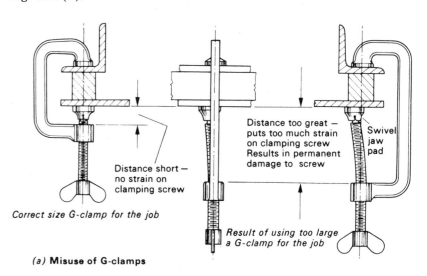

(a) Misuse of G-clamps

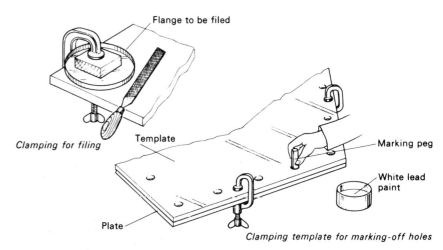

(b) Applications of use of G-clamps for work-holding

Fig 8.16 Use of G-clamps

8.7 Hand tools used for clamping (specific)

Special G clamps are available which are a great improvement on the ordinary G clamp. These special clamps are capable of performing many clamping operations more efficiently and quickly than by other means. They are fitted with a short, shielded screw which will outlast many ordinary clamp screws, the fine-pitched thread provides a very powerful grip. *Unlike the ordinary G clamp, there is no long screw to bend or damage.* Details of these clamps are given in Fig. 8.17, whilst typical applications are shown in Fig. 8.18.

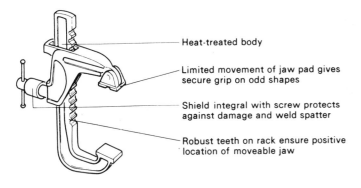

- Heat-treated body
- Limited movement of jaw pad gives secure grip on odd shapes
- Shield integral with screw protects against damage and weld spatter
- Robust teeth on rack ensure positive location of moveable jaw

TYPE	CAPACITY (mm)	THROAT DEPTH TO PAD CENTRE (mm)	RATING (kN)	REMARKS
Standard duty	0-152 0-305	60	11.34	For all normal clamping duties. Toolrooms, light and medium welding, sheet metal fabrication, welding box and die holding. Frame manufactured in heat-treated pearlite iron; bright cadmium-plated finish; weld spatter and rust resistant
L.M. deep throat	0-228	121	3.63	For light duty use in woodworking, pattern making, plastic fabricating, sheet metal work, corebox clamping etc. Pad tilt control screw to permit tip holding of small projections. Manufactured in heat-treated L.M. 10 Aluminium
Medium duty	0-228 0-457	90	10.10	For the heavier work, deeper jaw. Machine shop, medium heavy welding, and plate fabrication. Manufactured from heat-treated spheroidal graphite iron
Deep throat medium	0-228 0-457	203	9.08	For special uses and awkward positions. Metal fabrication, pattern making, plastic and fibre-glass moulding. Manufactured from heat-treated spheroidal graphite iron
Heavy duty	0-152 0-305 0-457 0-609 0-914 0-1372	114	27.23	The ultimate in hand applied clamping where extreme pressure or length is required. Frame forged steel, heat treated

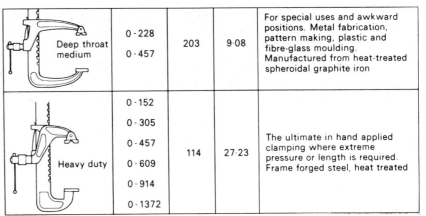

Fig. 8.17 Special G-clamps and applications

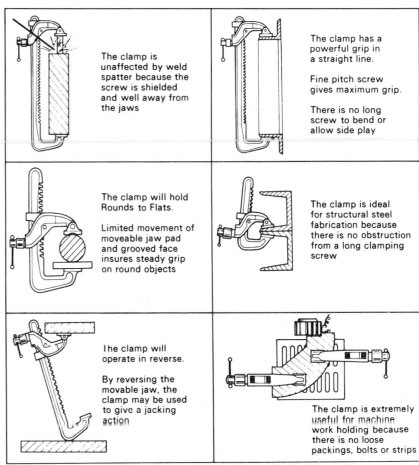

The clamp is unaffected by weld spatter because the screw is shielded and well away from the jaws

The clamp has a powerful grip in a straight line.

Fine pitch screw gives maximum grip.

There is no long screw to bend or allow side play

The clamp will hold Rounds to Flats.

Limited movement of moveable jaw pad and grooved face insures steady grip on round objects

The clamp is ideal for structural steel fabrication because there is no obstruction from a long clamping screw

The clamp will operate in reverse.

By reversing the movable jaw, the clamp may be used to give a jacking action

The clamp is extremely useful for machine work holding because there is no loose packings, bolts or strips

Fig. 8.18 The versatility of special G-clamps (rack clamps)

For all types of work where a large capacity is required, use is made of bar clamps, and these are available with capacities up to 2·4 m. Like the special G clamps they are simple and quick to use by just sliding the movable jaw to the nearest notch and tightening the screw. A range of tools is illustrated in Fig. 8.19.

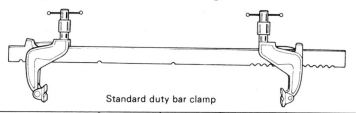

Standard duty bar clamp

TYPE	MAXIMUM CLAMPING LENGTH (mm)	CAPACITY (mm)	RATING (kN)	REMARKS
Standard duty	457, 609, 914, 1219, 1524, 1829	140 minimum to indicated length	9·1	For all types of work where a large capacity is required. The bar is manufactured from HIGH TENSILE STEEL
L.M. deep throat	457, 609, 914, 1219, 1524, 1829	152 minimum to indicated length	3·6	With all bar clamps, multiples of jaws may be used on a bar to give positive component location
Medium duty	609, 914, 1219, 1524, 1829	216 minimum to indicated length	18·2	Principally for use in medium weight fabrication shops. The bar is manufactured from HEAT-TREATED HIGH TENSILE STEEL
Deep throat medium	609, 914, 1219, 1524, 1829	254 minimum to indicated length	9·1	
Heavy duty	1829, 2134, 2438	305 minimum to indicated length	27·25	The ultimate in heavy duty clamping. HEAT-TREATED HIGH TENSILE BAR The jaws on all bar clamps may be reversed to give opening action
	Bar connector Standard duty, medium duty and heavy duty connection pieces may be used to couple any two bars for greater overall length			

Fig. 8.19 Range of bar clamps and applications

Tee-slot clamps are used for clamping the work on all tee-slotted machine tables, or welding positioners. The basic principle of a very efficient type of Tee-slot clamp is shown in Fig. 8.20.

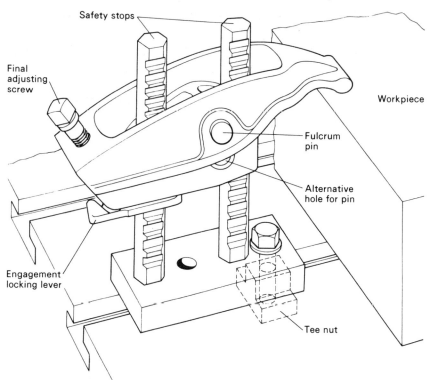

Fig. 8.20 Basic details of Tee-slot clamps

METHOD OF OPERATION

The required height is obtained by releasing the locking lever, pulling the head assembly rearwards out of engagement with the column teeth, and raising or lowering as required.

The head is then engaged by pushing forward, returning it to the column teeth and closing the engagement locking lever.

Final adjustment is made with the screw provided for the purpose.

The head is made of cast steel, the column and base of carbon steel and the adjustment screw of high tensile steel.

HOLDING POWER

There are two ranges of Tee-slot clamps available:
1. Working force   22·68 kN

Ultimate force 36·3 kN
2. Working force 36·3 kN
Ultimate force 113·4 kN

The smaller type are available in three heights, 102 mm, 152 mm and 203 mm.

The larger type is made in eight heights from 152 mm to 1 219 mm in increments of 152 mm.

8.8 Use of clamping devices for riveting

The toggle-action hand vices described in 8.6 are an invaluable aid for holding work for riveting, and are used in conjunction with special 'locating pins' or 'skin pins'. A typical application is shown in Fig. 8.21.

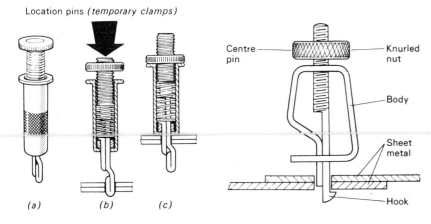

(a) Method of operation

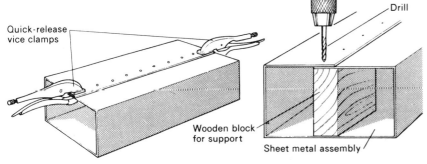

(b) Work clamped for drilling

Work of this nature must be supported to prevent sagging under the drill pressure which would result in misalignment of the rivet holes

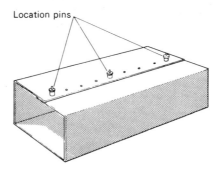

After drilling the holes are deburred and the assembly clamped with the aid of location pins ready for riveting

(c) Work clamped for riveting

Fig. 8.21 Workholding for riveting

METHOD OF OPERATION

1. Unscrew the knurled nut.
2. Depress the centre pin and insert in the hole in the sheets.
3. Tighten the knurled nut and draw the centre pin upwards clamping the sheets between the body and the hook.

Location pins are removed by simply unscrewing the knurled nut and depressing the centre pin.

Location pins are an essential aid for sheet metal assembly, particularly in the aircraft industry. The pins give accurate location, positive grip, no distortion of the holes, rapid one-hand operation, and additional clamping pressure by means of the knurled nut if required. They can be removed as rapidly as they are inserted, they are suitable for holes of 2·4 mm, 3·2 mm and 4 mm diameter, and cut assembly time considerably in sheet metal assembly work.

8.9 Use of soft iron wire and spring pegs

Many soldered or brazed lap joints require to be held in position to prevent the mating surfaces from moving when heated. Such joints only require light clamping, and this is achieved with the aid of soft iron wire or simple spring pegs. Applications of this simple but effective method of work holding are illustrated in Fig. 8.22.

177

operations. They are used to clamp plates, sections and plate formers on to the bending floor. The bending floor or bending table consists of a large cast iron slab which has numerous location holes in which the 'dogs' and 'pins' are inserted.

The basic principle of the use of these devices is illustrated by two practical applications: bevelling steel section (Fig. 8.23(a)); and the hot bending of angle section (Fig. 8.23(c)).

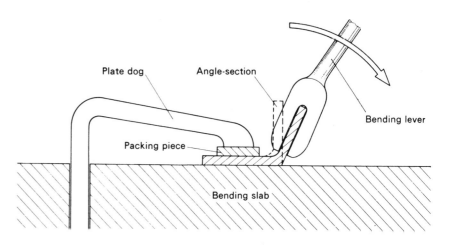

(a) Bevelling angle-section

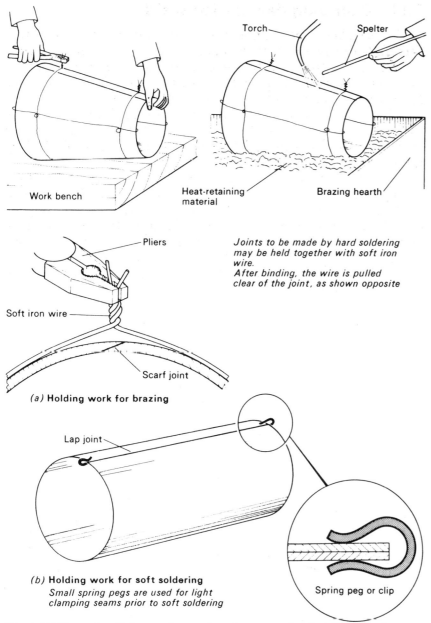

Fig. 8.22 Use of soft iron wire and spring pegs for holding work

8.10 Use of plate dogs and pins

'Plate dogs' and 'pins' are used extensively in the plate working and steel fabrication workshops for a variety of forming and straightening

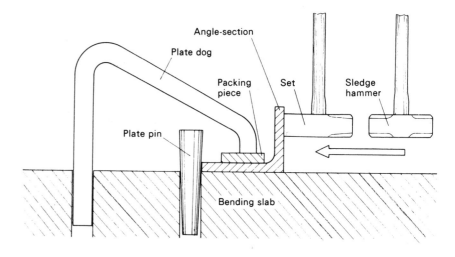

(b) Cold straightening angle-section

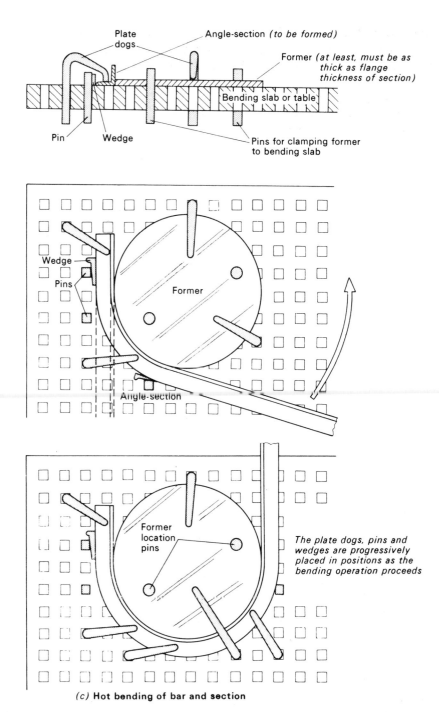

Fig. 8.23 Applications of slate-dogs and pins for workholding

(c) Hot bending of bar and section

8.11 Clamping devices for weld

In the assembly of parts prior to welding, clamps of many types are needed. G clamps are most commonly used, but *care must be taken to prevent 'spatter' from damaging the threads, otherwise the useful life of an ordinary G clamp will be extremely short.* This can be done by coating the threads with 'anti-spatter compound'.

Because of the damage which can be caused by 'spatter', the special G clamps (Fig. 8.17) have a very definite advantage over ordinary G clamps by the fact that they are fitted with a shield which protects the screw against damage and spatter.

Many of the clamps used for holding work for welding are quick-acting and can be rapidly adjusted for various thicknesses of workpiece. Quick-acting clamps are essential where work has to be assembled in welding jigs for batch production. Ordinary screw-type clamping devices would be uneconomical and time consuming. Quick-acting clamps are generally of the 'toggle-action' type or 'cam-operated'. Typical examples of these quick-acting clamping devices are shown in Fig. 8.24.

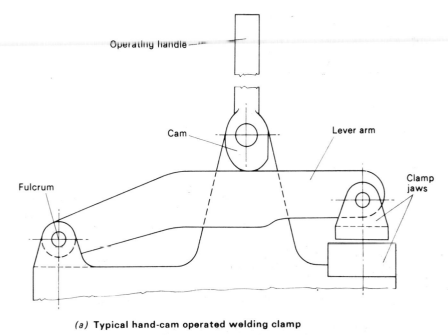

(a) Typical hand-cam operated welding clamp

Quick-acting clamps are essential for holding work in welding jigs

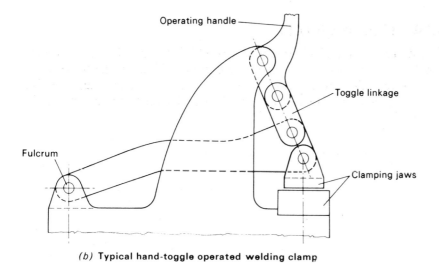

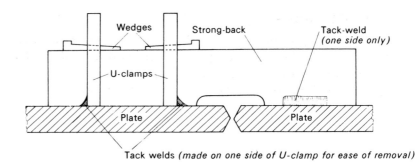

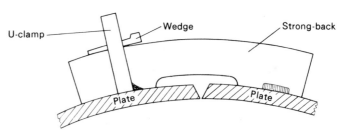

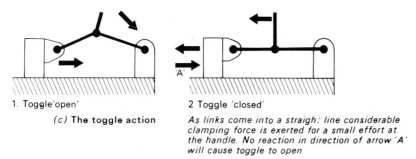

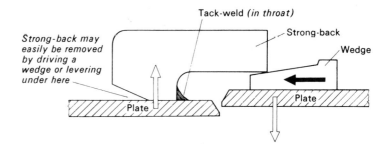

Fig. 8.24 Quick-acting clamping devices

Various other clamping devices in common use for holding work prior to welding include, wedges, clamps and angle bars, cleats, U clamps or bridges, strong-backs, jack-clamps and, chain and bar. These will now be discussed in detail.

8.12 Use of strong-backs

When parts are set up prior to welding, it is important that the plate edges should be correctly aligned. The root faces should be aligned and the root gaps uniform. Many devices may be used to achieve this alignment, one of which is called a 'strong-back'.

Basically, a strong-back consists of rigid piece of plate or angle-section which is tack-welded in position on one side of the joint and used in conjunction with a wedge or a bolt and dog, as shown in Fig. 8.25.

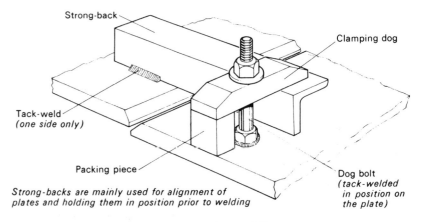

Fig. 8.25 The use of strong-backs for welding

8.13 Use of cleats

Angle cleats are used to push or draw the plate edges together when assembling long and circular weld seams. When used for pulling the joint edges together, they are often referred to as 'draw-lugs'.

Applications showing the basic principle involved are illustrated in Fig. 8.26.

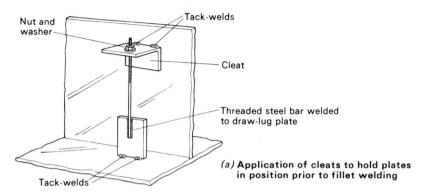

(a) Application of cleats to hold plates in position prior to fillet welding

When cleats are used as an aid to holding work for welding, they may be easily removed by tapping with a hammer to bend them over and break the tack-welds as shown by the arrows

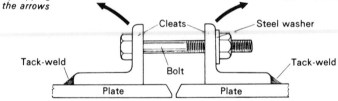

(b) Application of cleats as draw-lugs for closing a butt weld

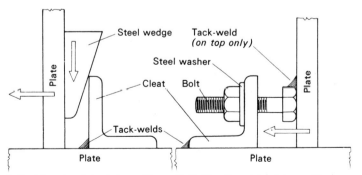

1. Pushing the plates into position 2. Pulling the plates into position

(c) Application of cleats for assembling plates in correct position for fillet welding

Fig. 8.26 The use of cleats for holding work for welding

8.14 Use of glands

Glands are simple rigid bridge pieces which are usually used in conjunction with cleats as an aid to assembly. Whereas cleats are used to maintain the correct uniform gap between the joint edges of the plates, glands provide the means of lateral alignment, as shown in Fig. 8.27.

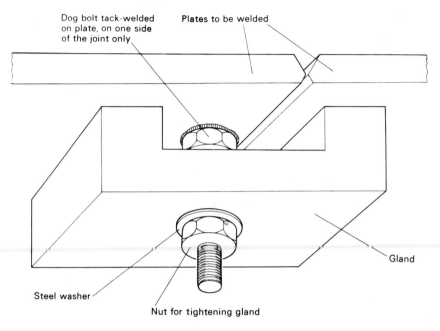

Fig. 8.27 Use of glands

Dog bolts or gland screws are positioned on one side of the joint and tack-welded to the plate.

The glands (bridging pieces) are placed in position and held by means of washers and nuts, as shown in the diagram.

The gland nuts are tightened and adjusted with a spanner until the desired alignment of the joint to be welded has been achieved. On completion of the welding operation, these useful assemblies are dismantled, and the tack-bolts removed.

8.15 Use of dogs and wood blocks

An effective method of clamping a stiffener, such as an angle bar, in position on the plate prior to welding, is to make use of a clamping dog and wood blocks. A bolt is tack-welded adjacent to the required location of the stiffener, and the assembly is held in position as shown in Fig. 8.28. *On completion of the welding, the bolt is removed by breaking the tack welds, and the local surface cleaned up with an angle grinder — this applies to any assembly aid which is tack-welded on to the workpiece.*

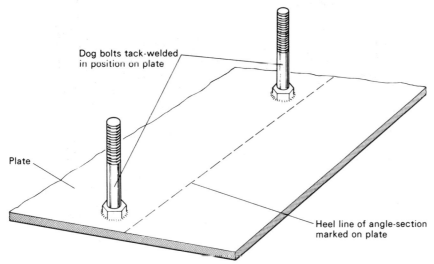

Dog bolts tack-welded in position adjacent to required position of stiffener to be welded to plate

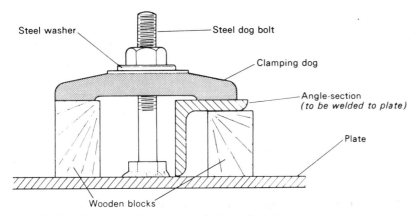

Fig. 8.28 Application of the use of clamping dogs

8.16 Use of bridge pieces

Bridge pieces are commonly referred to as U clamps, and are used together with wedges for clamping sections and plates together, as shown in Fig. 8.29.

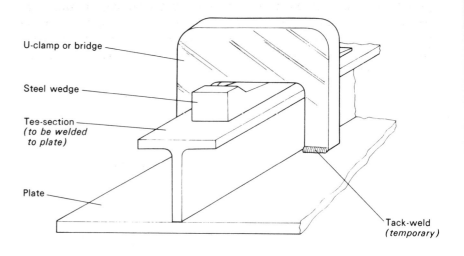

(a) **Application of clamping bridge**

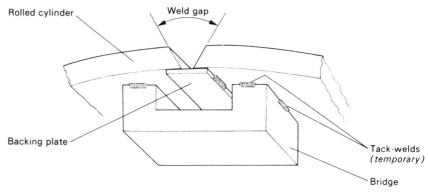

(b) **Application of assembly bridge**

Assembly bridges are a useful holding aid for maintaining the weld gap during welding.

They are used mainly on cylindrical work where backing straps are required

Fig. 8.29 Use of bridge pieces

8.17 Use of chain and bar

Difficulty is often experienced when welding longitudinal seams on cylindrical vessels which have been incorrectly rolled, causing considerable misalignment of the joint edges. This difficulty may be overcome by the use of a chain and bar, as illustrated in Fig. 8.30. This effective method may also be applied to flat plates which are out of alignment.

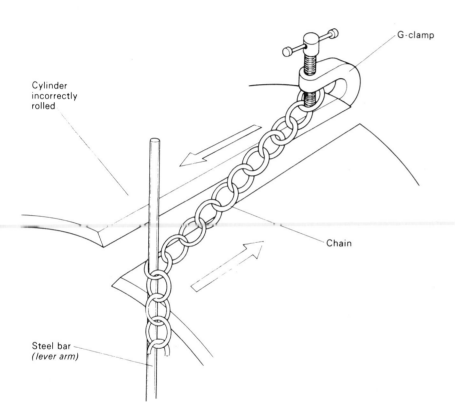

Fig. 8.30 Use of chain and bar

8.18 Use of jack clamps

Jack clamps are used for the attachment and alignment of heavy plate. An example of its use is shown in Fig. 8.31(a). It will be seen that it makes use of the *principle of moments* as described in *Basic Engineering*, Section 7.6. Figure 8.31(b) shows how this principle is applied to the jack clamp.

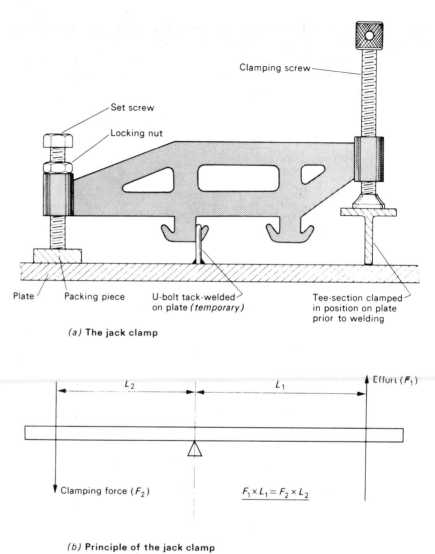

(a) The jack clamp

(b) Principle of the jack clamp

$$F_1 \times L_1 = F_2 \times L_2$$

Fig. 8.31 Use of jack clamp for holding work

8.19 Use of spiders

Large cylindrical vessels are often made up of two or more sections of rolled plate. The sections are then welded together circumferentially. It is essential that each section is truly circular to ensure correct alignment of the joint to be welded.

183

In order to provide correct alignment of the circumferential joints, 'spiders' are fitted at both ends of each section to ensure concentricity. Figure 7.32 illustrates the use of spiders for this purpose.

The spider assemblies are positioned both ends of the rolled and welded steel cylinder with the aid of suitable lifting tackle.

Steel packing pieces are placed between the set screws, on the ends of the legs, and the cylinder wall.

The set screws are tightened and adjusted in sequence until the ends of the cylinder are truly circular.

Once the contour has been checked by means of a suitable template, location plates are tack-welded to the legs and the inside cylinder wall.

8.20 General information on the use of clamping aids (welding)

When tack-welding clamping aids such as cleats, bridges, strong-backs and dogs, the tack welds should be small, but sufficient to perform their temporary function. They should be made so that they can be easily removed after they have served their purpose.

These clamping devices should not be used on hardenable steels because there is a tendency for local hard spots to develop as a result of the small tack welds.

For shop assemblies, a large, flat, rigid base plate with T-slots is generally used for bolting-down purposes. Such tables or plates are ideal for the use of Tee-slot clamps. It is common practice to use angle plates and simple framework for holding components in position prior to welding.

On site, a number of the clamping devices discussed in this chapter are extremely useful for manipulating plates and sections into the correct position and holding them there for welding.

Caliper clamps and magnetic clamps, as shown in Fig. 8.33, are also useful where components have to be welded together at an angle.

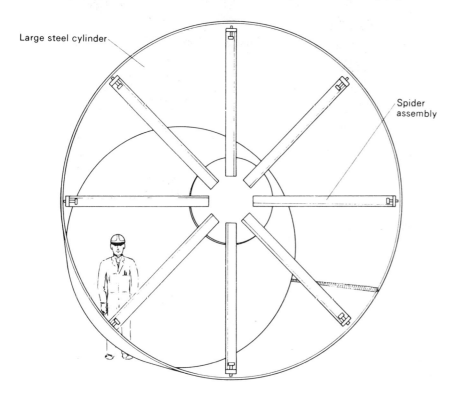

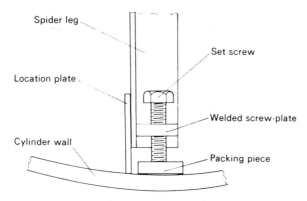

Fig. 8.32 Use of spider assemblies for holding work

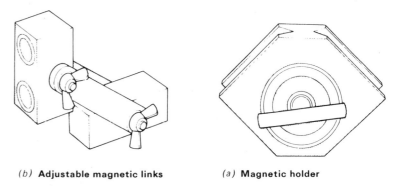

(b) Adjustable magnetic links   (a) Magnetic holder

Fig. 8.33 Magnetic clamps for welding

9 Workshop operations

This is not what the planning sheet meant by the component being chucked

9.1 Basic fabrication procedure

Fabrication engineering involves a multiplicity of operations and the use of a diverse range of tools and machines.

One can appreciate the immense variety and volume of work carried out in the fabrication industry by considering the very wide range of materials with which the fabricator works. A selection of the market forms of supply of materials used in the industry is listed in Table 9.1. Tables 9.2 and 9.3 list the classifications of hot- and cold-rolled materials respectively.

Table 9.1 Market forms of supply

MARKET FORM OF SUPPLY	REMARKS
Sheet metal	Produced in widths up to 1·83 m, in thicknesses ranging from 0·762 mm to 3·25 mm in FERROUS and NON-FERROUS materials, and in a range of qualities and finishes. May be coated or uncoated.
Plates	These are HOT-ROLLED in a wide range of sizes and thicknesses, and are usually supplied with square edges. Thicknesses in common use range between 3·25 mm and 51 mm. Heavy plates are available in widths up to 3·66 m and thicknesses from 9·5 mm to 51 mm or more, and light plates in thicknesses from 3·25 mm to 9·5 mm. Steel plate is the flat material used for ship's hulls, bridge building, large tanks and vessels such as gas holders. It is also used to make structural members and pipes of large diameter.
Flats	Are narrow plates up to 508 mm width and 25 mm thickness, with rolled edges.
Slabs	Are thick plates with rolled edges and are commonly used for the base plates of heavy stanchions.
Sections	In general, 'sections' are structural shapes used for making framework and bridge building. They include CHANNELS and ANGLES, TEES, UNIVERSAL BEAMS and COLUMNS, ROLLED STEEL JOISTS, RAILS, BARS (are usually straight lengths, which are produced in a wide variety of cross-sections, such as round, square, flat, hexagonal, etc., and in a wide range of sizes). Considerable use is made of ROUND, SQUARE and RECTANGULAR HOLLOW TUBES.
Wire rod and wire	Wire rods range from 6·35 to 12·7 mm in diameter. Wire is produced from wire rod by wire drawing. *Some grades of wire can be reduced by over 90% of the starting cross-section to produce over 10 times the original length of the rod.*

185

Table 9.2 Classification of cold-rolled sheets

CLASSIFICATION OF COLD-REDUCED STEEL SHEETS

MATERIAL TYPE	CLASSIFICATION	REMARKS
En2A/1	CR1	A non-ageing (stabilised) steel for exceptionally severe draws and/or pressings where it is necessary to avoid the appearance of stretcher strains and loss of ductility due to strain ageing.
En2A/1	CR2	For severely drawn parts.
En2A	CR3	For general deep drawing.
En2	CR4	For moderate drawing and simple bending.

FINISHES FOR COLD-REDUCED SHEETS

FINISH	CLASSIFICATION	DESCRIPTION
Cold-reduced general purpose	CR/GP	A commercially flat sheet, whilst reasonably free from surface defects, should not be used where finish is of prime importance.
Cold-reduced full finish	CR/FF	A commercially flat 'skin passed' high grade sheet with a minimum of surface blemishes suitable where a high standard of surface finish is desirable. The surface usually has a 'matt finish' and should be used on outside or visible parts and is suitable for any kind of high grade painting.
Cold reduced vitreous enamelling	CR/VE	A commercially flat sheet of special quality low metalloid steel with a surface specially prepared for vitreous enamelling.

The above table is based on B.S. 1449: Part 1B.

Table 9.3 Classification of hot-rolled materials

CLASSIFICATION OF HOT-ROLLED SHEETS

MATERIAL TYPE	CLASSIFICATION	REMARKS
En2A/1	HR1	For severely drawn parts.
En2A/1	HR2	For severely drawn parts.
En2A	HR3	For general deep drawing.
En2	HR4	For moderate drawing and simple bending.

FINISHES FOR HOT-ROLLED SHEETS

FINISH	CLASSIFICATION	DESCRIPTION
Hot rolled	HR	A commercially flat mill finish sheet.
Hot rolled pickled	HRP	A commercially flat pickled sheet significantly free from scale.
Hot rolled vitreous enamelled	HR/VE	A commercial flat sheet of special quality, low metalloid steel with a surface specially prepared for vitreous enamelling.

The above information is based on B.S. 1449: Part 1B.

QUALITIES OF STEEL

MATERIAL TYPE	DUCTILITY OR TENSILE STRENGTH	FINISH	
		HOT ROLLED	HOT ROLLED PICKLED
En2A/1	Extra deep drawing quality (killed steel)	HR11	HRP11
En21/1	Extra deep drawing quality.	HR12	HRP12
En2A	Deep drawing quality.	HR13	HRP13
En2A	Flanging or drawing quality.	HR14	HRP14
En2	Commercial quality.	HR15	HRP15

CLASSIFICATION OF HOT-ROLLED PLATES

MATERIAL TYPE	CLASSIFICATION	REMARKS
En2A/1	HR11	For severely drawn parts.
En2A/1	HR12	For severely drawn parts.
En2A	HR13	For moderately drawn parts.
En2A	HR14	Flanging or drawing quality steel.
En2	HR15	Commercial quality steel.

The above information is based on B.S. 1449: Part 1A.

The production of fabricated components or structures involves five principal operational stages:
1. MEASURING AND MARKING-OUT (see Chapter 5);
2. CUTTING TO SIZE (see Chapter 7);
3. FORMING TO REQUIRED DIMENSIONS (see Chapter 10);
4. JOINING AND ASSEMBLY (see Chapter 9, *Basic Engineering*);
5. SURFACE FINISHING FOR PROTECTION AND DECORATIVE PURPOSES (see Chapter 6, *Basic Engineering*).

In this chapter a number of examples of fabrication work will be used to illustrate a combination of basic fabrication procedures.

9.2 The need for planning operations

A sheet metal article may be made from one or from several pieces of metal. If the article is made from more than one piece it has to be fabricated to required dimensions by means of hand tools, by machine operations, or by a combination of both. Most components and assemblies involve a sequence of operations which must be planned in order to produce the finished article as economically as possible. *This can be best achieved by ensuring that each operation is performed, where possible, with equipment designed specifically for the purpose.*

The sequence of operations may vary slightly with the individual craftsman and the equipment available in his fabrication workshop.

If a sheet-metal article requires a number of joints to be made during the course of its fabrication, the sequence of operations should be planned to ensure that access is possible.

Generally, operations on sheet and plate commence with the guillotine. It is considered good practice to make a 'trim cut' on standard sized sheets or plates to ensure one datum edge from which other cuts are made. This reduces the size of the flat material, usually in length, and this must be taken into consideration when marking out blank sizes for cutting in order to obtain a maximum yield. Intelligent marking-out on the flat material is essential in order to avoid considerable scrap or 'off-cut' material. Economy in the use of materials should be the prime consideration of all craftsmen engaged in fabrication. The economic arrangement of patterns has been discussed in Chapter 5 and serves to illustrate that with planned marking-out and by careful cutting, a maximum number of blanks can be obtained. This means:
1. A REDUCED NUMBER OF STANDARD SHEETS ARE REQUIRED;
2. OFF-CUT MATERIAL IS KEPT TO A MINIMUM — resulting in a reduction in material cost per part.

9.3 Forming a 'pittsburgh lock' (use of folding machine)

The 'Pittsburgh lock' or 'lock seam' is used extensively in ductwork shops. These seams are normally produced by special 'lock-forming machines'. They can, however, be successfully produced on a folding machine provided the correct sequence of operations are performed. The need for adopting a planned sequence of operations and the necessary skill required to make a Pittsburgh lock can be appreciated by practising with scrap metal or off-cuts. Figure 9.1 shows the lock and the layout for forming the lock.

A pencil should be used for marking the fold lines because a scriber would cut through the protective zinc coating on the surface of the sheet.

Assembling the lock seam

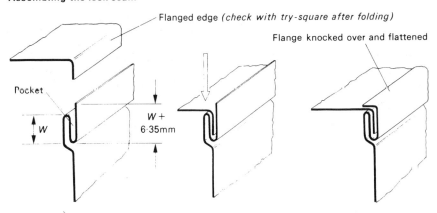

Flange inserted in pocket *(may have to be forced in by striking lightly with a hammer)*

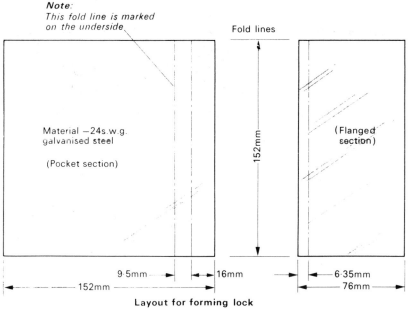

Tools and equipment required	
Guillotine, bench shears or straight snips	The guillotine produces the best straight line cutting
Engineer's rule and pencil	For marking out
Folding machine (with adjustable folding beam)	For forming the seam
Hammer and mallet	For adjusting and straightening the lock assembly and final flattening of flange

Fig 9.1 The Pittsburgh lock

Figure 9.2 shows the sequence of operations which may be adopted for making the lock, using two off-cuts of 24 s.w.g. (0·56 mm) galvanised steel sheet. The first operation is to cut two pieces of sheet metal to the required sizes. The fold lines are then marked in accordance with the dimensions provided on the drawing (Fig. 9.1). *The width of the flanged edge is normally made slightly less than the depth of the pocket.* The allowance for the pocket is equal to twice the width of the pocket plus an allowance for knocking over. In the example given, this allowance is W + W + 6·35 mm. Lock seams are used as longitudinal corner seams for various shaped ducts, and the necessary allowances have to be added to the pattern or layout.

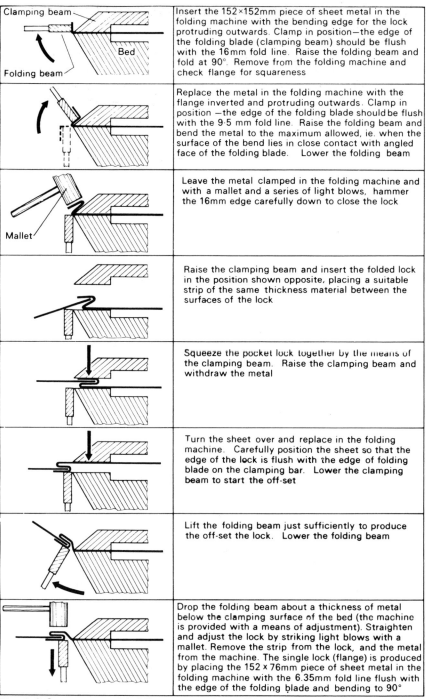

Fig 9.2 Sequence of folding operations (Pittsburgh lock)

9.4 Transferring paper or metal patterns on to sheet metal

It is most important when using patterns or templates, to position them in the proper manner on the metal to be used in order to avoid unnecessary scrap. Templates must be held in position in order to restrain any tendency to move during the marking out operation. These two factors for successful marking out with patterns or templates are illustrated in Fig. 9.3.

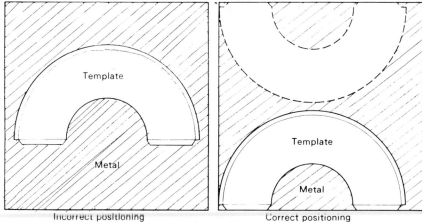

(a) Positioning of template on material to avoid unnecessary waste

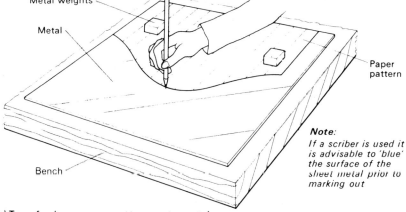

(b) Transferring a paper pattern on to metal

(i) Place the sheet of metal to be used on the surface of the bench and position the paper to avoid waste
(ii) To prevent the paper from creeping hold it in position with metal weights
(iii) With a hard sharp pencil or scriber scribe the outline of the pattern on the sheet metal
(iv) Remove the weights and the pattern and cut the metal to the outline scribed upon it using universal hand shears removing all burrs with a suitable file

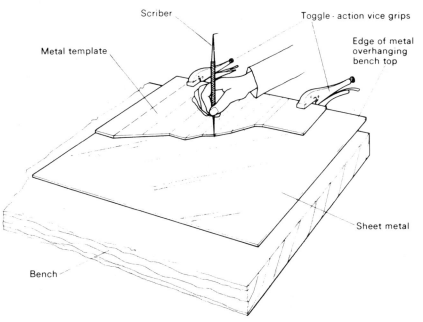

(c) Transferring a metal template on to sheet metal
(i) Place the sheet of metal to be used on the bench with one of its squared sides slightly overhanging
(ii) Position the metal template in position as shown, and clamp it securely with vice grips to restrain any movement
(iii) Scribe the outline of the template on the sheet metal, using a sharp scriber
(iv) Release vice grips to remove template and cut the sheet metal to the outline scribed upon it with a suitable pair of hand shears. Remove all burrs with a file

Fig 9.3 Methods of transferring patterns to metal

9.5 The use of notched corners (sheet metal work)

A common requirement in light sheet metal work is the notched corner, as in the manufacture of simple folded boxes, trays or electronic chassis. Notching is an essential operation where light-gauge fabrications include wired edges, self-secured joints, lap joints and welded corner seams. The term 'notching' is used to describe, in simple form, *the removal of metal from the edges of sheet metal blanks or patterns prior to carrying out any forming operations.*

Considerable thought must be given to marking-out for notching because good notching is of prime importance where the finished article is to have a neat appearance. There is nothing more unsightly than overlaps, bulges and gaps which result from not allowing for notching, or bad notching, on the initial blank.

It is advisable to make rough sketches of junctions which require notching before commencing to mark-out the pattern. This simple exercise will enable the craftsman to determine where metal removal is necessary, and decide upon the sequence of operations necessary to produce the article.

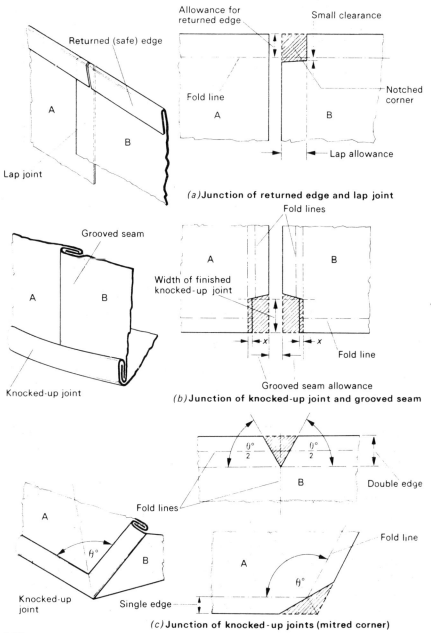

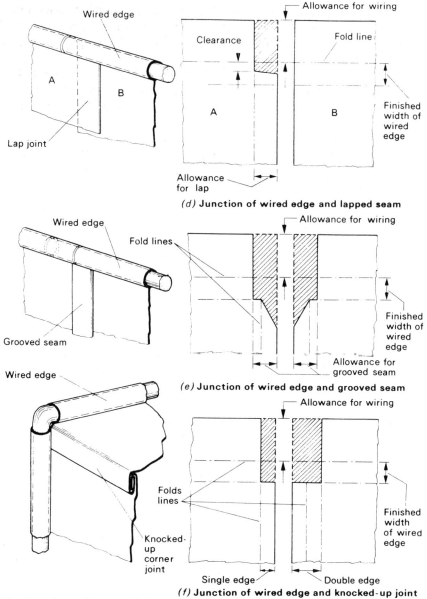

Fig 9.4 Junctions which require notching prior to forming

Note: In order to avoid gaps on assembly, Fig. 9.4(b), it is not advisable to notch to the full width of the groove seam allowance. Make a cut with snips to the inside corner of the notch and leave about 3 mm of the seam allowance, as shown at x. This material will overlap when the grooved seam is made, and a single cut with the snips will ensure a perfect butt of the notched edges.

Figure 9.4 contains sketches of typical junctions encountered in light gauge fabrications. These show the correct approach to the problems of marking-out for notching before forming operations are commenced.

Usually, the notching of corners on sheet metal blanks to be formed into boxes is limited by the depth of the box because of the large amount of scrap resulting from this operation. Consider the following examples:

(a) A rectangular tray is required to be fabricated by square notching the corners, folding the sides up square and welding the vertical corner seams, the depth of notch is to be 20 mm.
TOTAL AREA OF METAL REMOVED BY NOTCHING
$= 4 \times 20 \text{ mm} \times 20 \text{ mm}$
$= \underline{1\ 600 \text{ mm}^2}$

(b) A rectangular box is to be fabricated by the same method, and the depth of notch is to be 152 mm.
TOTAL AREA OF METAL REMOVED BY NOTCHING
$= 4 \times 152 \text{ mm} \times 152 \text{ mm}$
$= \underline{92\ 416 \text{ mm}^2}$

By comparison, it can be seen that although the depth of the box is roughly seven times greater than the depth of the tray, the area of scrap metal produced by notching is nearly fifty-eight times greater.

9.6 Method of fabricating a metal pan with wired edge and riveted corners

Figure 9.5 gives details of a rectangular pan to be manufactured from light gauge sheet metal together with the required layout. Table 9.4 lists the tools and equipment required.

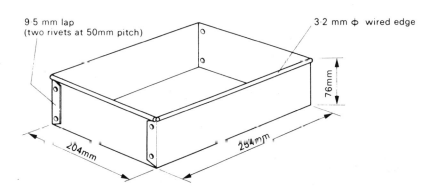

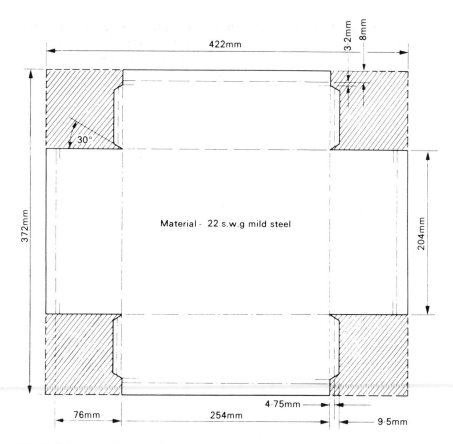

Fig 9.5 Layout for forming a pan

Table 9.4 Tools and equipment required to make a rectangular pan

TOOLS AND EQUIPMENT REQUIRED	REMARKS
Steel rule Straight edge Flat steel square Dividers Jenny odd-legs Bevel Scriber	These are used for marking-out the blank.
Centre punch Nipple punch Hammer	Required for marking the positions of the rivet holes. The hammer is also used for riveting.

191

Guillotine Universal snips	For cutting out the blank and notching the corners.
Rivet set	For the riveting operation.
Folding machine	For bending up the sides.
Bench Stakes Mallet	These are required for completing the bending operations by hand and for wiring the edge.
Cutting pliers	For cutting the wire and holding it in position during the wiring operation.
Bench vice	For bending the wire if a frame is used, and for holding the hatchet stake when throwing the flange off for the wired edge.
Tinman's hand-lever punch	May be used for punching the rivet holes in the corner flanges on the blank.
Drilling machine or portable drill	For drilling the rivet holes on assembly.

The following sequence of operations may be adopted:
1. Mark out the overall dimensions for the blank on a suitable size sheet of metal. If a standard size sheet is used, square the sheet by making a trim-cut on the guillotine and mark the overall length and width of the required blank from the datum edges. Check the rectangular outline for squareness by measuring the diagonals.
2. Cut out the blank on the guillotine and remove all sharp edges or burrs with a suitable file.
   Remember that handling sheet metal can be dangerous so 'ALWAYS REMOVE BURRS AFTER ANY CUTTING OPERATION'.
3. Mark off the allowance for the riveted flanges. These are normally on each end of the long sides of the pan.
4. Mark the centre lines for the rivets.
   Operations 3 and 4 may be performed after square notching the corners.
5. Notch the corners. Care must be taken, when cutting with the snips, not to 'over-cut' in the corners. The notching of corners can be performed much quicker and more efficiently on a notching machine. Most machines are capable of making square notches up to 102 mm in depth in sheet metal up to 1·6 mm thickness.
6. Mark out the clearance notches at each end of the rivet flanges. These are normally made at an angle of 30° as shown in Fig. 9.4, and may be marked with the aid of a bevel.
7. Mark the positions for the rivets on the flanges and centre punch.
8. Drill or punch the rivet holes. *If a drill is used the blank must be supported on a block of wood and the whole assembly securely clamped to the drilling table.*

Figure 9.6 illustrates a 'tinman's hand-lever punch' which may be employed for punching holes in sheet metal.

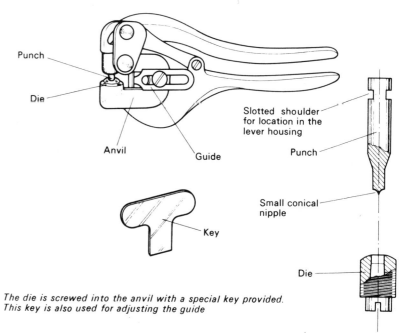

The die is screwed into the anvil with a special key provided. This key is also used for adjusting the guide

Fig 9.6 Tinman's hand-lever punch

Note: Hand-lever punches are supplied with a range of punches and dies suitable for punching small diameter holes in sheet metal. No distortion of the sheet occurs, and the holes are practically free from burrs — one advantage over drilling. The punch is located exactly in centre punch marks on the surface of the metal by 'feel' rather than by 'sighting'. The punch is designed with a small conical nipple for this purpose.

9. Bend the long sides up first in the folding machine to an angle of about 45° and flatten back as shown in Fig. 9.7(a). This provides a crease line for final bending by hand over a hatchet stake after the other two sides have been folded up.
10. Bend up the ends in the folding machine to 90°. This operation will also bend up the rivet flanges as shown in Fig. 9.7(b).
11. Complete the bending operations over a hatchet stake as shown in Fig. 9.7(c).
12. During operation 13, the flanges have to be knocked back slightly to accommodate the ends. This is rectified by bending the laps over a suitable stake with a mallet as shown in Fig. 9.7(d).
13. Support each corner, in turn, on a suitable anvil or bench stake and mark the centres of the rivet holes through the holes in the flange with a nipple punch.
14. Drill the top hole in each corner first and insert a rivet to maintain correct location before drilling the bottom holes. During this drilling operation the job must be properly supported on a block of wood and securely clamped.
15. Deburr the holes. The assembly is now ready for riveting.
16. Rivet the corners. The method of riveting is shown in Fig. 9.8.

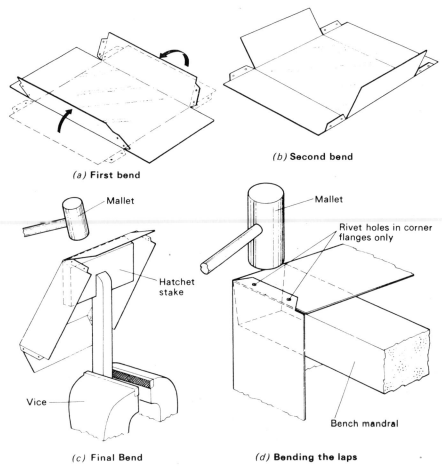

Fig 9.7 Bending the sides of the pan

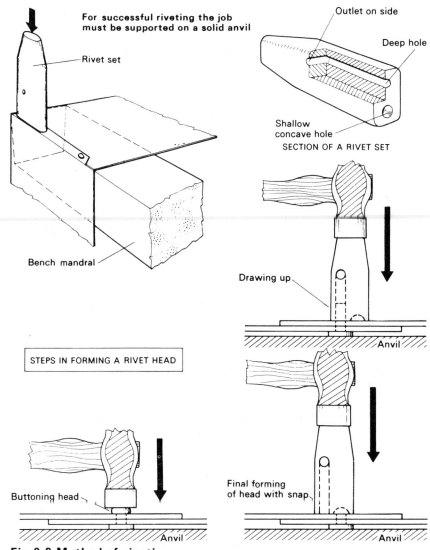

Fig 9.8 Method of riveting

17. Bend the edges for the wire over a hatchet stake, using a mallet as shown in Fig. 9.9(a).
18. Cut the wire to the required length and make a frame as shown in Fig. 9.9(b). The frame may be bent to shape in the bench vice. A great many craftsmen prefer to apply the wire direct to the job, thus dispensing with the operation of forming the frame. *In either case the ends of the wire should terminate about the middle of one of the short sides of the pan.*

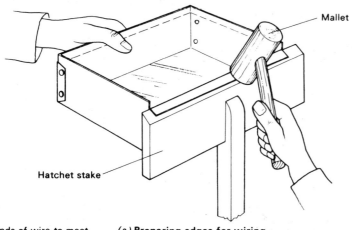

(a) **Preparing edges for wiring**

Never use hand shears for cutting wire — use wire cutting pliers

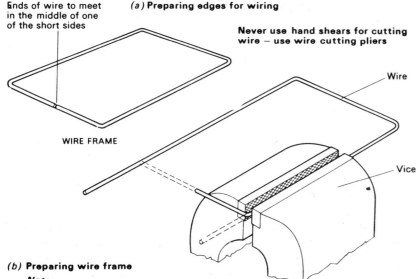

(b) **Preparing wire frame**

Note:
The wire frame may be bent in the jaws of a vice, the corner being closed with a hammer.
Cut off any surplus wire so that the ends of the wire just butt together. File the ends of the wire smooth to prevent marking of the bead during the wiring operation

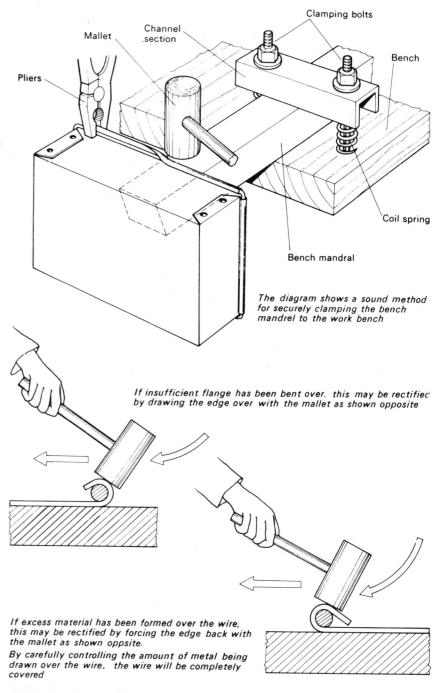

The diagram shows a sound method for securely clamping the bench mandrel to the work bench

If insufficient flange has been bent over, this may be rectified by drawing the edge over with the mallet as shown opposite

If excess material has been formed over the wire, this may be rectified by forcing the edge back with the mallet as shown oppsite.
By carefully controlling the amount of metal being drawn over the wire, the wire will be completely covered

(c) **Commencing to wire the edge of the pan**

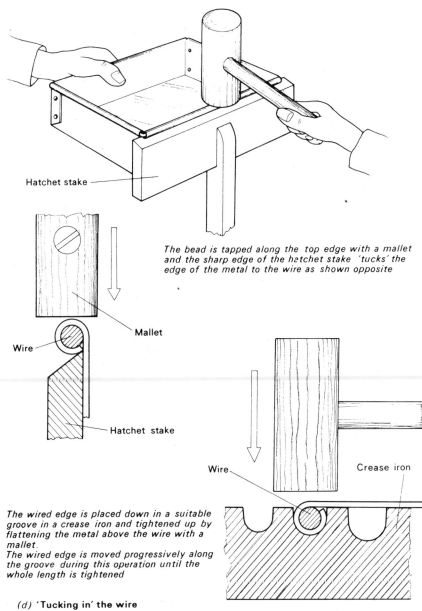

Fig 9.9 Wiring the edge of the pan

19. Insert the wire and form the metal over the wire using a mallet and a suitable bench stake as shown in Fig. 9.9(c).
20. Complete the wiring by 'tucking-in' over a hatchet stake using a mallet as shown in Fig. 9.9(d).

9.7 Faults when wiring an edge

In most fabrication shops wiring is essentially a hand operation. As a general rule the allowance for a wired edge is 2½ times the diameter of the wire. The flange or edge is folded up just over 90° and knocked over the wire with a mallet. A skilled craftsman can control the operation of forming the metal over the wire by pulling over or knocking back the metal with the mallet. Figure 9.10 illustrates common faults which may occur when wiring a straight edge.

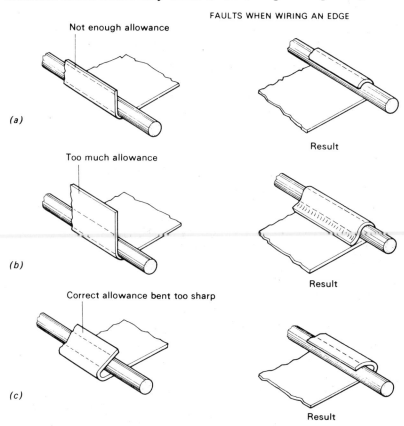

Fig 9.10 Typical faults when wiring a straight-edge

Figure 9.11 shows the faults that may occur when wiring a curved edge. The conical jug shown in Fig. 9.11(a) has a wired edge. This edge, like all contoured edges must be wired in the flat. This type of wiring operation can prove difficult if the correct technique is not adopted. Figure 9.11(b) and (c) illustrate the incorrect and correct methods of performing the operation of wiring a curved surface in the flat.

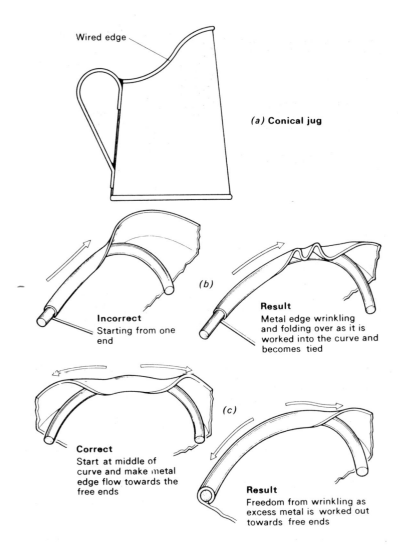

Fig 9.11 Typical fault when wiring a curved edge

9.8 Methods of wiring cylinders and cones

The two methods of producing a wired edge on curved surfaces are:
1. WIRING IN THE 'FLAT' BEFORE FORMING BY ROLLING;
2. WIRING THE EDGE AFTER FORMING BY ROLLING.

The method employed is optional, but the choice may be influenced by certain factors which will now be explained.

Wiring before rolling

On externally wired edges, the decision whether to wire in the 'flat' or wire after rolling will be influenced by the fact that when a cylinder is rolled after wiring, the inside diameter of the wired edge will be somewhat less than the required inside diameter of the cylinder. This is explained in Fig. 9.12.

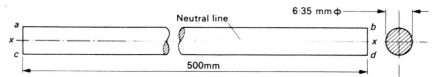

If a piece of 6·35mm diameter wire 500mm in length is rolled to form a ring it will have a **mean diameter** of 159mm. Likewise, when sheet of 1·6mm thick metal of the same length is rolled to form a cylinder, it wills have a **mean diameter** of 159mm

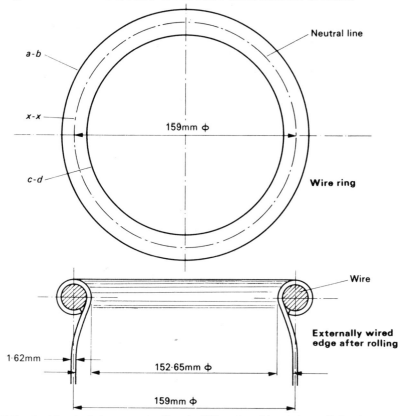

If the sheet is wired along on edge (i.e. the 500mm length) and then rolled to form a cylinder, the above effect will result. It will be seen that both sheet metal and wire forms assume a common **mean diameter** thus causing a considerable constriction as illustrated above

Fig 9.12 The effect of wiring before rolling

It should now be appreciated that *wiring before rolling is unsuitable if the cylinder is to have a constant inside diameter.*

The constriction in diameter caused by the wire can be minimised by light lubrication of the wire prior to wiring. This lubrication tends to assist the slight movement of the wire inside the bead during the rolling operation. Because the cylinder resists the constricting force of the wire, thus preventing it from attaining its smallest diameter, a gap results where the two ends of the wire should butt together. This unavoidable gap produces no real problem provided that the ends of the wire are slightly chamfered with a file to prevent them marking the bead when 'truing-up' with a mallet.

On internally wired edges, rolling after wiring produces the opposite effect to that obtained on externally wired edges of cylinders. *The wired edge tends to become slightly larger in diameter than that of the cylinder.* The process of internally wiring before rolling is used where it is not essential to maintain a constant outside diameter on cylindrical articles.

Wiring after rolling

If a cylinder is required to have a constant diameter then the wiring operation must be carried out after the rolling operation.

Figure 9.13 illustrates the operation of rolling after wiring.

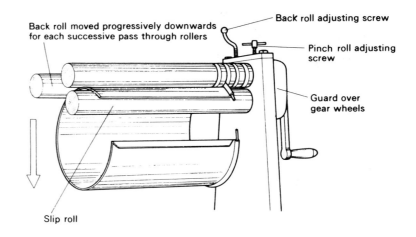

Fig 9.13 Rolling after wiring

Note: It is common practice to insert a narrow strip of metal between the hooked flanges for the grooved seam to prevent them from being flattened during the rolling operation. *The material being rolled should be lightly gripped between the pinch rolls.*

SLIP ROLLS are used for rolling cylinders after wiring. The cylinder must be rolled over the slip roll. Care must be taken before inserting the metal in the rolls to ensure that:
(a) The wired edge rests in the correct groove in one of the 'pinch rolls';
(b) The machine is checked to determine whether it 'rolls up' or 'rolls down'.

The slip roll enables the cylinder to be withdrawn easily from the machine on completion of the rolling. The sequence of operations for wiring after rolling is shown in Fig. 9.14.

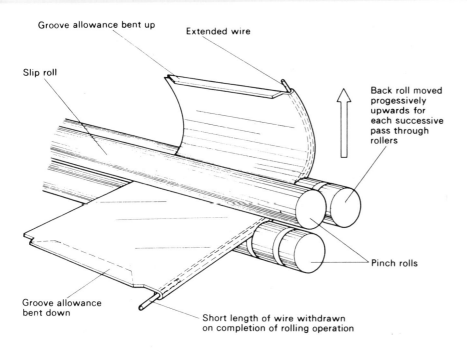

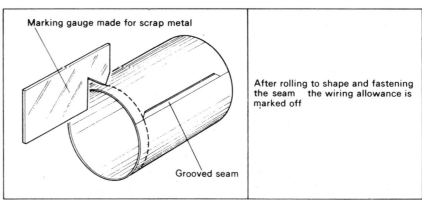

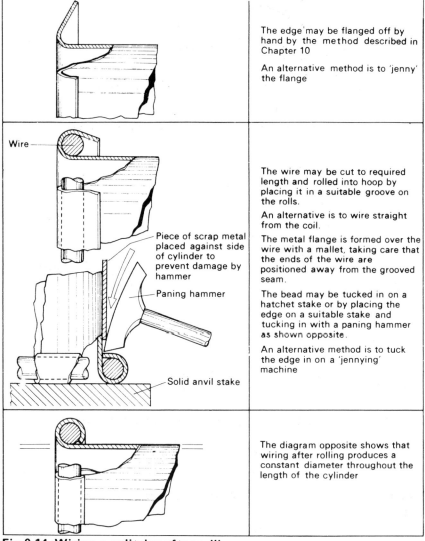

Fig 9.14 Wiring a cylinder after rolling

9.9 Method of fabricating a domestic funnel (tinplate)

Figure 9.15 gives details of a domestic funnel which is to be fabricated from tinplate. A funnel has been selected for the purpose of showing the sequence of operations and the combined use of hand tools and machines necessary for the construction of a tapering article with a grooved seam and a wired edge.

When making a funnel, the normal procedure, as is the case with all tapering articles, is to wire the edge after it has been formed to the required shape and fastened together.

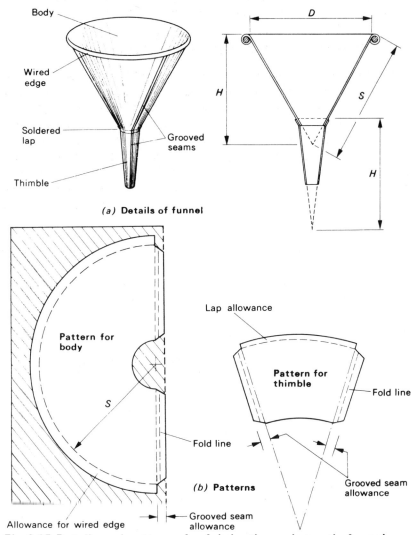

Fig 9.15 Details and patterns for fabricating a domestic funnel

The funnel shown in Fig. 9.15(a) is made in two parts, a BODY and a THIMBLE. Each are frustums of a right cone, the patterns for which are developed by the RADIAL LINE METHOD which is explained in Chapter 2. By making the diameter equal to the slant height S the development of the body will be a quadrant, as shown in Fig. 9.15(b). It is common practice to make the distance H from the base of the cones to the apex constant. The patterns may be marked out on paper and transferred on to the tinplate, or marked out directly on to the sheet. The appropriate allowances for the seams and for notching are made in the usual manner.

9.10 Breaking the grain of the metal

The first operation when working with tinplate should be to 'break the grain' of the metal to prevent ridges forming on its surface. This consists of rolling the piece of tin plate backwards and forwards a few times in the rolls, reversing the bending each time. This process will ensure that the article to be formed by rolling will have a smooth surface free from kinks or ridges. If this 'breaking' operation is not carried out, a cylindrical or conical article, with only one rolling, develops ridges around the shaped body which, if not seen, may be more easily felt by passing the palm of the hand over the rolled surface.

It is sound practice always to break the grain, especially on metals which have been 'cold reduced' before commencing forming operations by rolling.

Once the breaking operation is completed the pattern or blank should be rolled out flat in readiness for forming operations.

9.11 Forming the funnel

Figure 9.16 shows typical workshop operations for forming the body and thimble for the funnel, and Table 9.5 lists the tools and equipment required.

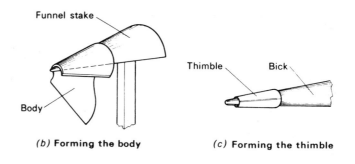
(b) Forming the body (c) Forming the thimble

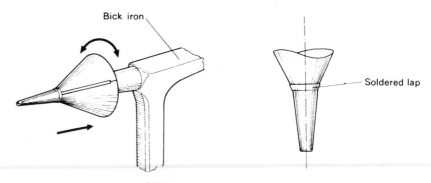
(d) Finishing the funnel

Fig 9.16 Operations for forming a funnel

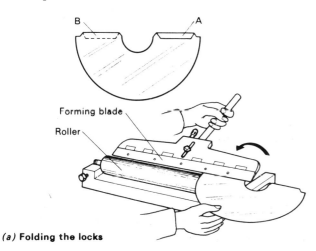
(a) Folding the locks

Table 9.5 Tools and equipment required to make a funnel

TOOLS AND EQUIPMENT REQUIRED	REMARKS
Steel rule Scriber Dividers	These are used for marking out.
Guillotine Bench shears Universal snips	For cutting out the blanks and developed shapes. Straight snips may be used for straight line and external curve cutting, but bent pattern snips will be required for cutting internal curves. Universal snips will perform both these operation.
Rolling machine	For BREAKING-IN the TINPLATE.

Flat bed folding machine	For forming the locks for the grooved seams. An alternative method for performing this operation is to throw the edges over a hatchet stake using a mallet.
Funnel stake Bick iron Mallet Stretching hammer	Used for forming the body of the funnel. Used for forming the thimble. For stretching lap circumference of thimble.
Grooving tool Hammer	For fastening the seams.
Soldering stove Soldering iron	For soft soldering the inside of the seams and the lap joint between body and thimble.
Cutting pliers	For cutting wire for the wired edge.
Jennying machine	Two operations are required: 1. Turning the wire allowance. 2. 'Tucking' the edge over the wire.

The tapered flange on the thimble for the lap can be produced by a light stretching of the edge on a suitable stake using a stretching hammer.

1. Fold the locks for the grooved seam one up and one down as shown at A and B in Fig. 9.16(a) on both the body and thimble.
   A suitable machine for this operation is the BENCH ROLL TYPE FOLDER.
   The edge to be folded is pushed under the forming blade until it butts against the back guide plate. The sliding back gauge plate is adjusted backwards or forwards until the fold line is aligned with the edge of the forming blade. The edge is folded right over by pulling the operating handle upwards. In this case the metal blank is positioned at one end of the machine, as shown in the diagram. One edge is folded and then the blank is turned over and inserted in the same position, thus ensuring that one edge is turned up and the other one down.
2. The body may be formed to shape over a funnel stake. This operation consists of bending the body of the funnel by hand with a sliding motion over the stake. A mallet is used for final forming to shape after grooving the seams (Fig. 9.16(b)).
3. The thimble is formed in a similar manner over a long tapering bick iron. It is formed roughly to shape, the grooved seams are interlocked and the thimble is driven on to the bick iron to hold the interlock tight before completing the grooved seam (Fig. 9.16(c)).
4. The body and thimble are joined by soft soldering. A bick iron is used to flange the joint as shown in Fig. 9.16(d).

An alternative method of attaching the thimble to the body is to make the allowance for the soldered lap on the thimble. The patterns must be adjusted accordingly. In this case the small end of the body for the funnel is reamed out on the bick iron as shown above. This operation will set the lap edge to the correct angle to allow the thimble to be forced in.

Note: Once the body and the thimble have been formed to the required shape, the seams are grooved with a grooving tool and soft soldered on the inside. When forming the thimble, it is advisable to over bend with the locks overlapping. Then when the thimble is forced on to the tapering bick iron the interlocking edge will be held securely in position ready for the grooving operation.

The next operation is to cut the required length of wire and roll it into a hoop on the rolls. The wire allowance is flanged up on the jenny, the wire inserted and the metal closed over the wire with a suitable pair of rolls on the jenny, making sure that the butt joint on the hoop is positioned away from the grooved seam on the body.

9.12 Forming operations on a universal jennying machine

Figure 9.17 shows the types of wheels or rolls which are used on a universal jennying machine together with the various operations which they perform.

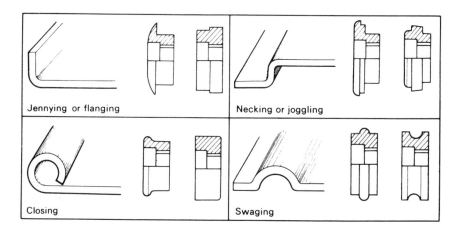

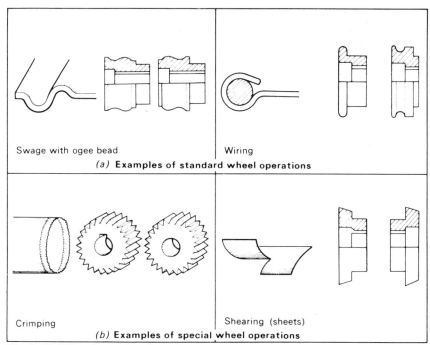

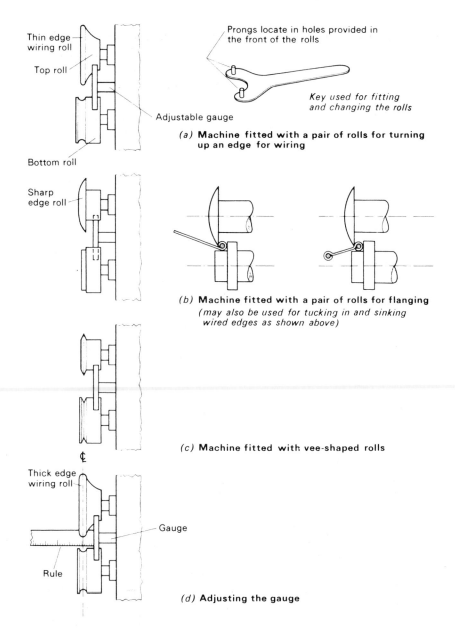

Fig 9.17 Wheels and rollers for jennying machines

Figure 9.18 illustrates a number of operations for preparing edges with the universal jenny machine. In Fig. 9.18(a) the working faces of the rolls are radiused and are used for forming narrow edges on circular and irregular articles. In Fig. 9.18(b) the upper roll has a sharp edge and is used for turning single edges on curved work and discs. These rolls are also used for tucking in wired edges.

The 'vee'-shaped rolls shown in Fig. 9.18(c) are ideal for turning up a double edge on elbows for paned down or knocked up seams.

Precaution: When using sharp edged rolls for flanging operations, do not over-tighten the top roll. If the top roll is too tight the metal will be sheared. The top roll should be adjusted so as to afford a light grip on the metal between the rolls.

For a wired edge the allowance must be equal to 2½ times the diameter of the wire. The measurement is taken from the face of the gauge to the centre of the upper roll as shown in Fig. 9.18(d). The gauge is adjusted, usually by a knurled nut at the side of the machine. After setting to the required measurement the gauge is locked in position.

Fig 9.18 Edge preparation on a jennying machine

When flanging a disc, a small piece of metal folded (as shown in Fig. 9.19) should be used to prevent injury to the operator's hand. As an extra precaution remove all burrs from the metal disc before commencing the operation.

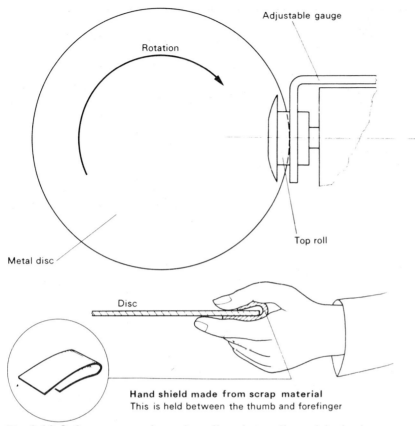

Fig 9.19 Safety precaution when flanging a disc with the jenny

9.13 Fly press operations

A fly press is basically a hand press where the ram is worked in the frame by the operation of a screw which is rotated by the operator turning a handle or 'fly'.

The types of fly press usually found in a fabrication shop are:
1. *Standard or c-frame* The main dimensions are the bed to guides, centre to back, and the screw diameter.
2. *Tall type* These are available with a range of bed to guide dimensions — maximum about 360 mm.
3. *Deep back* These have a range of centre to back dimensions which are greater than the standard models — maximum about 306 mm.
4. *Bar type* The solid bed is omitted and provision is made to fit a bar on which the work may be supported. Mainly for work on cylinders which pass over the bar.

The deep back type and the bar fly press are illustrated in Fig. 9.20.

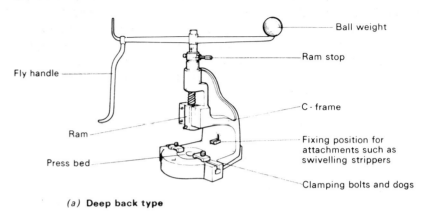

(a) **Deep back type**

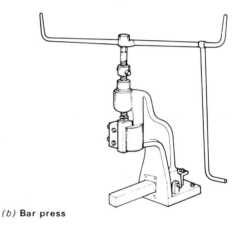

(b) **Bar press**

Fig 9.20 Types of fly press in common use

Hand screw presses are generally selected dimensionally, but the pressure exerted is directly related to the screw diameter. As a general rough guide, the rating in tonnes* of a fly press is twice the screw diameter. The force can be increased by fitting ball weights to the fly arms. *In the interest of safety, take care when operating a fly press because the fly arm and the weights can cause injury, not only to the operator but to those in close proximity.*

Notching, piercing and punching, light pressing, bending and riveting are all operations which can be performed on a fly press.

*Note: The force, in kilo-newtons exerted by the press on the work piece is equal to ten times the rating in tonnes mass.

Figure 9.21 shows details of a typical sub-chassis in 18 s.w.g. mild steel sheet. All the operations necessary to form this chassis from the blank can be easily performed on a fly press fitted with appropriate tooling. This type of sheet-metal work is normally done in batches, which means that the fly press is set up for one specific operation and the batch is run through before resetting for the next operation. Alternatively more than one fly press may be set up to reduce waiting time due to resetting operations.

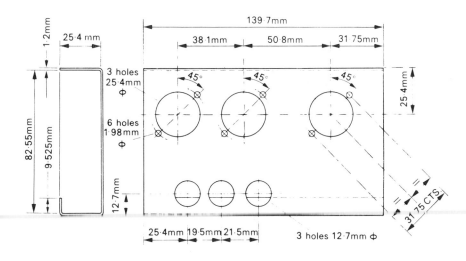

Material: En2/CR/FF. 18 s.w.g.

Fig 9.21 Details of a typical sub-chassis

The sequence of operations may be as follows:
1. A template is used to mark the positions of all the holes. Such a template is usually marked out on 10 s.w.g. mild steel plate on a surface table using a vernier height gauge and an angle plate. Small pilot holes are drilled, and once the template has been passed by inspection these are opened out with the correct size drill to suit the diameter of a nipple punch. The template is provided with location buttons to give an accurate location for the blanks. Figure 9.22 shows the template positioned over the blank ready for transferring the hole positions with a nipple punch. The use of such a template is a fool proof system which not only provides identical hole positions on each blank, but dispenses with the use of guides and locations having to be set up on the press.

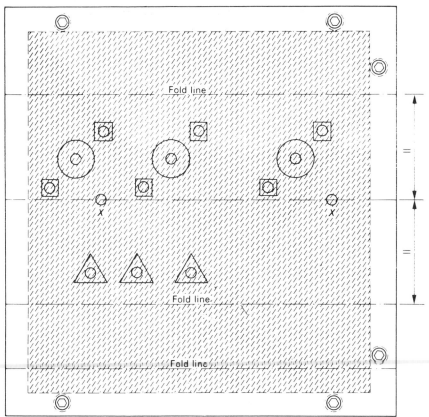

The master template is placed over the blank (shown shaded) and is accurately located by means of location buttons. The master template may be used for transferring hole positions to the blank using a nipple punch or, where large diameter holes are indicated, the pilot holes are punched through the template and the blank in one operation. Location holes (sometimes termed 'tooling holes') as shown at X may be provided on the template for more accurate location for bending

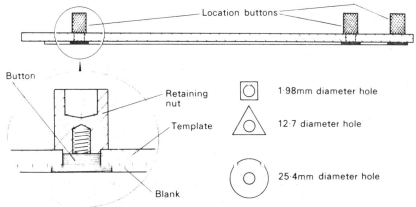

Fig 9.22 Marking hole centres (master template)

2. The fly press is set up for punching the holes and the ram-stop is set to give optimum punching conditions for ease of stripping. A universal bolster outfit is used extensively on fly-press work of this nature. The bolster frame is a heavy casting which has a central hole bored to receive the dies and standard holders for the smaller dies. It is also provided with two pairs of holes drilled and bored for adjustable gauges through bosses cast on the sides. In addition four tapped holes are provided in the top face of the bolster for securing special attachments in place of the gauges mentioned.

Two types of stripper are available with the bolster outfit. One is an automatic stripper for attachment to the punch unit, and this removes the part quickly on the return stroke of the ram. This type of stripper is supplied with three plates which have different apperture sizes to suit the varying punches. The plates are easily removed for interchanging with another member without dismantling the unit from the punch holder.

The other type of stripper consists of two swivel heads which are located in a bracket secured to the rear of the press column, and the operator can rotate them clear from the region of the punch in order to make the component easier to load and remove. With this type of stripper adequate support is applied close to where the stripping action takes place, i.e. where it is really needed.

The punches used with a universal bolster outfit range from 2·4 to 95 mm diameter and to accommodate this range of punches the bolster is supplied with five insertable die holders. *The main advantage of the bolster outfit is that all interchangeable sets of punches, dies and die holders will assume concentric alignment with each other without having to reset the unit.*

The normal procedure when punching a series of holes in a blank is to punch all the smallest diameter holes first. The smaller diameter punches are provided with a centre point which accurately locates in the dot made by the nipple punch. Larger diameter punches are provided with a pilot pin for location. Large diameter holes are first punched to the diameter of the location pin and then re-punched to the required diameter.

3. The bolster unit is removed from the flypress by releasing the clamping arrangement which is usually 'Tee'-bolts and dogs. This is replaced by a standard bending tool consisting of a 'vee' block mounted on the press bed. Bending small articles is an operation which can take up considerable time if hand methods are utilised,

and the use of the bending tool illustrated in Fig. 9.23 can solve many problems of this nature.

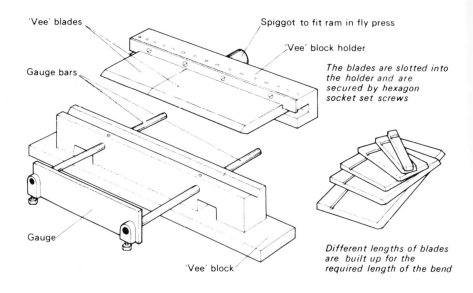

Fig 9.23 Bending tool for small sheet metal components

This tool is basically a greatly reduced version of the massive press-brake tools used on that class of machine and is ideally suitable for the hundreds of different brackets and bend components, corner strips and the edges of small trays or panels. The top 'vee' blades are made in varying lengths, as Fig. 9.23 shows, and by sliding them into a groove machined through the blade holder they may be built up for various lengths of bend. With this composite arrangement, it is possible to insert the blades at different intervals enabling two or more bends to be made along a sheet of material.

4. The ram-stop is set to control the downwards movement of the 'vee' blade, thus ensuring a constant bending angle, in this case 90°. A back stop mounted on the 'vee' block itself is set to provide the location for a specific bend. The blank is pushed up to this stop with one hand and the press is operated with the other. Provision is made on the top of the ram screw for positioning the fly handle at a suitable station for this two-handed operation. Figure 9.24 shows details of the bending operation

(a) with the use of a back stop;
(b) with a pair of tooling holes in the blank which are placed on a centre-line parallel to the fold required.

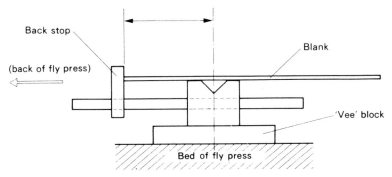

(a) Position of fold by back stop

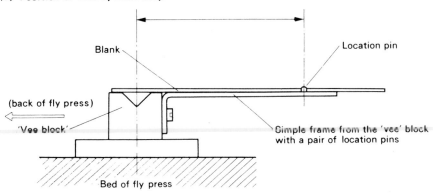

(b) Position of fold by tooling hole and pin

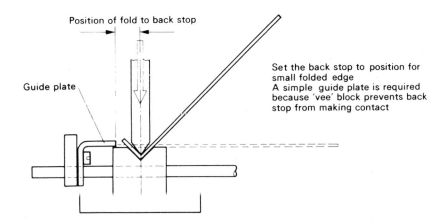

Set the back stop to position for small folded edge
A simple guide plate is required because 'vee' block prevents back stop from making contact

(c) Make the small fold first

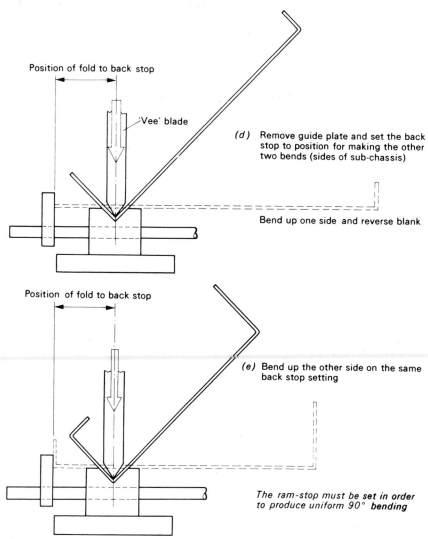

(d) Remove guide plate and set the back stop to position for making the other two bends (sides of sub-chassis)

Bend up one side and reverse blank

(e) Bend up the other side on the same back stop setting

The ram-stop must be set in order to produce uniform 90° bending

The general allowance on the blank for fly press bending is inside dimensions plus half a thickness of material for each bend

Fig 9.24 Bending the sub-chassis on the fly press

Note: Producing a fold in a blank stiffens the part considerably. If the component has been designed to support a certain weight over a span and has to be manufactured from 16 s.w.g. mild steel (1·6 mm) to give the required stiffness, then by the addition of a fold along its length it could be manufactured from considerably lighter material, for example 20 s.w.g. (0·9 mm) and still maintain its original stiffness.

Every sheet metal workshop encounters examples where it becomes necessary to cut and bend sheets to the form of boxes or shallow trays, and in order to obtain this shape the blank needs notching at each corner prior to undertaking the bending operations. Figure 9.25 illustrates the following useful cutting tools which can be used with a universal bolster on the fly press:

Corner notching tool   The die is inserted in the bolster frame and the punch is aligned by the dovetail leg operating in a corresponding groove — this useful feature makes incorrect setting impossible;

Corner radius tool   Which can remove four different radii according to the set of the tools. The four legs on the punch extend into the die apperture and so ensure proper alignment. The die is attached to the bolster frame by screws which locate in the four tapped holes provided on its top face. This tool is used primarily for cutting the corners of sheets and panels;

Cropping and end-rounding tool   Enables the simultaneous cropping and rounding of the ends of strip components such as links. The narrow strip locates between the guides and the end location is secured from the cross bar shown in the diagram. The complete die assembly is secured to the bolster frame by locating it in the recess.

Rectangular and square punches   The production of rectangles and squares in several sheets of material is a lengthy and tedious operation if press tools are not employed for the task. A range of punches and dies may be used with the bolster unit, and may be set to produce single holes, several appertures in different positions in the component, notch the sheet edges, or produce 'L' of 'T' shaped cuts.

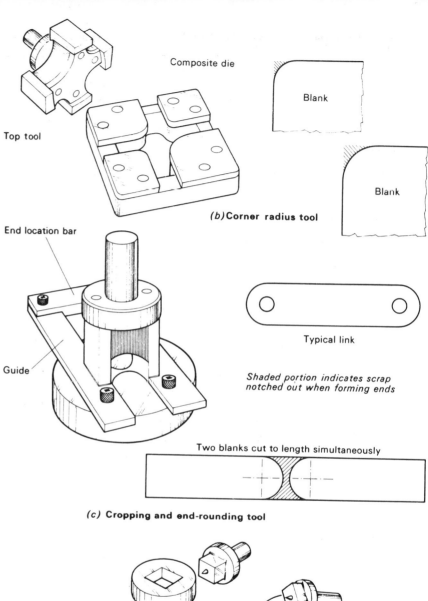

(b) Corner radius tool

(c) Cropping and end-rounding tool

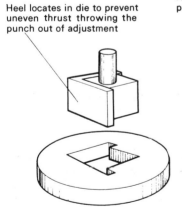

(a) Corner notching tool

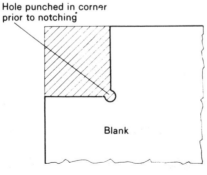

To produce flat surfaces with a notched and folded chassis, holes are required to prevent the metal tearing at the corners. If the chasis is to be welded this hole is not so important

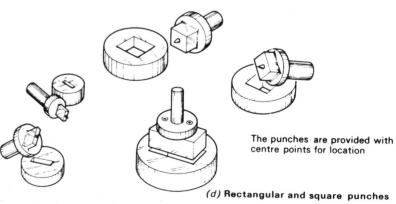

(d) Rectangular and square punches

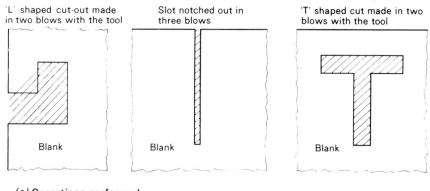

(e) Operations performed

Fig 9.25 Typical cutting tools used on the fly press

9.14 Setting the guillotine

Treadle and power guillotines are fitted with FRONT and SIDE GAUGES. These usually consist of a FIXED SIDE GAUGE, sometimes referred to as the 'SQUARING GUIDE', and a FLAT BAR FRONT GAUGE.

On some machines the side gauge can be extended for wide sheets, and may be graduated in millimetres.

The front gauge is adjustable across the bed or table, and further along extension bars (arms fitted to the front of the machine). Figure 9.26(a) shows a plan view of the side and front gauges. Figures 9.26(b) and 9.26(c) show how the gauges are set.

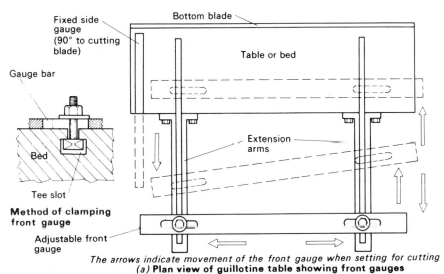

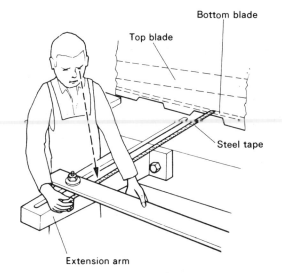

(b) Setting the front gauge - using a rule

(c) Setting the front gauge – using a steel tape

Fig 9.26 Front and side gauges (guillotines)

The bulk of cutting performed on the guillotine is when the sheet or plate is located against a BACK GAUGE, and there are several types of these. The simplest, and usual standard type, consists of an ANGLE GAUGE BAR. Back gauges are mounted on an attachment fixed to the movable cutting beam and move up and down with it. Figure 9.27 gives basic details of gauges and guides used on guillotines.

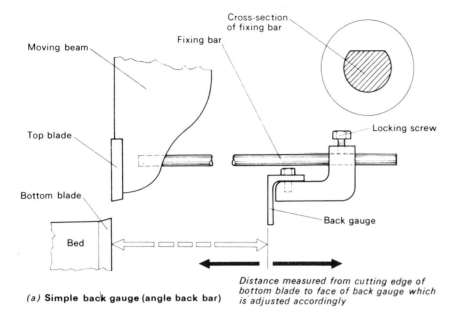

Fig 9.27 Details for setting back gauges (guillotines)

Where precision cutting is required, the simple back gauge will take a while to set in order to achieve the exact size register.

On batch or bulk cutting operations where the back gauge has to be moved constantly a more elaborate type of back gauge is necessary to enable the settings to be carried out more quickly, reducing both the actual time taken to move the gauge and then the time to set it to a dead size.

Even the simple type of back gauge can have an attachment for final fine adjustment, as shown above. The guide is set in an approximate position and locking lever B tightened. The final setting to correct register is achieved by slackening lever A, and turning the hand wheel on the fine adjustment screw. The back gauge is then locked in position by tightening locking lever A.

9.15 Operations on the guillotine

The fixed side gauge is used for positioning the material. To square off two adjacent sides of a sheet or plate, a trim-cut of approximately 6 mm is made on one edge. The second edge is then sheared at 90° to the first by holding the trimmed edge firmly against the side gauge which is normally located on the left-hand side of the table.

After marking off from the trimmed edge (DATUM EDGES), further straight cuts are made by sighting each scribed line or witness mark, in turn, on the edge of the fixed bottom cutting member of the guillotine.

If a number of identical blanks are required to be cut, the material is located against the side gauge and the back guide which is set at the necessary distance from the back edge of the bottom blade. The front guide is used when the material has to be cut to close tolerances. It is normal practice to set the back guide about 6 mm over the required dimension for the cut, and the front guide to the exact measurement. The material is passed over the front guide, through the gap between the blades to locate on the back guide, and the initial cuts made off the back guide. The cut blanks are then located on the front guide and finally cut to required dimension. Taper cutting is performed by setting the front guide at an angle.

Simple guide plates may be made up and clamped in position on the table by means of 'tee'-bolts. These are very useful improvised attachments which enable many bevel and mitre cutting operations to be easily performed on the guillotine. Figure 9.28 shows a typical gusset plate that can be produced on the guillotine shear.

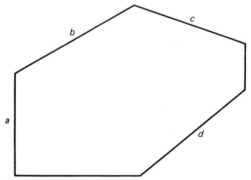

Fig 9.28 Typical gusset plate

The sequence of operations for cutting out the gusset plate is explained below and shown in Fig. 9.28. (The required gusset plate is shown shaded for all the sequence of operations.)

OPERATION 1

Set the back guide to cut the edge marked *a*. The length of flat is located against the side gauge and the back gauge, as shown in the diagram opposite.

OPERATION 2

Cut the number of blanks required.

Notes:
(a) If a suitable width of flat is not available, suitable widths may be cut from a standard plate. The back gauge would be set for this purpose.
(b) A simple bevelling plate may be flame cut and into it a series of location holes, as shown in D, E, F and G, drilled for clamping purposes. It will be seen that a pair of holes is used for each setting. The first blank is marked out and used as the pattern for all the settings. The bevelling plate is set for cutting edge *b* and is securely clamped to the table using the Tee slot on the left of the machine, the clamping bolts located at positions D and E.

OPERATION 3

The blanks are located in the bevel plate whilst the edge *b* is trimmed. The bevel plate is bolted to the guillotine through holes D and E.

OPERATION 4

Edge *c* is cut by setting the bevelling plate as shown opposite, using location holes F and E.

OPERATION 5

The bevelling plate is removed from its location with the 'tee' slot on the left-hand side of the machine, and is turned over and located in the required position using the 'tee' slot on the right-hand side for cutting edge *d*, as shown below. It is clamped securely in position using bolt holes D and E.

Note: Any number and variation of bevelled shapes may be cut on the guillotine using this simple set up. It is good practice to stamp some information on such bevelling plates. They can then be stored and used time and time again when a similar cutting operation is called for. On completion of the cutting, the gusset plates are marked off for the hole centres and punched or drilled.

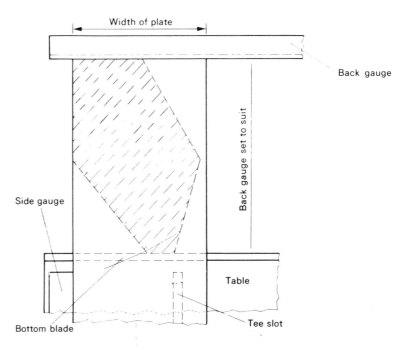

(a) Operation 1 and 2

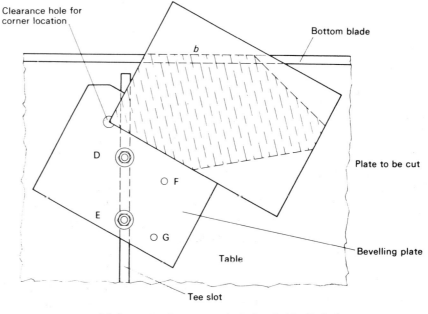

(b) Operation 3 Left-hand side (facing)

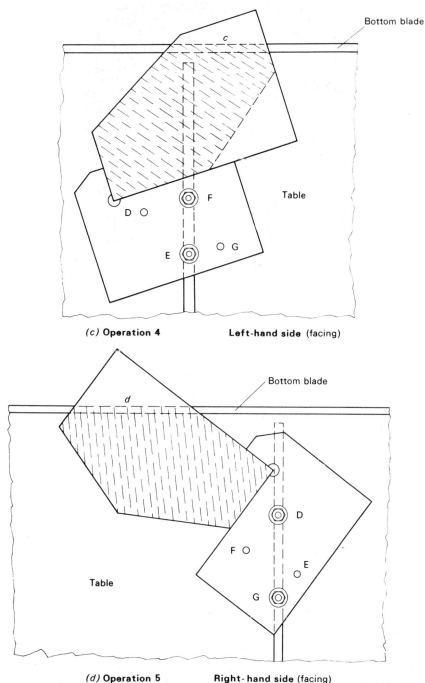

(c) Operation 4   **Left-hand side** (facing)

(d) Operation 5   **Right-hand side** (facing)

Fig 9.29 Sequence of operations for cutting gussets

9.16 Cutting apertures and circles (shearing machines)

Rotary shears or combined nibbler-shearing machines are used for straight or irregular cutting of metal sheets and plate.

The basic principle of these machines has been explained in Chapter 7.

Rotary shears may be used for cutting out appertures in panels on which the outline has been marked, and this operation requires considerable skill. Figure 9.30 shows a method of cutting out a rectangular apperture with radiused corners with the aid of a suitable guide. This operation requires less skill.

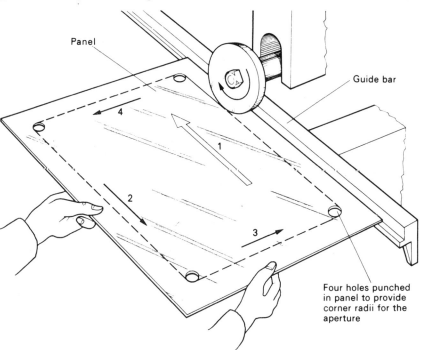

The panel is located against a guide bar and each cut is made in turn between a pair of holes

Fig 9.30 Straight line cutting operation on a rotary shearing machine

Circular flanges and discs can be easily produced on shearing machines, and this operation is explained in Fig. 9.31. As a general guide, when performing cutting operations on a combination shearing and nibbling machine:

1. Use shearing blades for straight line or large radius curves;
2. Use nibbling punch for cutting small radii circles or complex shapes.

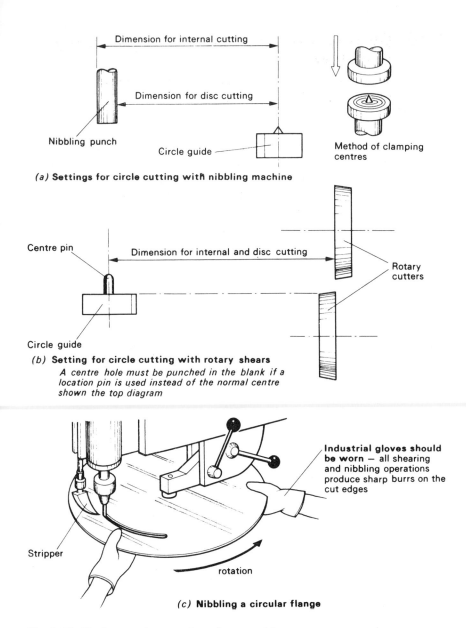

(a) Settings for circle cutting with nibbling machine

(b) Setting for circle cutting with rotary shears
A centre hole must be punched in the blank if a location pin is used instead of the normal centre shown the top diagram

(c) Nibbling a circular flange

Fig 9.31 Circle cutting on shearing machines

(a) Typical curve, cutting-out and nibbling shears

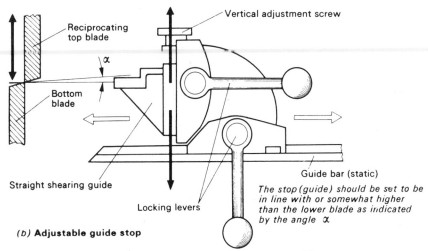

(b) Adjustable guide stop

The stop (guide) should be set to be in line with or somewhat higher than the lower blade as indicated by the angle α

Fig 9.32 Combination shearing and nibbling machine

9.17 Angle section work

Apart from constructional steelwork, angle sections in particular, and rolled sections in general, are often used in conjunction with sheet metal in various fabrications. In this section some of the operations involved in angle section work will be described, and these may be applied to 'tee', channel and similar sections.

Figure 9.32 shows a typical universal shearing and nibbling machine which, with interchangeable tools, is capable of other operations such as beading, folding, flanging and louvre cutting. The diagram also illustrates how various adjustments may be made on the guide stop which is a feature of these types of machine.

Cutting

The methods generally used for cutting angle sections are:
1. Shearing.
2. Use of abrasive wheel machines (see Chapter 7).
3. Cold sawing.
4. Notching — see under 'shearing' below.
5. Flame cutting. This process has been fully described in Chapter 7.

Figure 9.33 illustrates types of welded connections for rolled steel sections, all of which have to be prepared by cutting operations.

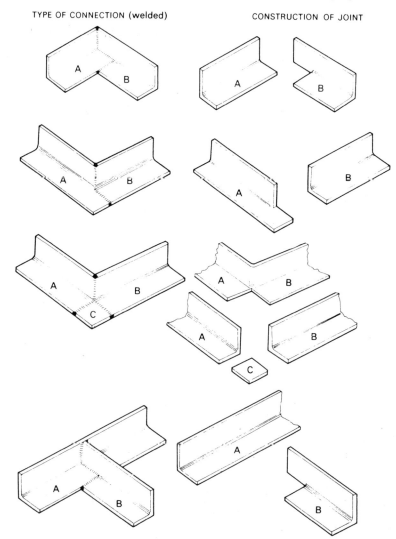

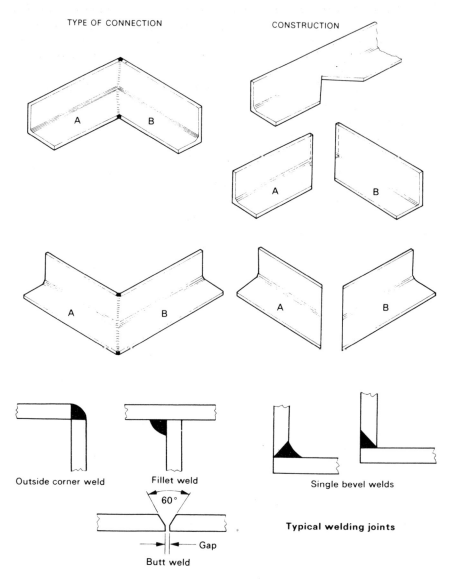

Fig 9.33 Types of welded connections for rolled steel sections

SHEARING

Cutting by shearing is quick and probably the most economical production method. The shearing of rolled steel sections is performed in dies designed to suit the section. The dies are mounted in a special shearing machine. This operation is commonly referred to as CROPPING.

With all cropping operations there is always some risk of distortion on one of the cut edges. This problem may be easily overcome by reinserting the portion with the distorted edge in the opposite direction and making a trim-out. Some allowance must be made for this in the initial marking-off or setting-up prior to cutting.

Hand-operated angle shearing machines are generally limited to sections up to 76 x 76 x 9·5 mm for, as will be appreciated, a considerable amount of manual effort is required to shear, for example, 51 x 51 x 6·35 mm section.

In most fabrication shops, cutting operations on rolled steel sections are carried out on power machines. Machines are available which perform a combination of cutting operations, such as punching, shearing and notching, the shearing operations including not only section shearing, but round and square bar cropping, and plate shearing. Angle section has to be notched in order to permit it to be bent, and most of the notches are of the 'vee'-notch or the square-notch type. Figure 9.34 shows a typical universal shearing machine, whilst Fig. 9.35 shows some of the operations performed on such a machine.

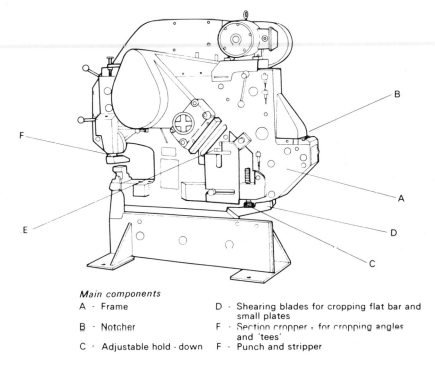

Main components
A - Frame
B - Notcher
C - Adjustable hold-down
D - Shearing blades for cropping flat bar and small plates
E - Section cropper, for cropping angles and 'tees'
F - Punch and stripper

Fig 9.34 Universal steel shearing machine

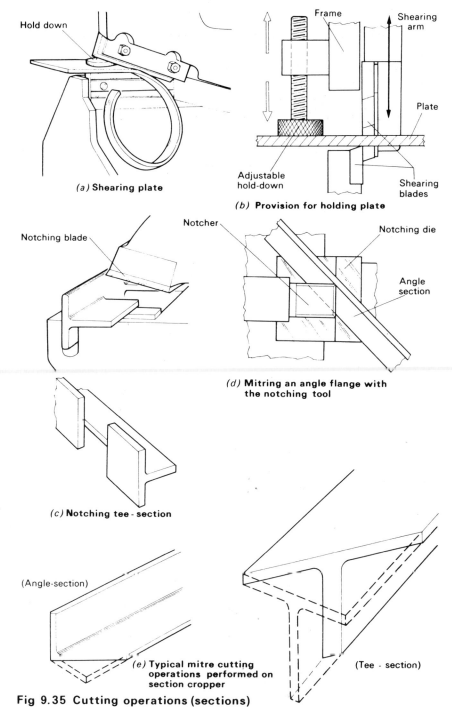

Fig 9.35 Cutting operations (sections)

213

ABRASIVE WHEEL CUTTING

In general, cutting operations using an abrasive wheel machine are confined to the lighter sections. Cutting may be carried out 'wet' or 'dry'. Wet cutting gives a rougher cut, but does give longer wheel life. Data for dry cutting is given in Table 9.6.

Table 9.6 Abrasive cutting data for some common sizes of angle section

ANGLE SECTION (Dimensions in mm)	APPROXIMATE CUTTING TIME IN MINUTES FROM FLOOR TO FLOOR	APPROXIMATE NUMBER OF CUTS PER WHEEL
76 x 76 x 9.5	2	110
51 x 51 x 6.35	1	300
38 x 38 x 6.35	0.5	400

Note: The above figures are typical for DRY CUTTING. Times for WET CUTTING will be slightly slower, but would give increased wheel life.

COLD SAWING

For cutting operations on the heavier sections, the cold saw is preferred to the abrasive cutting wheel.

Three types of high-speed cold sawing machines in general use in the fabrication industry are:

1. *Table type sawing machine* This is especially useful for cutting long bevels through the webs of joists and channels, in addition to making normal square cuts. The machine is similar to a woodworker's circular saw in that the blade centre is below the level of the work table, but the travel of the blade is provided for in a slot through the length of the table. The material to be cut must be securely clamped to the table during the cutting operation.

2. *Upright type sawing machine* With this type of machine, the circular blade is fed in a horizontal direction. Joists or other sections are placed with the widest part of the section upright. When bevels are required to be cut through the flanges of channels or joists, this machine may be used on a turntable and is fitted with a 'vee'-shaped side-block device which is specially designed for this purpose.

3. *Vertical-lift type sawing machine* This circular blade is fed vertically downwards into the section being cut. This machine is normally fitted with self-centring vices for square cutting, and the section to be cut is fixed with the narrowest part in the vertical position.

Power- or hand-operated holding clamps are fitted to cold sawing machines to secure the steel sections against the powerful thrust of the blade while they are being cut.

The blades used on cold saws consist basically of a body with segments made from hard tool steel rebated over and riveted to the body, the centre of which is reinforced and provided with a suitable hole for fixing to the machine. To ensure that the saw teeth are kept free from steel chips during the sawing operation, chip removers are fitted.

The blade must be adequately lubricated, and provision is made for a continuous flow of a soluble oil and water mixture on to the blade during the cutting operation.

Bending

For bending angle section to various angles after notching, the standard type of bar bender (see Chapter 10) is used.

Three examples of angle bending together with a sequence of operations are illustrated in Figs 9.36, 9.37 and 9.38.

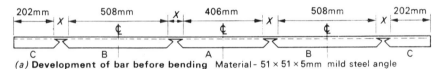
(a) Development of bar before bending   Material - 51 x 51 x 5mm mild steel angle

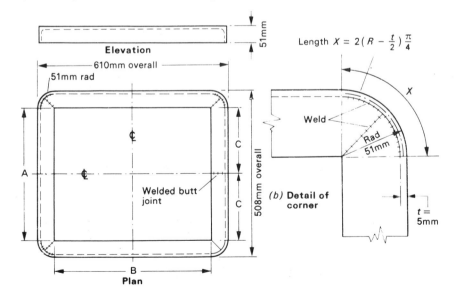

Length $X = 2\left(R - \dfrac{t}{2}\right)\dfrac{\pi}{4}$

(b) Detail of corner

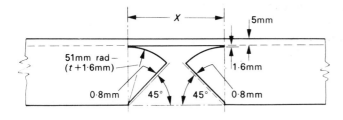

(c) Development of corner

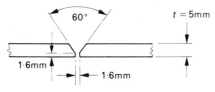

(d) Preparation for welding

Fig 9.36 Angle frame with rounded corners

Table 9.7 Sequence of operations for forming the frame

1. Draw a plan view of one corner showing the joint preparation as shown at (b).

2. Calculate the corner length, marked 'X', using the formula:
$$\text{LENGTH } X = 2\left(R - \frac{t}{2}\right)\frac{\pi}{4}$$

3. Calculate total length of angle section required.

4. On a suitable length of angle section, scribe a line at 90° to the heel in the centre of the section as shown in (a).

5. On each side of this centre line mark-off the notched corners as shown at (a), using a sheet metal template for the development of the corners as shown at (c). With a centre punch and hammer 'witness-mark' the lines to be cut.

6. Flame cut and prepare the joints for welding.

7. Bend the angle section around a suitable jig to form the corners.

8. Check the frame for dimensions and shape. Adjust if necessary.

9. Tack-weld mild steel straps diagonally to corners of frame, or clamp frame to suitable plate (jig) to prevent undue warping during welding.

10. Weld using metallic arc welding equipment.
    Allow to cool and check frame for size and adjust if necessary.

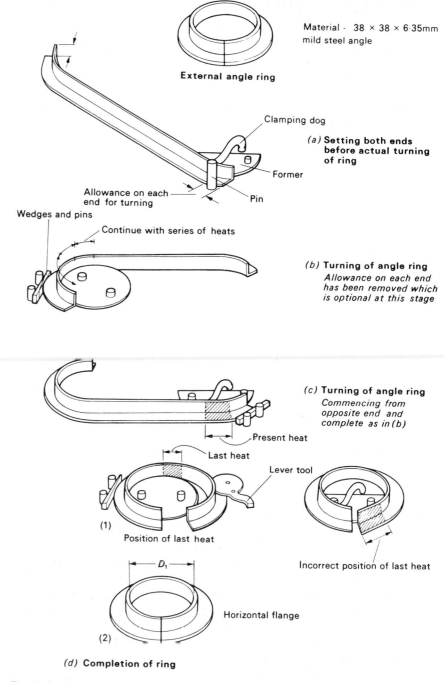

Fig 9.37 External angle ring

Table 9.8 Sequence of operations

1. Determine the length of the angle section required to make the ring by using the formula

$$(D_1 + T + \tfrac{1}{3}T)\,\pi$$

where D_1 = Diameter of heel,

T = Thickness of angle section,

π = 3·142.

Note: An additional allowance should be made on the length of the section to assist in bending both ends.

2. Heat the ends of the section uniformly in the fire to a bright red heat and then bend them on the bending table as shown at (a). (The bending allowance on the ends of the section may now be flame cut or left till the ends of the section are overlapping).

3. Heat the next section of the angle in the fire and bend on the bending table as shown in (b) until approximately the angle section has been bent.

4. Commencing from the other end of the angle section continue the bending as shown in (c) until the section is formed approximately to the shape shown at (d)(1).

5. With the last heat, complete the bending of the bar as shown at (d)(2).

6. Check the contour of the ring, and adjust if necessary.

7. Heat the ring in the fire, square-up the standing flange and round-up the angle section to correct diameter.

8. Flame cut the joint and prepare for tack-welding. Tack-weld and completely weld the joint.

Precautions:
1. Do not overheat the angle section.
2. When rounding-up and flattening the section use the 'fuller' and 'flattener' in preference to the hammer to avoid damaging the angle section.

Internal angle ring

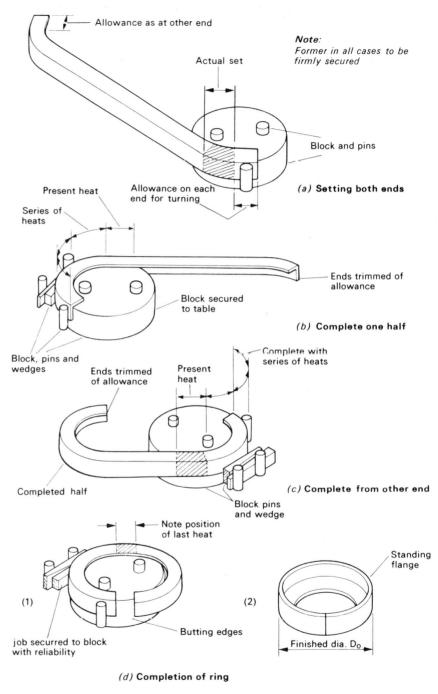

Fig 9.38 Internal angle ring

Table 9.9 Sequence of operations for forming the internal angle ring

The same sequence of operations may be adopted as shown in Fig. 9.31, but in this case determine the length of angle section required to form the ring by using the formula

$$(D_o - T - \tfrac{1}{3}T)\pi$$

where D_o = Diameter of the ring,

T = Thickness of angle section.

π = 3·142.

The methods used for clamping have been explained in Chapter 8.

9.18 Locating hole centres on large diameter flanges

The method of marking out hole centres on circular flanges has been explained in Chapter 5. Using a centre punch mark for datum centre, this operation is relatively easy when marking out on a blank before the flange is cut. However, very often in the fabrication industry the craftsman has to mark out hole centres for drilling on large diameter blank flanges. Figure 9.39 explains how this problem is overcome by the use of drilling jigs.

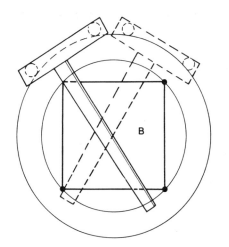

2. Locate centre using a centre square as shown at B

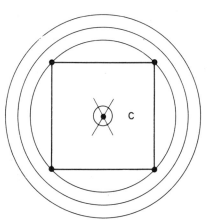

3. Centre punch the centre, and with trammels set to required radius, scribe the P.D.C. as shown at C

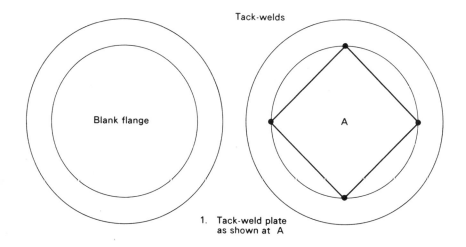

1. Tack-weld plate as shown at A

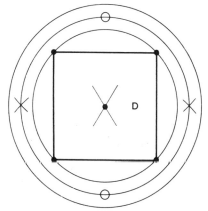

4. Using straight edge scriber and trammels mark four hole centre on the P.D.C. at 90° to each other and centre punch as at D

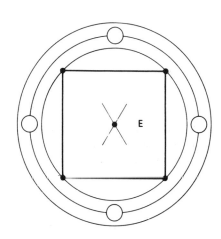

5. Drill four holes to suit bolt diameter as shown in E

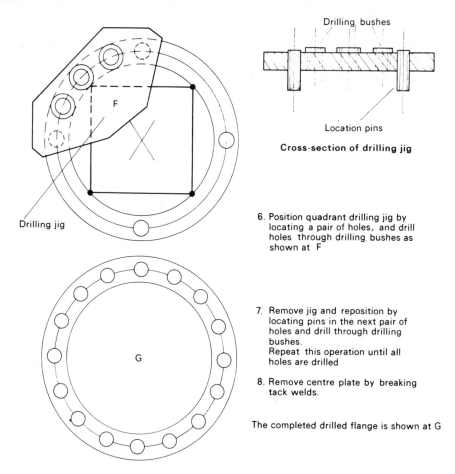

6. Position quadrant drilling jig by locating a pair of holes, and drill holes through drilling bushes as shown at F

7. Remove jig and reposition by locating pins in the next pair of holes and drill through drilling bushes. Repeat this operation until all holes are drilled

8. Remove centre plate by breaking tack welds.

The completed drilled flange is shown at G

Fig 9.39 Use of jig for drilling large flanges

9.19 Forging equipment

The principles of forging have already been introduced in 6.9 and 6.10. Some of the more practical aspects of this forming process will now be considered. The equipment required for hand forging is as follows:

1. *The hearth* Since the forging operations performed in the average workshop are rarely more ambitious than 'hot bending', the hearths are usually gas or oil-fired. These have the advantage of being clean in use and do not throw off abrasive dust that could settle on the slideways of nearby machines, causing excessive wear.

However, if the full range of forging operations are to be performed, then the traditional open-fire hearth, burning coke breeze or anthractite peas, is essential. Figure 9.40 shows a typical hearth and its blowing equipment.

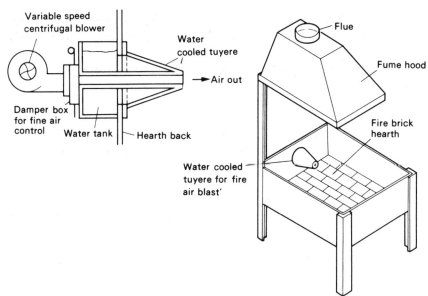

Fig. 9.40 Blacksmith's hearth

In skilled hands this type of hearth can be fired to produce a wide range of temperatures from a dull red heat for bending to white heat for forge welding. The area of the fire can also be controlled. Unfortunately, this type of hearth produces a lot of abrasive dust. Therefore, it is best kept in a separate shop, away from machine tools.

2. *The anvil* The anvil is used to support the work while it is hammered. Figure 9.41 shows a typical SINGLE BICK LONDON PATTERN anvil.

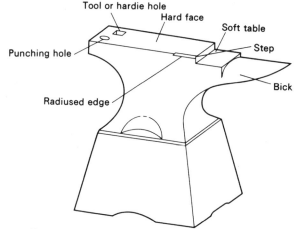

Fig. 9.41 The anvil

The anvil itself is made from a number of forgings that are forge welded together. The workface is usually made of a medium carbon steel and hardened.

The anvil is supported on an angle iron or malleable cast iron stand to bring it up to a convenient working height.

3. *Work-holding* Because of the high temperatures at which metals are forged, it is not possible to hold them in the hand. Tongs are used to hold the work and their mouths vary in shape to suit the work being held. A range of tongs is shown in Fig. 9.42.

4. *Hammers* These are used to form the workpiece, either by striking the metal directly or by striking forming tools (see 5). Various hammers used by the blacksmith are shown in Fig. 9.43.

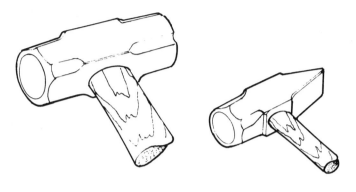

Fig. 9.43 Hammers

The smaller hand hammers are used by the blacksmith himself on light work and finishing operations. The larger sledge hammers are wielded by the blacksmith's assistant or STRIKER.

5. *Forming tools* Figure 9.44 shows a variety of forming tools used by the blacksmith. The use of these tools will be described later in the chapter.

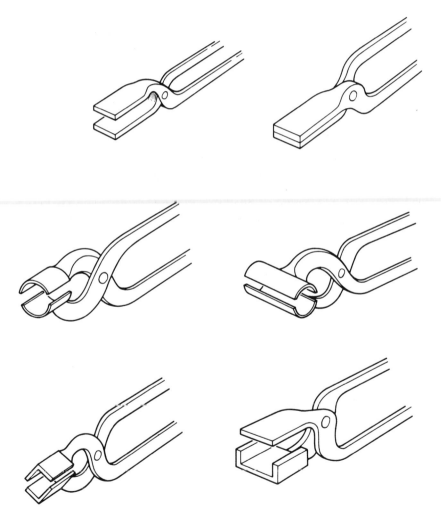

Fig. 9.42 Tongs

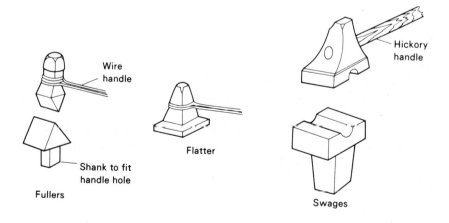

Fig. 9.44 Forming tools

9.20 Basic forging operations

The basic forging operations are:
1. Drawing down.
2. Upsetting or jumping up.
3. Punching or piercing.
4. Cutting with hot or cold chisels.
5. Swaging.
6. Bending and twisting.
7. Welding.

1. Drawing down This operation is used to reduce the thickness of a bar and to increase its length. The tools used are the FULLERS, and this operation is shown in Fig. 9.45; since the fullers leave a corrugated surface, the component is smoothed off with a FLATTER.

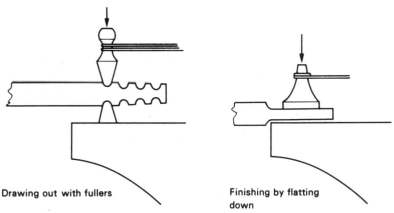

Fig. 9.45 Drawing down

2. Upsetting This operation is the reverse of drawing down and is used to increase the thickness of a bar and to reduce its length. Generally, the increase in thickness is only local as when forming a bolt head. Figure 9.46 shows the operation of upsetting and it will be seen that only the hammer is used.

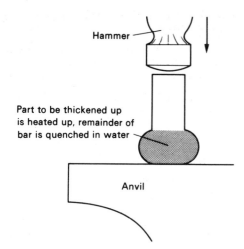

Fig. 9.46 Upsetting

3. Punching Holes are started by being punched out (round or square) at a size convenient for the piercing or slug to pass through the punch hole to the anvil.

The punched hole is then opened out to size using a drift. The advantages of drifting the hole to size are:
1. The surface finish in the hole is improved.
2. The metal round the hole is swelled out and so avoiding any weakening of the component.

Figure 9.47 shows the operations of punching and drifting to produce a hole.

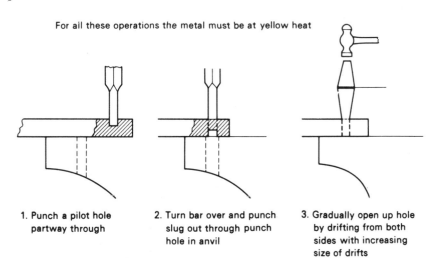

Fig. 9.47 Punching and drifting

4. *Cutting* Metal can be cut hot or cold, depending upon the equipment used. The cold chisel has already been discussed in *Basic Engineering*, Section 8.9. The hot chisel is more slender so that it can knife its way through the hot, plastic, metal. When cutting cold, the work is supported on the soft table of the anvil to avoid damaging the cutting edge as the chisel breaks through. The techniques of cutting have already been introduced in Chapter 7.

5. *Swaging* This is the operation of rounding a component. The component is broken down into an octagon with the flatter, and then rounded off between a pair of swages or between a single swage and the swage block, as shown in Fig. 9.48.

6. *Bending* This operation is shown in Fig. 9.49. The point to note when bending is that there is a tendency for the metal to thin out round the bend, causing weakness. This can be overcome by either upsetting the bar prior to bending, or forge welding an additional piece of metal, or GLUT, into the bend. The latter method forms a very strong corner.

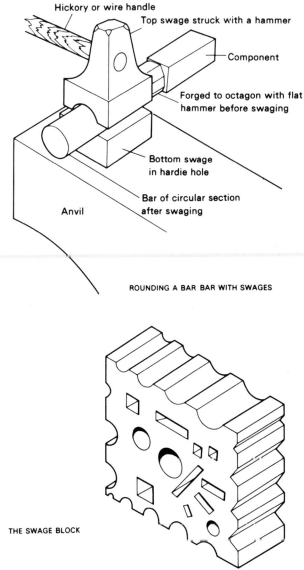

Fig. 9.48 Swaging

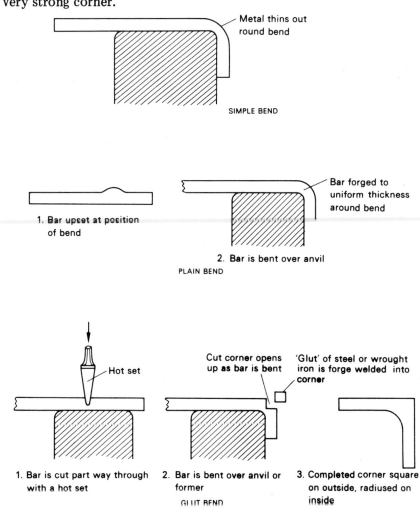

Fig. 9.49 Bending

7. Welding This is a difficult operation requiring great skill. When done correctly the resulting joint is superior in strength to either arc or oxy-acetylene welding. Forge welding is restricted to wrought iron and mild steel. A flux is necessary, and this may be a proprietary brand such as Laffite, or a natural material such as silver sand. The former is the easier to use. There are various types of weld, and two examples are given in Fig. 9.50.

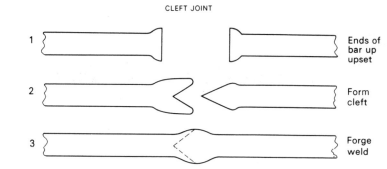

Fig. 9.50 Forge welding

STRAIGHT SCARF JOINT

Stages in making a scarf joint using Laffite welding plate

1 Upset each bar end

2 Scarf is formed on each bar end

3 Bars are heated to a dull red heat A piece of Laffite welding plate is placed in the joint The joint is lightly hammered until the bars are stuck together

4 Joined bars are reheated to a light welding heat

5 Bars are forged out flat at welding heat to complete joint

Stages in making a scarf joint using silver sand as a flux

1 Upset each bar end

2 Scarf is formed on each bar end

4 Bars are quickly withdrawn from the hearth-Tapped on anvil to remove dirt- Sprinkled with silver sand where weld is required

5 The bars are quickly positioned and hammered together to form the weld
(a) First blows—top, middle to drive out scale
(b) Second blows—top, thick end so that weld is complete with lower thin end before it is chilled by anvil
(c) Third blows—top, thin end

6 Take another heat and forges until joint is no longer visible

10 Fabrication processes

Fred was formed by a simple bending operation

This chapter is concerned with the plastic manipulation of sheet metal at room temperature. Figure 10.1 shows some typical cold-forming operations.

10.1 Folding or bending

The terms '*folding*' and '*bending*' are rather loosely used in industry because they are so similar. The difference between them is so slight that they are both carried out with the same purpose in view, which is to deflect the metal from one flat plane to another so that it stays there permanently.

If the deflection is sharp, and the radius small, the metal is said to be *folded*.

Should the curvature be large and the deflection cover a large area, it is called *bending*. In this respect the rolling of a hollow body, such as a cylinder, is called bending.

Folding or bending involves the deformation of a material along a straight line in two dimensions only.

10.2 The mechanics of bending

When a bending force is gradually applied to a workpiece under free bending conditions, the *first stage* of bending is elastic in character. This is because the TENSILE AND COMPRESSIVE stresses that are developed on opposite faces of the material are not sufficiently high to exceed the YIELD STRENGTH of the material. The movement or STRAIN which takes place as a result of this initial bending force is elastic only, and upon removal of the force the workpiece returns to its

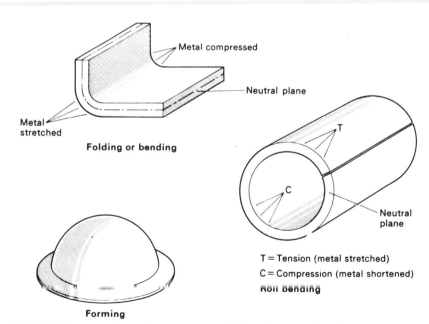

Fig.10.1 Comparison of common cold forming applications

original shape.

As the bending force is continued and gradually increased, the stress produced in the outermost fibres (on both the compression and tension sides) of the material eventually exceeds the yield strength.

Once the yield strength of the material has been exceeded, the movement (strain) which occurs is PLASTIC. This permanent strain occurs only in the outermost regions furthest from the NEUTRAL PLANE. Between the outermost fibres and the neutral plane there is a zone where the strain produced is elastic.

On release of the bending force, that portion adjacent to the neutral plane loses its elastic stress, whilst the outer portions, which have suffered plastic deformation, remain as a permanent set. *The elastic recovery of shape is known as 'SPRINGBACK'.*

Figure 10.2 illustrates the effects of a bending force on a material.

When bending sheet metal to an angle, the *inner fibres* of the bend are comp*ressed* and are given COMPRESSION STRESSES. The *outer fibres* of the bend are *stretched* and given TENSILE STRESSES.

Between the two stressed zones, which are in opposition to each other, lies a boundary which is a NEUTRAL PLANE. This boundary is termed the 'NEUTRAL AXIS' or 'NEUTRAL LINE' (see Fig. 10.3). See section 10.24 for fuller explanation of neutral line or axis.

The position of the neutral line will vary in different metals because of their differing properties, and also vary due to the thickness of the material and its physical condition. It is important to establish the position of the neutral line as it is required, in practice, for the purpose of calculating bend allowances (see **10.24**).

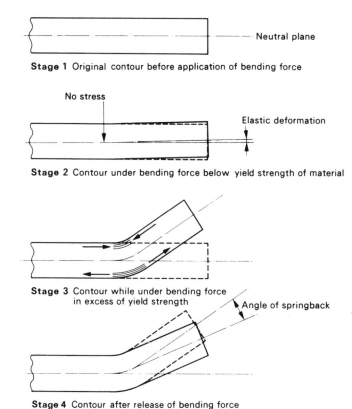

Fig.10.2 The effects of a bending force on a material

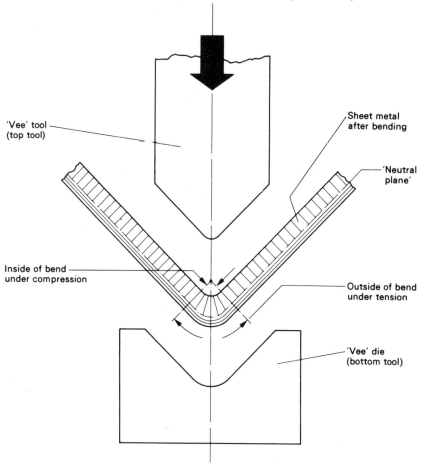

Fig.10.3 Bending action – pressure bending

10.3 Springback

During the bending of a material an unbalanced system of varying stresses is produced in the region of the bend. When the bending force is removed (on completion of the bending operation) this unbalanced system tends to bring itself to equilibrium. The bend tends to spring back, and any part of the elastic stress which remains in the material becomes RESIDUAL STRESS in the bend zone.

The amount of springback action to be expected will obviously vary because of the differing compositions and mechanical properties of the materials used in fabrication engineering. Some materials, because of their composition, can undergo more severe cold-working than others.

The severity of bending a specific material depends on two basic factors:
1. The radius of the bend.
2. The thickness of the material.

A 'tight' (small radius) bend causes greater cold deformation than a more generous bend in a material of the same thickness.

A thicker material develops more STRAIN HARDENING than is experienced in thinner material bent to the same inside radius.

The *'condition'* of the material upon which bending operations are to be performed, has an influence on the amount of springback likely to result. For example, using the same bend radius, a COLD-ROLLED NON-HEAT-TREATABLE aluminium alloy in the 'HALF-HARD' temper, or condition, will exhibit greater springback than the same alloy of equal thickness when in the 'FULLY ANNEALED' condition.

The limit to which free bending can be carried out is determined by:
1. The extent to which the material will stretch (ELONGATE) on the *tension side* (outside of bend).
2. The failure due to such COMPRESSIVE EFFECTS as *buckling*, *wrinkling* or *collapse* on the inside of the bend in respect of hollow sections (pipe bending).

10.4 Methods of compensating for springback

The clamping beam on a folding machine is specially designed to compensate for springback. This is illustrated in Fig. 10.4.

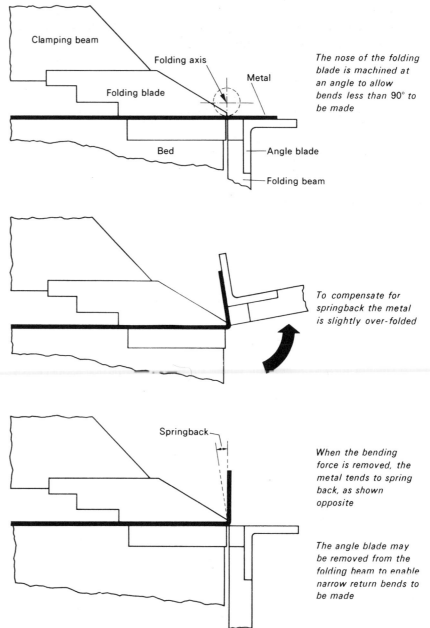

Fig 10.4 Allowing for springback on a folding machine

There are two methods of reducing springback when using a press brake, or a 'vee' tool in a fly press. These are illustrated in Fig. 10.5.

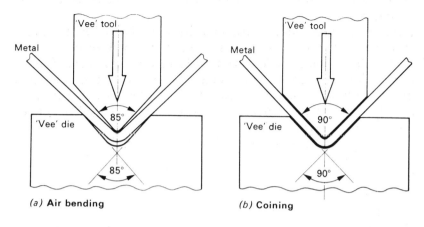

Fig.10.5 Two methods of pressure bending

10.5 Basic bending methods

There are two main machine methods of bending sheet metal:
1. Bending in folding machines.
2. Bending in pressure tools.

Because of the great force required to bend plate, the bending of platework is normally carried out with the aid of pressure tools in a press brake.

The bending of bar, angle and flat section may be carried out with simple bench-mounted bending machines, as shown in Fig. 10.6.

Air bending

This allows partial bending and various angles to be bent by THREE POINT LOADING. The three points are the two edges of the VEE DIE (bottom tool) and the nose of the VEE PUNCH (top tool).

During air bending, the sheet or plate retains its ELASTICITY. In this case the bending angle must be over-closed to compensate for the springback of the material after removal.

The bending tools are designed accordingly, both the top and bottom vees have an included angle of less than 90°. In general, the angle of these tools is 85°.

ADVANTAGES IN AIR BENDING

1. Less power required to bend the material.
2. Ability to bend heavy sheets and plates.
3. Ability to form various angles with the same tooling.

Coining

This type of bending can be compared with a deep-drawing operation. The nose of the vee tool crushes the natural air-bending radius on the inside of the bend. This COMPRESSION removes the ELASTICITY of the sheet or plate. THIS RESULTS IN THE BEND RETAINING THE EXACT ANGLES OF THE BENDING TOOLS.

Both tools have an included angle of 90°. The advantage of 'coining' is high angular accuracy.

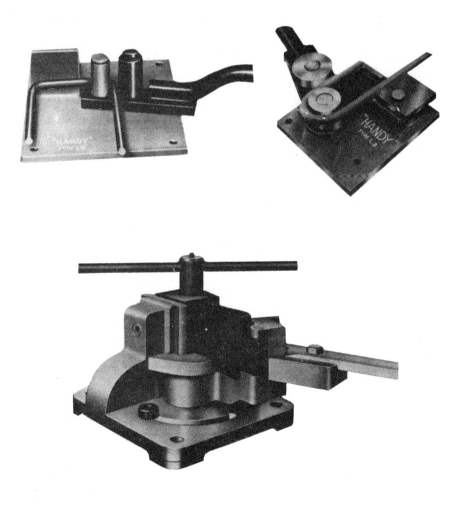

Fig.10.6 Simple bench mounted bending machines

The simple bar-bending machines shown in Fig. 10.6(a) have the following capacities:

Bends 1 rod of 9·53 mm dia.
Bends 2 rods simultaneously of 7·94 mm dia.
Bends 3 rods simultaneously of 6·35 mm dia.
Bends 4 rods simultaneously of 4·76 mm dia.
The diameter of the centre spindle is 25·4 mm.

The heavier machine shown in Fig. 10.6(b) has replaceable formers for bending the bar round. This enables bends of different radii to be made on the machine, which has the following capacities:

Bends rounds 12·7 to 19 mm dia. (supplied with 3 bending formers)
Bends rounds 12·7 to 25·4 mm dia. (supplied with 5 bending formers)
Bends rounds 12·7 to 31·75 mm dia. (supplied with 6 bending formers).

The angle and flat bar bending machine shown in Fig. 10.6(c) enables the material to be bent hot or cold.

The machine shown is supplied with a locking bar for the angle stop and swivel blade, an operating lever and double-edged folding blade. It is used for making bends in notched angle, flats, rounds and squares. Its capacity is increased when making hot bends.

10.6 Folding machines

The main specifications of folding machines are as follows:
1. The maximum length and thickness to be bent. For example, the capacity of the machine may be 1·5 m times 1·62 mm. This means that the machine is capable of folding a sheet of metal 1·5 m wide and of 1·62 mm (16 s.w.g.) thickness.
2. The lift and shape of the clamping beam.

The smallest width of bend is 8 to 10 times the metal thickness.

The minimum inside corner radius of the bend is 1½ times the metal thickness.

The THREE MAIN STEPS in folding work are:
1. *Clamping.* In clamping, the amount of lift of the clamping beam is important. It should be sufficient to allow the fitting and use of special clamping blades, or to give adequate clearance for previous folds.
2. *Folding.* Care must be taken to see that the folding beam will clear the work, particularly when making second or third folds.

Some folding machines are designed to fold radii above the minimum, either by the fitting of a radius bar or by adjustment of the folding beam.

3. *Removal of the work.* Care must be taken in folding to ensure that the work may be easily removed on completion of the final bend. The sequence of folding must be carefully studied. The lift of the clamping beam is important here. Some folding machines, known as 'UNIVERSAL FOLDERS' have a swing beam. The work may be completely folded around this beam, which is then swung out to allow removal of the work.

Some of the above points are illustrated in Fig. 10.7.

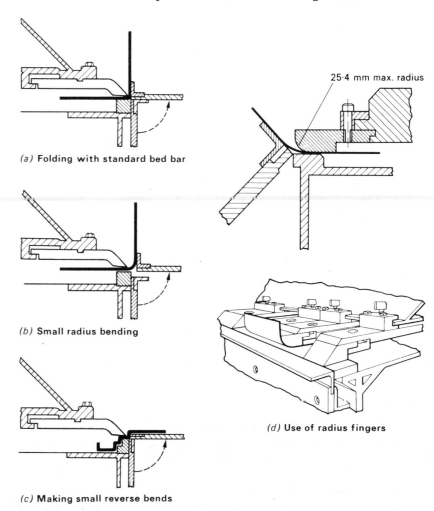

(a) Folding with standard bed bar

(b) Small radius bending

(c) Making small reverse bends

(d) Use of radius fingers

Fig. 10.7 The use of folding machines

Figure 10.7(a) shows a section of a 'box and pan' folding machine. It is fitted with a standard bed bar and fingers. The sheet metal is shown in position after completion of a right-angle bend, using a standard angle folding bar.

Figure 10.7(b) shows small radius bends being made. The folding beam is lowered, and the metal is clamped in the normal way. When the folding beam is raised, the gap between the nose of the folding blade and the face of the folding bar allows a larger radius to be made.

Figure 10.7(c) shows small return bends being made on this machine by using a specially stepped bed bar. Such a bar is very useful for moulded work. The clamping beam lifts high enough to allow that part of the metal on the inside of the beam to be withdrawn over the bar. In this case the standard folding bar has been substituted by a narrow blade, giving smaller face width to the folding beam.

Figure 10.7(d) shows the use of radius fingers with the standard angle folding bar. This allows radius bends up to a maximum of 25 mm radius to be made.

The fingers may be positioned where required on the clamping beam to allow short lengths to be folded.

By fitting a special narrow bar to the folding beam, it is possible to form reverse bends narrower than the face of the standard angle bar. This is shown in Fig. 10.8. This, of course, reduces the maximum gauge of sheet metal which would be normally folded with the standard blade.

The variety of bends and combinations of bends that can be made on the folding machine are shown in Fig. 10.9.

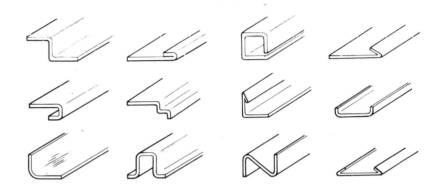

Fig.10.9 Examples of work produced on a folding machine

Some folding machines have provision for inserting a round mandrel in trunnion arms on the machine. Such is the case with UNIVERSAL SWING-BEAM FOLDERS, as shown in Fig. 10.10.

The amount of lift in the clamping beam is very important. It governs the maximum size of the mandrel used. A machine with a clamping beam lift of between 175 and 200 mm will allow a mandrel of 152 mm diameter to be used.

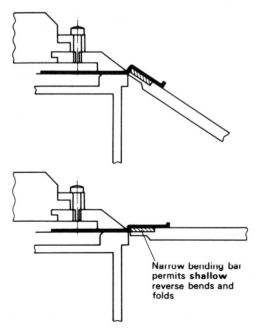

Fig.10.8 Use of a narrow bending bar

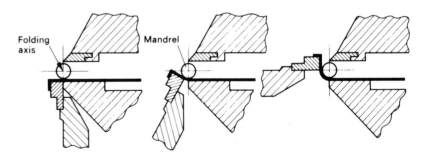

Fig.10.10 Use of a mandrel in a folding machine

The folding of shallow depth boxes and pans may be successfully performed on a universal folding machine, provided there is sufficient lift of the clamping beam. An angle clamping blade is attached to the clamping beam as shown in Fig. 10.11.

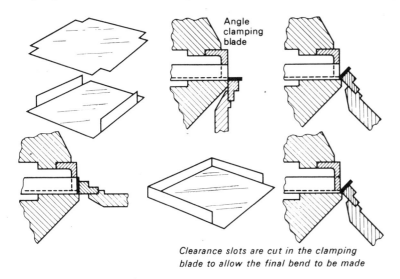

Clearance slots are cut in the clamping blade to allow the final bend to be made

Fig 10.11 Use of an angle clamping blade in a folding machine

Figure 10.12 shows three types of sheet metal folding machines in common use in the light fabrication industry.

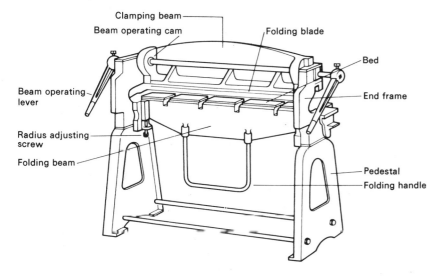

This machine allows sheet metal sections to be completely folded around the beam. The beam is then swung out to permit removal of the work

Fig.10.12 Types of folding machine

10.7 Bending in pressure tools

The simplest types of machines which may be used for pressure bending are:
1. A fly press fitted with a 'Vee' punch and a 'Vee' die. This arrangement is ideal for bending small sheet metal components such as brackets or electronic chassis.
2. A hand-operated angle bender. This is limited to short lengths of sheet metal angle or 'Zed' section. An angle bender is shown in Fig. 10.13.

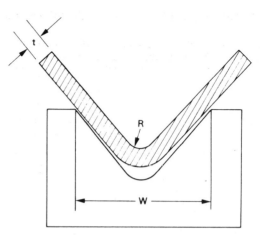

Fig. 10.14 Die ratio

Standard air bend
$W = 8t \cdot R \simeq t$

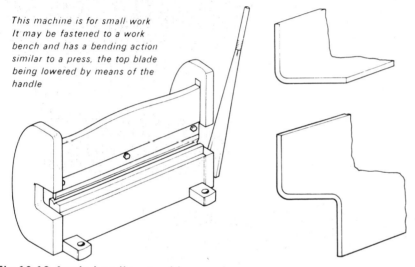

This machine is for small work. It may be fastened to a work bench and has a bending action similar to a press, the top blade being lowered by means of the handle

Fig.10.13 Angle bending machine

By far the greatest amount of pressure bending is done on a press brake. The larger capacity machines are capable of bending heavy plate.

Whether a simple machine or a press brake is used, the principle of pressure bending is the same. THE METAL IS FORMED BETWEEN THE TOP AND BOTTOM 'VEE' TOOLS UNDER THE APPLICATION OF A FORCE.

'Vee' die openings. The force required to make simple rightangle bends depends upon the size of the 'Vee' opening in the bottom tool. Figure 10.14 illustrates the relationship between the 'Vee' opening and the thickness of metal to be bent: R = radius; t = thickness; w = width at die opening.

Table 10.1 Comparison of 'vee' die ratios

METAL THICKNESS		FORCE TONNES/METRE Required to produce 90° 'air bends' in mild steel (Tensile strength 450 N/m^2 using a die ratio of:)		
s.w.g.	mm.	8:1	12:1	16:1
20	0·9	6·8	4·1	3·0
18	1·2	9·1	5·8	4·1
16	1·62	12·2	7·5	5·4
14	2·0	14·9	9·5	6·8
12	2·64	19·6	12·2	8·8
	3·2	23·7	14·6	10·5
	4·8	35·2	22·0	15·9
	6·4	47·4	29·5	21·3
	8·0	58·9	36·6	26·8
	9·5	70·8	44·0	31·8
	11·0	82·6	51·5	37·3
	12·7	94·5	58·9	42·7
	14·3	115·3	66·0	48·1
	15·9	118·2	73·5	53·2
	17·5	129·7	80·9	58·6
	19·0	141·9	88·1	64·0
	20·4	153·5	95·5	69·1
	22·2	165·2	102·6	74·5
	23·8	177·1	110·1	79·9
	25·4	189·3	117·5	85·3

Table 10.2 Bending forces required for metals other than mild steel

Select pressure from the above table and **multiply** by the appropriate figure in table 10.2 below.

MATERIAL	MULTIPLY BY:
Stainless steel	1·5
Aluminium — soft temper	0·25
Aluminium — hard temper	0·4
Aluminium alloy — heat treated	1·2
Brass — soft temper	0·8

10.8 Principle of the press brake

Press brakes are designed to bend to a rated CAPACITY based on a *'Die ratio'* of 8:1 which is accepted as ideal conditions. Figure 10.14 shows the meaning of *Die ratio*. This is recommended for use with a standard 'Vee' die for 90° 'air bends', and gives an INSIDE RADIUS approximately equal to the THICKNESS of the metal.

Different thicknesses of plate formed over the same die will have the same inside radius, but the force or load required for bending will vary considerably.

If the die opening is less than 8 times the metal thickness, fracturing on the outside of the bend may occur. However, it is possible to produce satisfactory bends in light gauge sheet metal using a die opening of 6 times the metal thickness, but this requires a greater pressure.

For 'HIGH TENSILE' plates and plates above 9·525 mm thickness, it is recommended that the die opening be increased to 10 to 12 times the metal thickness. This considerably reduces the bending load required.

The pressure required for bending is in direct relation to the tensile strength of the material. For materials other than MILD STEEL the capacity would have to be DECREASED or INCREASED accordingly. A long machine will bend plates thicker than the rated capacity but only over shorter lengths than the rated capacity, providing the plates are appreciably shorter than the maximum tool or die length.

Thinner plates than the rated capacity can usually be bent over the full length of the dies but the MAXIMUM WIDTH OF FLANGE IS DETERMINED BY THE DEPTH OF GAP. A standard bend has an inside radius approximately equal to the thickness of the metal. If this radius is not important and a slightly larger radius would be quite satisfactory then, in many cases, a larger die opening could be employed and the machine will be able to bend plates thicker than the rated capacity over the rated length.

10.9 Types of press brakes

Press brakes are usually MECHANICAL or ELECTRO-HYDRAULIC. A press brake is really a wide ram press, and as such can be used for an extremely wide range of pressing work, if fitted with suitable tooling.

Their capacities are usually given in either or both, pressure exerted, or actual maximum work done based on W = 8T. [See Fig. 10.14].

Figure 10.15 illustrates two types of press brake.

(a) Mechanical press brake
This machine has a capacity of 76 tonnes, and is capable of bending a 2.44 metre length of 4 mm steel plate. Some of the smaller capacity press brakes are available with a swing-out bending beam

(b) Down-stroking hydraulic press brake
This machine has a capacity of 500 tonnes. Details of the dead stop mechanism are illustrated on the following page

Fig. 10.15 Press brakes

75 tonnes, medium-duty machines between 75 tonnes and 150 tonnes, and heavy-duty machines between 150 tonnes and 500 tonnes. Some of the very large machines have table length of 5·5 m. Figure 10.16(*a–h*) illustrates the versatility of the press brake. Note that the tonne is a unit of mass equal to 1 000 kg. Therefore the load on the ram of a 25 tonne machine would be equivalent to a mass of 25 × 1 000 = 250 000 kg. Such a load would exert a force on the work piece of 25 000 × 9·81 newtons. That is 245 250 N or 2·5 MN. Similarly, a 152 tonne machine would exert a maximum force on the work piece of 15·2 MN.

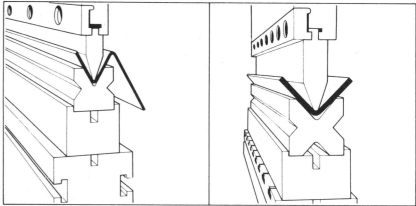

(a) **Four-way dies**

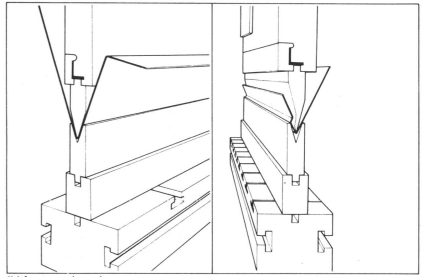

(b) **Acute angle tools**

Press brakes may be 'DOWN-STROKING' or 'UP-STROKING'. With a down-stroking press the ram brings the top tool down to the bottom fixed tool. An up-stroking press brake is one in which the ram pushes the bottom tool up to the fixed top tool. Many hydraulic press brakes are up-stroking.

Some smaller press brakes are available with a swing-out bending beam. Light-duty machines are rated between 25 tonnes and

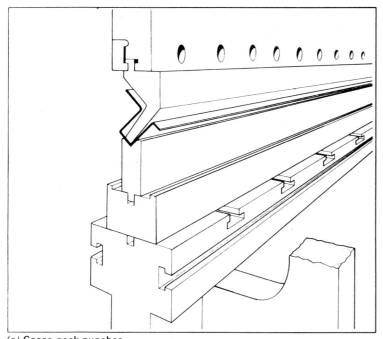
(c) Goose-neck punches

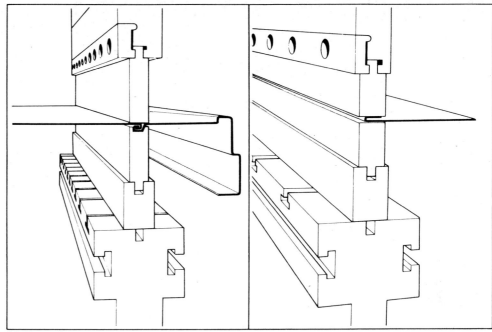
(d) Flattening

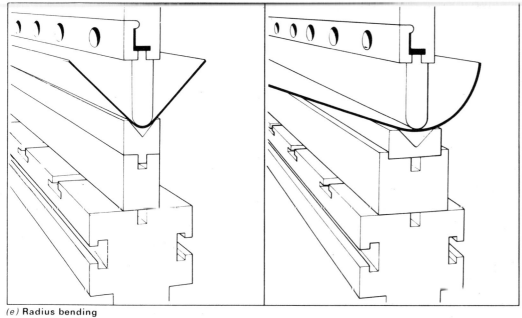
(e) Radius bending

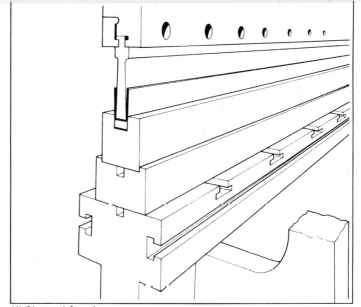
(f) Channel forming

233

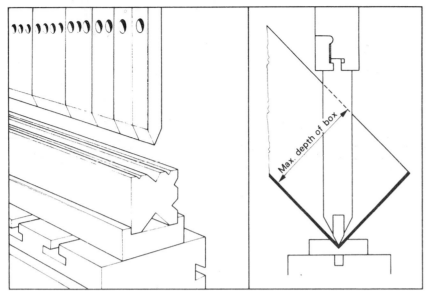

(g) Box making

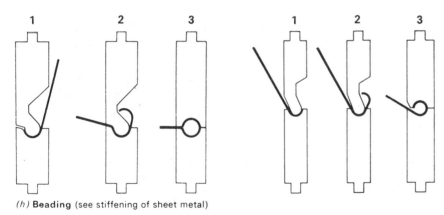

(h) Beading (see stiffening of sheet metal)

Fig. 10.16 Versatility of pressure bending

These tools are available for any angle, but if the female die is less than 35° the sheet tends to stick to the die.

Acute angle dies may be set to bend 90° by adjusting the height of the ram.

The Goose-neck punch—Fig. 10.16(c). When making a number of bends on the same component, clearance for previous bends has to be considered. Goose-neck punches are specially designed for the above purpose. These tools are very versatile, enabling a variety of sheet metal sections to be formed.

The bending force for MILD STEEL is given in table 10.1, whilst the bending force for other materials is given in table 10.2.

Flattening (planishing) tools—Fig. 10.16(d). Flattening tools of various forms may either be used in pairs for flattening a returned edge, or hem, on the edge of sheet metal or in conjunction with a formed male or female die (as illustrated for closing a countersunk grooved seam in sheet metal work).

Radius bending—Fig. 10.16(e). A radius bend is best formed in a pair of suitable tools. The radius on the male punch is usually slightly less than that required to allow for 'springback' in the material. A large radius can be produced by simply adjusting the height of the ram and progressively feeding the sheet through the tools.

Channel dies—Fig. 10.16(f). Channel dies are made with 'pressure pads' so that the metal is held against the face of the male die during the forming operations. As a general rule, channel dies are only successful on sheet metal up to and including 2·64 mm thickness.

A channel in heavy gauge metal is best made in a 'Vee' die with a 'Goose-neck' type of male punch.

Boxmaking—Fig. 10.16(g). Male punches for box making must be as deep as possible. Most standard machines are fitted with box dies which will form a sheet metal box 170 mm deep. If deeper boxes are required, the machine must be provided with greater die space and longer male dies. **For each extra 25 mm of die space the depth of the box is increased by 17 mm.**

Beading—Fig. 10.16(h). Three operations are necessary to form a bead on the edge of sheet metal.

10.10 Roll bending

In the fabrication industry there is a need to bend not only circular bodies of normal diameter range and length for tanks, boilers and pressure vessels, but also cylinders or pipes of small diameter and long length, in the maximum thickness of plate possible. In addition the need to bend conical sections has also to be catered for.

Machines used to bend sheet metal or plate into cylindrical or

Interchangeable four-way dies—Fig. 10.16(a). The interchangeable female dies are used for bending medium and heavy plate. They are provided with 85° openings on each of the four faces.

Male punches for use with four-way dies are usually made with a 60° angle.

Acute angle dies—Fig. 10.16(b). Acute angle dies have many uses and, if used in conjunction with flattening dies, a variety of seams and hems may be produced on sheet metal.

part cylindrical forms with either parallel or conical sides are called 'BENDING ROLLS'.

Bending rolls for sheet metal are made in various sizes from the bench type for tin plate work to the larger pedestal types which are suitable for general sheet metal work.

Figure 10.17 shows the main features of a bending rolls suitable for bending sheet metal.

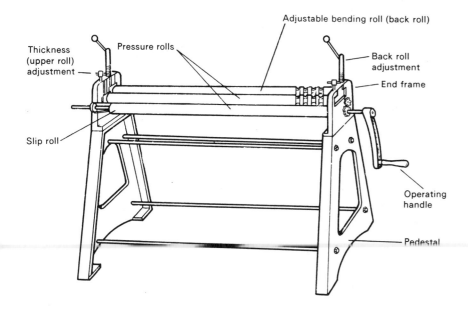

Fig.10.17 Sheet metal bending rolls

The basic type of rolls used in sheet metal work is known as the 'PINCH-TYPE ROLLERS'. These machines have two front rollers which lightly grip (PINCH) and draw the sheet through, and a 'free roller' at the rear to 'set' the metal to the desired radius.

There are two kinds of pinch-type rollers, basic details of which are illustrated in Fig. 10.18.

Roll-up type—Fig. 10.18(a). These machines have adjustment in a vertical direction on the top or bottom pinch roll, and in an upward direction on the back roller. This type will roll any size of curvature above the size of the top roll.

Roll-down type—Fig. 10.18(b). These machines have adjustment in a vertical direction on top or bottom pinch roll, and in a downward direction on the back roller. This type will not roll more curvature than will pass beneath the pedestal frame of the machine.

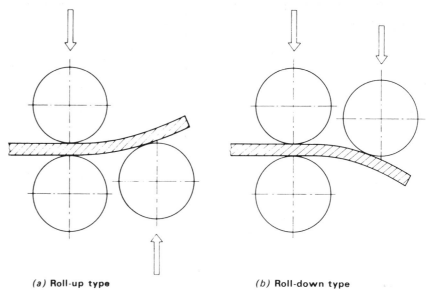

Fig. 10.18 Basic arrangement of the rolls in 'pinch-type rolls

As a general rule, the minimum diameter which can be rolled on a rolling machine is in the order of 1½ to 2 times the diameter of the roll round which it is being rolled.

Most machines roll the metal in an upward direction because this does not restrict the size of the cylinder or curve to be rolled.

The IDENTIFICATION of either kind of pinch-type rollers can easily be determined by visual inspection, as follows:

Where pinch-type rollers have wiring or beading grooves, if these grooves are in the top and back rolls, the machine is designed to roll down. When the grooves are in the bottom and back rolls, then the machine is designed to roll up.

10.11 The rolling of plate

The majority of machines in general use for the roll bending of plate are of the horizontal type, although considerable use is made of vertical plate rolls.

Rolls used for platework are more robust than those used for sheet metal work, and are normally power operated. Whereas pinch-type rolls are suitable for bending light plate, 'PYRAMID TYPE ROLLERS' are used for medium or heavy plate bending. Pyramid-type rolls, as the name suggests, have three rolls arranged in pyramid fashion as shown in Fig. 10.19.

Most plate-rolling machines are provided with longitudinal grooves along the lower rolls to assist in gripping the plate. These grooves are useful for initial alignment of the plate.

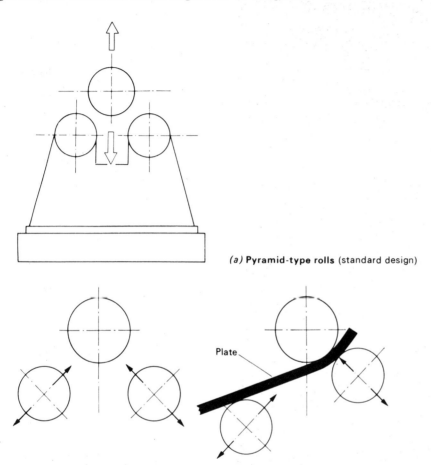

(a) Pyramid-type rolls (standard design)

(b) Pyramid-type rolls with adjustable bottom rollers

Fig. 10.19 Pyramid-type rolling machines

Figure 10.19(a) shows the basic arrangement of the rollers in a standard pyramid-type rolls. The top roll is adjustable up or down, and may be 'slipped' to allow removal of the work when completely rolled around it, as in the case with cylinders (see **10.12**).

Figure 10.19(b) shows the rolls set for 'pinch bending'. One great advantage of this type of machine is that for heavy plate the bottom roll centres are wide, and the load on the top roller consequently reduced. Being mounted in inclined slideways, the bottom roller centres are automatically reduced as the rollers are adjusted up to the work on thin plate or to small diameters, thereby providing a slightly increased capacity at the top and bottom ends of plate range.

Figure 10.20 shows a 'four-in-one' universal pyramid/pinch-type rolling machine. These machines are capable of performing all the roll bending operations normally carried out in fabrication workshops. In the hands of a capable operator it is a universal machine for all types of roll bending in light or heavy platework.

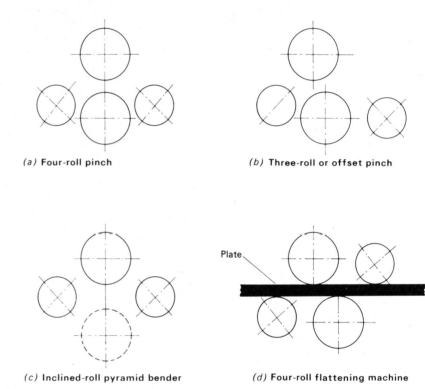

(a) Four-roll pinch

(b) Three-roll or offset pinch

(c) Inclined-roll pyramid bender

(d) Four-roll flattening machine

Fig. 10.20 The four-in-one universal pyramid/pinch-type rolling machine showing alternative settings of the rollers

10.12 Slip rolls

When rolling complete cylinders, the finished cylinder is left round the roll, so provision has to be made for its removal. Rolls with this provision are referred to as 'SLIP ROLLS'. With most sheet metal rolling machines, the roll around which the cylinder formed is made to slip out sideways. The slip roll on the powered plate-rolling machines usually slips upwards for the removal of the cylinder. Figure 10.21 shows a plate-rolling machine with a slip roll.

(a)

Fig. 10.21 Bending a pipe line section on a plate rolling machine

(b) Sequence of operations for rolling a steel pipe line section of the four-roller twin initial pinch hydraulic bending machine illustrated in figure (a).

These machines are available in the following sizes:
Plate width 380mm to 4·6m
Plate thickness 1·62mm to 76mm (cold)

10.13 Cone rolling

A certain amount of cone rolling may be carried out on both hand- and power-operated rolls. To enable this to be done provision must be made to adjust the curving roll to a suitable angle in the horizontal plane to the pinch rolls, or in the pyramid rolls to the other two.

10.14 Angle ring-bending rolls

These machines may be hand-operated by suitable gearing, or power-operated. They are used for the cold-bending of channel, angle and tee bar rings. The rollers may either be horizontal or vertical, as shown in Fig. 10.22.

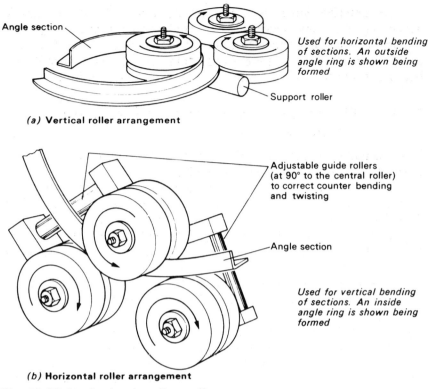

(a) Vertical roller arrangement

(b) Horizontal roller arrangement

Fig. 10.22 Angle ring bending rolls

An angle ring-bending rolls consists of three rollers arranged in triangular formation. Each roller can be split into two sections to take the flat flanges of angles or channels as they are being bent.

When bending an outside ring, the flat flange of the angle is adjusted in the slots of the two rollers, and for an inside ring the flat flange is adjusted in the slot of the single central roller. 'Tee' section may also be formed on these machines. Pressure is exerted, during rolling, by a screw arrangement which moves the single central roller towards the gap between the other two rollers.

10.15 Forming sheet metal

The expression 'forming metal' usually means shaping or bending the material in three dimensions. This is much more difficult than simple bending in two dimensions, for some part of the metal must be STRETCHED or SHRUNK, or both.

Consider a FLANGE to be 'thrown' on a curved surface, for example a cylinder, as illustrated in Fig. 10.23(a). It will be apparent that the edge of the cylinder, after externally flanging, has a greater circumference than it had before the flange was thrown.

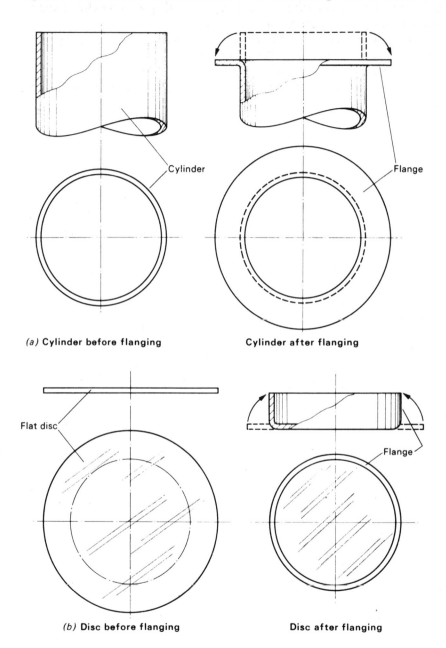

(a) Cylinder before flanging   Cylinder after flanging

(b) Disc before flanging   Disc after flanging

Fig.10.23 Comparision of flanging methods

Now consider a FLANGE to be worked up around the edge of a flat metal disc, as shown in Fig. 10.23(*b*). Here it will be appreciated that the edge of the disc, after flanging, will have a smaller circumference than it had before flanging.

In case (*a*) the metal has been STRETCHED, whilst in case (*b*) the metal has been SHRUNK or COMPRESSED.

By way of contrast, the effects of increasing or decreasing the surface area of one flange of an angle strip are illustrated in Fig. 10.24.

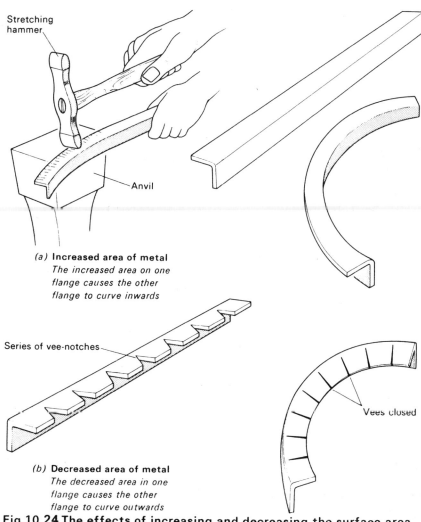

(a) **Increased area of metal**
The increased area on one flange causes the other flange to curve inwards

(b) **Decreased area of metal**
The decreased area in one flange causes the other flange to curve outwards

Fig. 10.24 The effects of increasing and decreasing the surface area of one flange on an angle strip

In practice metal is not generally removed by the simple expedient of cutting 'Vee' slots. The surface area is reduced by compressing or shrinking the surplus metal. It is much more difficult to produce an internally curved flange than an externally curved one, because it is much easier to stretch metal than to shrink it.

10.16 The main aspects of forming sheet metal by hand

It is essential for the craftsman to possess a thorough basic knowledge of the properties of the materials which he has to use.

This enables him to understand, and even forecast the behaviour of materials under applied forces, and thus be in control of the desired direction of their flow during a particular forming process.

Failure to understand these points will, inevitably, result in much valuable material, time, and effort being spent producing faulty shapes, stressed areas, splitting and other undesirable features.

The methods employed for shaping work by hand are similar for most materials, the main difference being concerned with such factors as:

(*a*) The FORCE with which the blow strikes the metal;
(*b*) The DIRECTION in which the FORCE or blow is applied.

It should be appreciated that if, for example, a piece of aluminium and a piece of mild steel sheet of equivalent thicknesses are struck with blows of equal force, the aluminium — being softer and more malleable and ductile — will be deformed to a much greater extent than the mild steel.

Since the shaping of sheet metals by hand is essentially a HAMMERING PROCESS, it is important to consider the types of blow which can be struck on sheet metal, and that each has its own field of application for particular purposes.

Types of blow used on sheet metal

1. *A solid blow* Where the metal is struck solidly over a solid steel head or anvil. A solid blow will stretch the metal. Figures 10.25 and 10.26 illustrate a typical application of the solid blow.

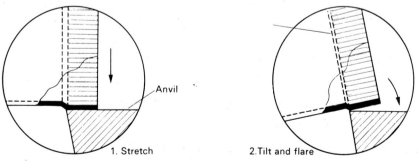

Fig. 10.25 Increasing circumference by stretching

A ring appears during the initial stretching — this acts as a guide line for throwing a constant width of flange

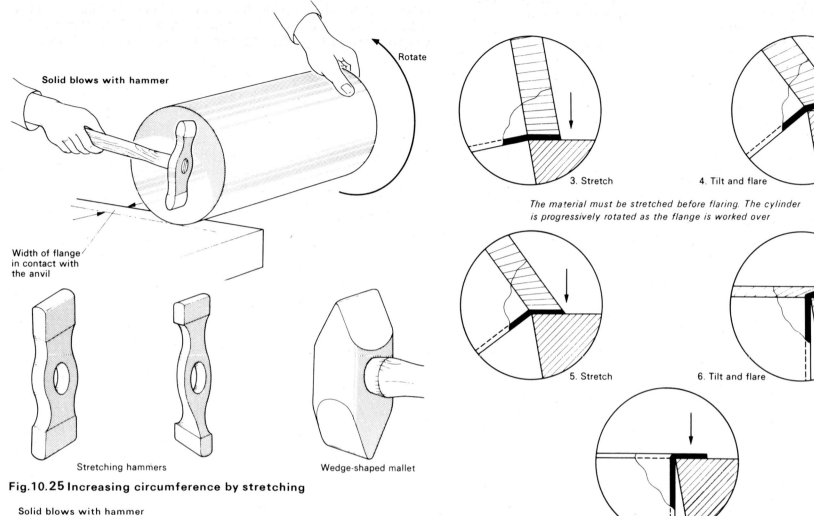

The material must be stretched before flaring. The cylinder is progressively rotated as the flange is worked over

Fig. 10.26 Sequence of operations for producing an external flange on a cylindrical body

2. *An elastic blow* Where either the head or the tool (or both) are made of a resilient material such as wood. An elastic blow will form sheet metal without unduly stretching it, and can be used to advantage to thicken the metal when shrinking it. The use of an elastic blow is shown in Fig. 10.27.

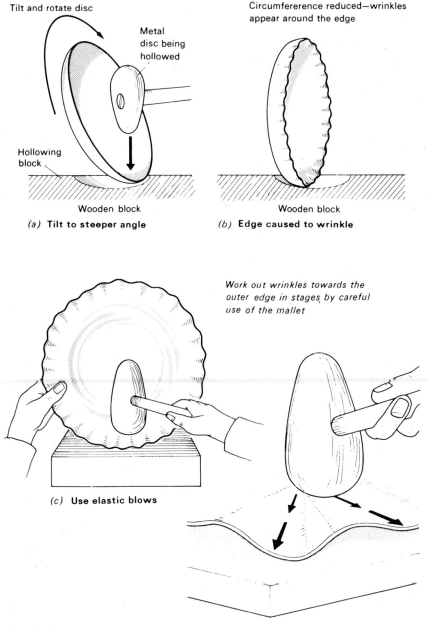

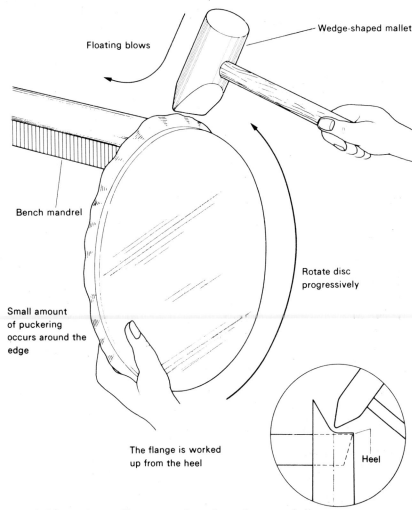

process. It is delivered while the metal is held over a suitable head or stake, hitting it 'off the solid', thus forming an indentation at the point of impact. Fig. 10.28 illustrates the use of floating blows.

Fig. 10.28 Raising a flange on the edge of a metal disc

Fig. 10.27 Shrinking the edge when hollowing

3. *A floating blow* Where the head or the anvil is not directly under the hammer. The floating blow is one which is used to control the direction in which the metal is required to flow during the forming

10.17 The use of bench tools for forming sheet metal

The craftsman often finds it necessary, when suitable machines are not available, to resort to the use of various types of metal anvils when bending or forming sheet metal articles.

These anvils are commonly referred to as 'STAKES' and are designed to perform many types of operations for which machines are not readily available or readily adaptable. Some stakes are made of forged mild steel faced with cast steel. The better-class stakes are made either of cast iron or cast steel. All are sold by weight because of their variety of size and shape.

A stake used for sheet metal work consists basically of a shank and a head or horn. The shanks are generally standard with respect to their taper. They are designed to fit into a tapered bench socket. The heads and horns are available in a great variety of shapes and sizes. Their working faces are machined or ground to shape.

The more common stakes and some of their uses are illustrated in Fig. 10.29(a—r).

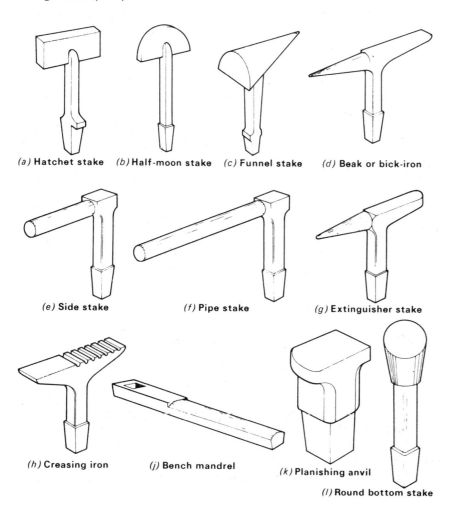

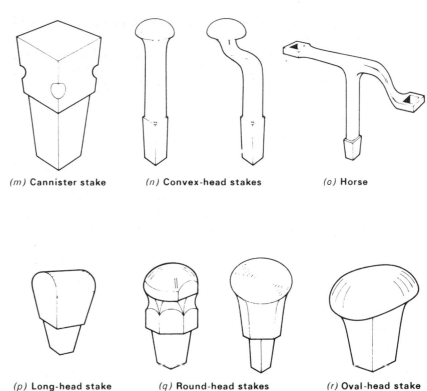

Fig.10.29 Typical bench stakes

1. *Hatchet stake* — Fig. 10.29(a). The hatchet stake has a sharp, straight edge, bevelled along one side. It is very useful for making sharp bends, folding the edges of sheet metal, forming boxes and pans by hand, 'tucking-in' wired edges, and seaming.

2. *Half-moon stake* — Fig. 10.29(b). This stake has a sharp edge in the form of an arc of a circle, bevelled along one side. It is used for throwing up flanges on metal discs or contoured blanks, preparatory to wiring and seaming. It is also used for 'tucking-in' wired edges on contoured work.

3. *Funnel stake* — Fig. 10.29(c). As the name implies, it is used when shaping and seaming funnels, and tapered articles with part conical corners, such as 'square-to-rounds'.

4. *Beak- or bick-iron* — Fig. 10.29(d). This stake has two horns, one of which is tapered, the other a rectangular shaped anvil. The thick tapered horn or 'beak' is used when making spouts and sharp tapering

articles. The anvil may be used for squaring corners, seaming and light riveting.

5. *Side stake* — Fig. 10.29(*e*). A side stake has one horn which is not tapered. It is more robust than the bick-iron and can withstand considerable hammering. Its main uses are forming, riveting and seaming pipe work, and making knocked-up joints on small cylindrical articles. It may also be used when forming tapered work of short proportions.

6. *Pipe stake* — Fig. 10.29(*f*). This is a much longer version of the side stake and, as the name implies, is useful when forming and seaming pipes.

7. *Extinguisher stake* — Fig. 10.29(*g*). Is very similar to the bick-iron, in that it has a round and tapered horn at one end and a rectangular shaped horn at the other. Some contain a number of grooving slots on the working surface of the rectangular horn. These are useful when creasing metal and bending wire. The tapered horn is used for forming, riveting, or seaming small tapered articles. It is also useful when forming wrinkles or puckers prior to 'raising'.

8. *Creasing iron* — Fig. 10.29(*h*). This has two rectangular shaped horns, one of which is plain. The other horn contains a series of grooving slots of various sizes. The grooves are used when 'sinking' a bead on a straight edge of a flat sheet — i.e. reversing wired edges. These irons are also used when making small diameter tubes with thin-gauge metal.

9. *Bench mandrel* — Fig. 10.29(*j*). This is firmly fixed to the bench by means of strap clamps which may be quickly released allowing the mandrel to be reversed or adjusted for length of overhang. It is double ended — the rounded end is used for riveting and seaming pipes, while the flat end is used when seaming corners of pans, boxes, square or rectangular ducting, and riveting. It has a square tapered hole in the flat end for receiving the shanks of other stakes and heads. Bench mandrels are available in four sizes ranging from 20 to 114 kg in mass.

10. *'Planishing anvils* — Fig. 10.29(*k*). Used for planishing all types of flat and shaped work, they are highly polished on their working surfaces. The one illustrated is called a 'TINSMITH'S ANVIL', and is used when planishing flat surfaces.

11. *Round bottom stake* — Fig. 10.29(*l*). These stakes are available in various diameters and have flat working surfaces. They are used when forming the bases of cylindrical work, and for squaring knocked-up seams.

12. *Canister stake* — Fig. 10.29(*m*). This has a square and flat working surface. Its main use is for working in corners, and squaring-up seams when working with square or rectangular articles.

13. *Convex-head stakes* — Fig. 10.29(*n*). Used when forming or shaping double-contoured and spherical work. It is usually available in two patterns — with a straight shank or with an off-set or cranked shank.

14. *Horse* — Fig. 10.29(*o*). This adaptable stake is really a double-ended support. At the end of each arm — one of which is usually cranked downwards for clearance purposes — there is a square hole for the reception of a wide variety of heads. Four of these heads are illustrated:

A long head — Fig. 10.29(*p*). This is used when making knocked-up joints on cylindrical articles, and for flanging;

Two types of *round head* — Fig. 10.29(*q*) which are used when raising;

An *oval head* — Fig. 10.29(*r*) which is oval in shape and has a slightly convex working surface. It sometimes has a straight edge at one end.

The condition of the stake has much to do with the workmanship of the finished article. Therefore great care must be taken when using them.

If a stake has been roughened by centre punch marks or is chisel marked, such marks will be impressed upon the surface of the workpiece and spoil its appearance.

A stake should never be used to back up work directly when centre-punching or cutting with a cold chisel.

A mallet should be used wherever possible when shaping sheet metal, and when a hammer is used care must be exercised to avoid making 'half-moons' on the surface of the stake.

Bench tools which have been abused and damaged as a result, should be reconditioned immediately. Regular maintenance will avoid the risk of marking the surface of the workpiece, as such marks in the metal cannot be removed.

10.18 'Hollowing' and 'raising'

Hollowing or raising are both methods employed for the purpose of forming sheet metal into hollow or double-curvature work by hand. As a general guide with regard to the choice of method employed:

HOLLOWING is employed where the desired shape or contour is to be only slightly domed or hollowed.

RAISING is always employed where shapes or contours of much greater depth are required.

Figures 10.30 to 10.35 inclusive show the comparison between the two processes.

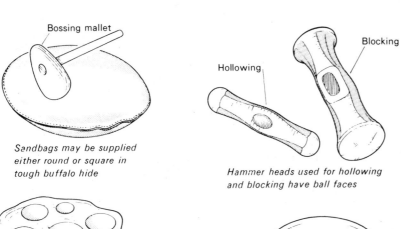

Fig. 10.30 The basic tools for hollowing

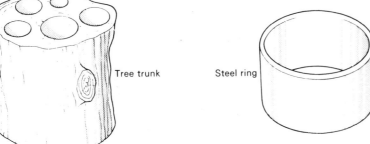

(a) Shallow depth
This would be formed by hollowing

(b) Too deep for hollowing
This would be formed by raising

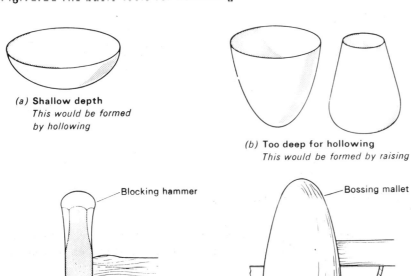

(c) Depth limitations

Fig. 10.31 Comparison of hollowing and raising

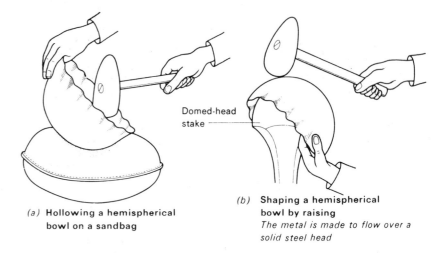

(a) Hollowing a hemispherical bowl on a sandbag

(b) Shaping a hemispherical bowl by raising
The metal is made to flow over a solid steel head

Fig. 10.32 The two basic methods of forming a bowl

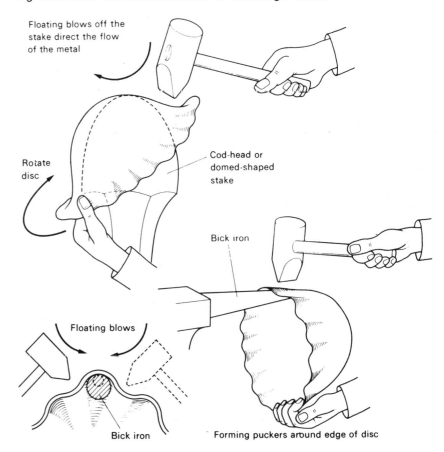

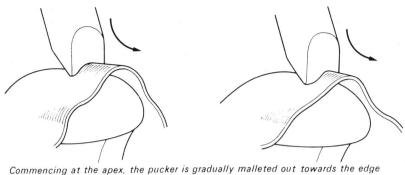

Commencing at the apex, the pucker is gradually malleted out towards the edge

Fig.10.33 Taking-in surplus metal when raising

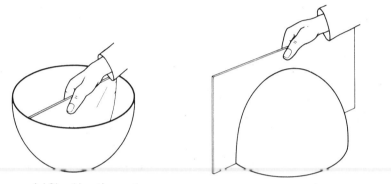

(a) **Checking the contour**
Checked with the aid of inside and outside templates

(b) **Double-curvature work**
Produced by either hollowing or raising is placed on a suitable stake and lightly malleted all over in order to remove any high spots on the surface

(c) **Overlapping blows**
These are made with the planishing hammer directly over the point of contact of the workpiece with the surface of the stake

Fig.10.34 Finishing processes for double curvature work

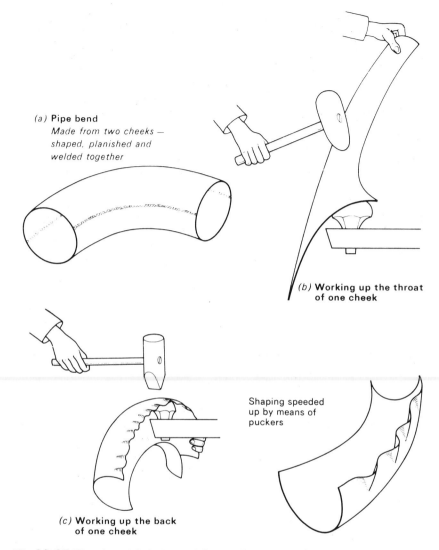

(a) **Pipe bend**
Made from two cheeks – shaped, planished and welded together

(b) **Working up the throat of one cheek**

(c) **Working up the back of one cheek**

Fig.10.35 Pipe bend fabricated from sheet metal

Hollowing is a process whereby sheet metal is beaten into a small indentation. The metal being formed is stretched and has its original thickness reduced.

Raising is a process whereby sheet metal is beaten and induced to flow into the required shape by application of carefully controlled 'floating' blows struck whilst the metal is slightly off the head or former being used. The metal being formed is compressed upon itself and has its original thickness increased.

245

10.19 Introduction to metal spinning

Most sheet metals lend themselves readily to cold-forming by spinning. There will, of course, be some variations in performance, and this is due to the individual characteristic mechanical properties and work-hardening behaviours. However, the *basic principles* of spinning are identical regardless of the particular metal being formed.

Metal spinning is a process whereby sheet metal discs are pressed or rolled to specific shapes by forcing the metal to flow over a suitable mandrel or chuck of the required form, usually by hand tools whilst chuck and metal are rotating together.

Causing flat sheet metal to alter its shape by PLASTIC FLOW under pressure manually applied through the LEVERAGE of spinning tools, calls for skill that can only be acquired through training and experience. The manner in which such shape-changing is accomplished is illustrated in the line diagram, Fig. 10.36.

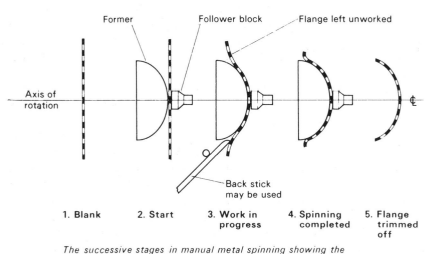

Fig.10.36 Metal spinning

10.20 Spinning lathes

The general design of a spinning lathe follows closely that of an ordinary machine shop lathe. There are, however, several necessary and important differences:

1. *Headstock and spindle* The HEADSTOCK and SPINDLE must be sturdily built to be able to withstand the considerable force which is applied during the spinning operation.

The HEADSTOCK is unlike that of a machinist's lathe in that the sheet metal disc (*B*) from which the article is to be shaped is not held in a CHUCK, but is held by FRICTION between a FORMER and a FOLLOWER.

2. *Former and follower* The FORMER (*A*) is solidly fixed to the headstock spindle and turns with it (see below). The FOLLOWER (*C*) is a block of metal or hardwood introduced between the job and the nose of the tailstock barrel, as shown in Fig. 10.37. Various diameters of follower block are used, and each is provided with a CENTRE location.

3. *Live centre* A special 'LIVE CENTRE' (*D*) is employed which is capable of rotating freely without friction under the great amount of end-thrust which is unavoidable in metal spinning.

4. *Tailstock* The TAILSTOCK must have provision for rapid advance and withdrawal of the BARREL. Due to the severe deformation of the metal being spun, it is necessary, with certain metals and alloys, that the work should be frequently ANNEALED, and for this purpose the work must be easily removed from the lathe. An ordinary lathe tailstock, in which the barrel is fed forward and backward by a screw operated from a handwheel, would render this operation far too slow. A specially designed tailstock makes this operation a relatively simple task. A quick-action LOCKING LEVER, when slackened, disengages the screw and allows the tailstock barrel to slide with ease, enabling it to be rapidly advanced or withdrawn.

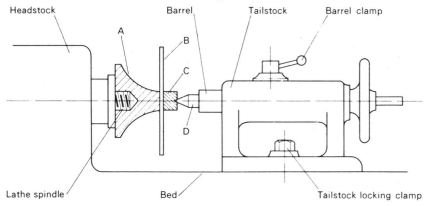

Fig.10.37 The basic features of the spinning lathe

In Fig. 10.37 pressure is applied by means of a TAILSTOCK HANDWHEEL 'D'. Sufficient pressure must be applied in order to hold the *former* 'A', the *work* 'B', and the *follower block* 'C' so tightly that they will revolve as one. The special 'LIVE CENTRE' is shown at 'E'.

Formers

Sometimes referred to as 'mandrels' or 'chucks', upon which the final contour or shape of the spun article is dependent, are usually made of hardwood (for short runs), steel or cast iron (for long runs). Wooden formers are generally made from MAHOGANY or LIGNUM VITAE. These hardwoods have *high strength* and *resistance to wear*. Small formers are made in one piece, i.e. from a single section of hardwood. For economy, and also to lessen the effects of shrinkage, some large formers are made from laminated hardwood blocks.

10.21 The spinning process

The hand-spinning process which is the most commonly used, except when considerable quantities of components are mass produced, is performed with the aid of a number of uniquely designed tools which have their actual working surfaces of special shapes according to the nature of the work, or part being spun. Some of these hand-spinning tools are illustrated in Fig. 10.38.

Hand-spinning tools are not standardised and may have a wide variety of shapes — some craftsmen shape their own forming tools.

A hand-forming tool consists of two parts:

(a) TOOL BIT. This is approximately 300 – 450 mm in length and usually forged to shape from high speed steel (round bar) and hardened. It has a 'tang' which fits into a handle.

(b) A WOODEN HANDLE. This approximately 600 mm in length.

The 'bit', when securely fitted, projects from the handle for a distance of 200 mm.

THE AVERAGE OVERALL LENGTH OF A HAND-FORMING TOOL IS BETWEEN 750 AND 850 mm.

The most common forming tool used is the COMBINATION BALL AND POINT. Its range of usefulness is large on account of the variety of shapes that may be utilised merely by rotating the tool in different directions.

A BALL TOOL is used for finishing curves.

The HOOK TOOL is shaped for use on inside work.

The bulk of spinning operations involve starting the work and bringing it approximately to the shape of the former, after which 'smoothing' or 'planishing' tools are used to remove the spinning marks and produce a smooth finish.

The FISH-TAIL planishing tool is one which is commonly used for finishing the work. It is also a very useful tool for sharpening any radii in the contour.

These hand tools are used in conjunction with a TEE-REST and FULCRUM PIN. The manner in which the fulcrum pin is advanced as the spinning operation progresses is extremely simple, as will be seen by studying the accompanying diagrams (Figs. 39–42). The tee-rest is accommodated in the adjustable TOOL-REST HOLDER which is clamped to the bed of the lathe. The tool-rest provides a wide range of adjustment in six directions, and a further fine adjustment can be made by releasing a clamp bolt and swivelling the tee-rest. All these features are illustrated in Fig. 10.39. The action of the spinning tool is shown in Fig. 10.40. (See 'Principle of moments'; *Basic Engineering*, 7.6).

When commencing spinning operations, the initial strokes are made outwards towards the edge of the disc being spun. In order to speed up the operation and also to avoid possible thinning of the metal, strokes are also made in the opposite direction, i.e. inward strokes.

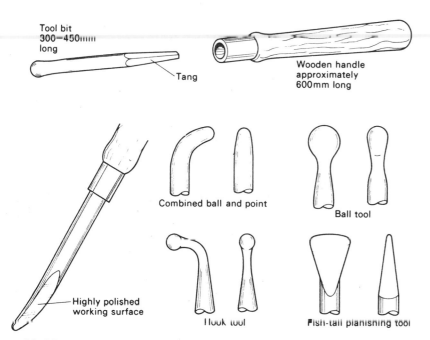

Fig. 10.38 Hand forming tools for metal spinning

'B' represents the working portion of the fulcrum pin. This varies in length between 75 and 100mm.
'C' represents the portion of the pin which is a free fit in the holes provided in the tee-rest.
It is smaller diameter than 'B' and is approximately 40 to 50mm in length

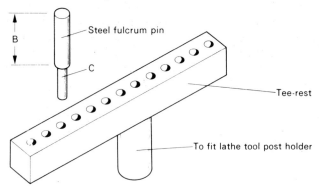

Simple tee-rest and fulcrum pin

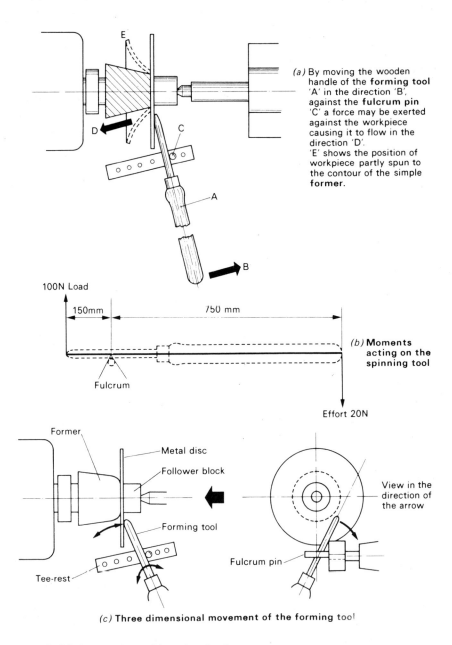

(a) By moving the wooden handle of the forming tool 'A' in the direction 'B', against the fulcrum pin 'C' a force may be exerted against the workpiece causing it to flow in the direction 'D'.
'E' shows the position of workpiece partly spun to the contour of the simple former.

(b) Moments acting on the spinning tool

(c) Three dimensional movement of the forming tool

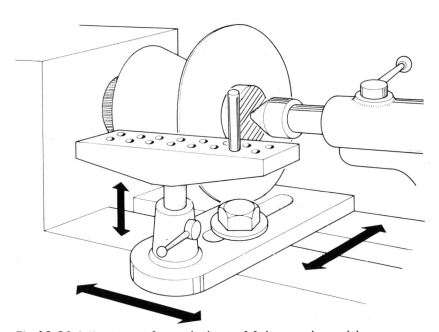

Fig.10.39 Adjustment for variations of fulcrum pin position

Fig.10.40 The action of hand spinning

In either case it is important not to dwell in any one position on the workpiece so as to cause excessive WORK-HARDENING. The tee-rest and the position of the fulcrum pin are reset as the work progresses, and both the forming tool and the outer surface of the metal disc are LUBRICATED frequently.

The process of spinning HARDENS THE METAL *by reason of the* COMPRESSION AND STRETCHING *which it undergoes whilst being shaped to the contour of the former.*

Figure 10.41 illustrates the metal-spinning process.

When spinning fairly large diameter discs, particularly when the shape of the former does not lend any support, the use of a 'BACK-STICK' is necessary.

Back-sticks, as their name implies, are always positioned at the back of the metal disc immediately opposite the forming tool. They are used to prevent wrinkles forming and, in the case of thin gauge metal, to hold and stiffen the edge and prevent it from collapsing. Figure 10.42 shows that a back-stick is used in the same manner as a forming tool, but on the opposite side of the disc. Pressure is applied in the direction indicated, and the work revolves between the back-stick and the forming tool, two fulcrum pins being used.

The tee-rest and holder may be adjusted to a variety of positions as work progresses

Key
A—Former B—Sheet metal disc C—Follower
D—Fulcrum pin

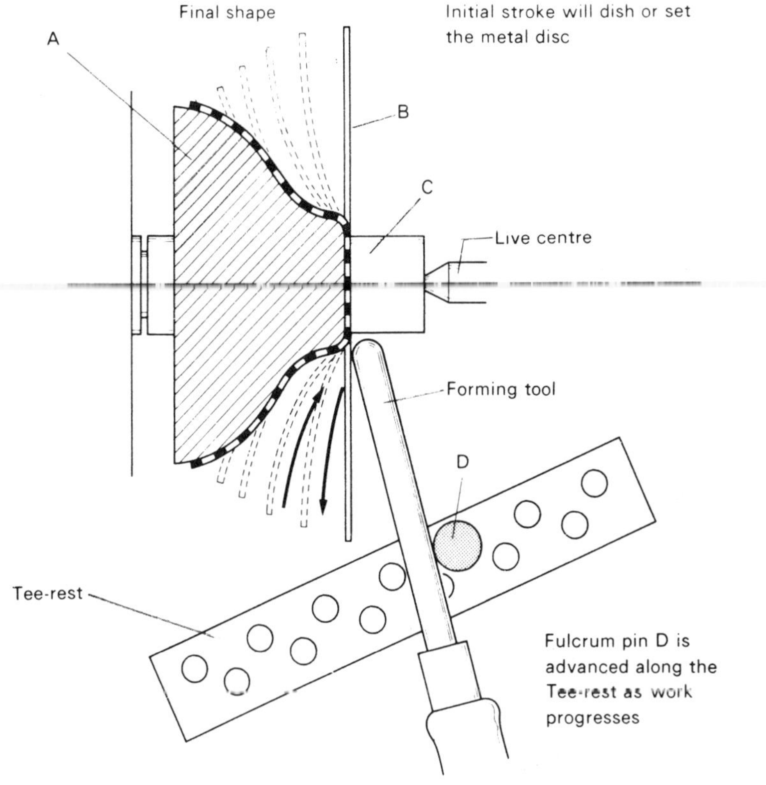

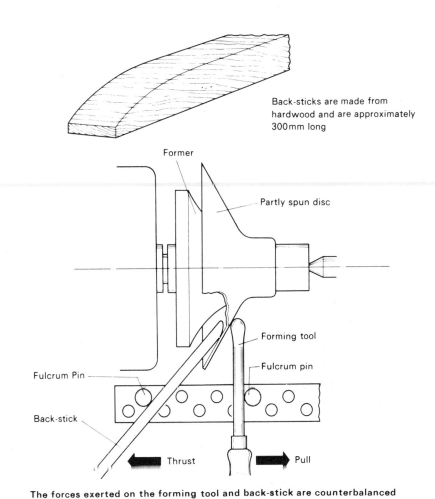

Fig.10.41 The metal spinning process

The forces exerted on the forming tool and back-stick are counterbalanced

Fig.10.42 The use of the back-stick

10.22 Lubrication for spinning

It is necessary that the FRICTION between the nose of the forming tool and the work be reduced to a minimum in order to prevent: excessive HEAT being developed; scratching or cutting of the metal surface; and possible damage to the tool.

LUBRICANTS are used for this purpose. ANY LACK OF LUBRICANT WILL RESULT IN DAMAGE TO THE WORKPIECE AND THE TOOL.

Lubricants must be frequently applied to the surface of the work and the tools during the spinning operations. It is important to use the correct type of lubricant. It must be sufficiently ADHESIVE to cling to the metal disc when it is revolving at high speeds.

In hand spinning, TALLOW and INDUSTRIAL SOAP, or a mixture of both, are generally used as lubricants.

10.23 Spindle speeds for spinning

Spindle speeds for metal spinning are very important, and these will depend on a number of factors, some of which are:
(a) The DUCTILITY of the material being spun;
(b) Whether the material is FERROUS or NON-FERROUS;
(c) The DIAMETER of the metal disc being spun;
(d) The THICKNESS of the material being spun;
(e) The SHAPE of the former;
(f) The type of TOOL SHAPE.

The drive to the spindle is usually by a two-speed motor on to a three-step cone pulley, giving SIX POSSIBLE SPINDLE SPEEDS. As a general rule: MILD STEEL requires the slowest spindle speeds.

BRASS requires about twice the speeds suitable for mild steel.
ALUMINIUM should be spun at about the same speeds as brass or slightly higher.

10.24 Bend allowances for sheet metal

When sheet metals are bent through angles of 90° the material on the outside surfaces becomes STRETCHED, whilst that on the inside surfaces of the bends is COMPRESSED. It is therefore necessary to make an allowance for these effects when developing a template or when marking out a blank sheet for bending.

Figure 10.43 illustrates the importance of the 'NEUTRAL LINE'.

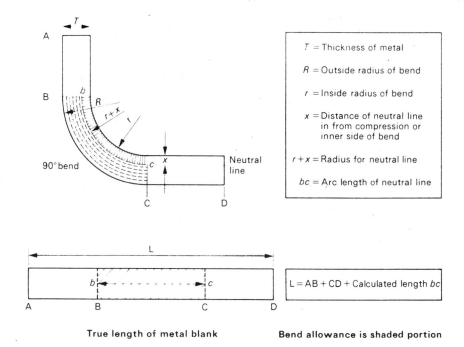

Fig.10.43 Bend allowances for sheet metal

An enlarged cross-section of a 90° bend in sheet metal is shown in figure 10.43.

THE NEUTRAL LINE IS AN IMAGINERY CURVE SOMEWHERE INSIDE THE METAL IN THE BEND. ITS POSITION DOES NOT ALTER FROM THE ORIGINAL FLAT LENGTH DURING BENDING.

Because there is a slight difference between the amount of COMPRESSIVE STRAIN and the amount of TENSILE STRAIN, the NEUTRAL LINE lies in a position nearer the inside of the bend.

For the purpose of calculating the allowance for a bend in sheet metal the neutral line curve is regarded as an arc of a circle whose radius is equal to the sum of inside bend radius and the distance of the neutral line in from the inside of the bend.

The true length of the sheet metal blank is never equal to the sum of the inside, or outside, dimensions of the bent metal.

The precise position of the neutral line inside the bend depends upon the number of factors which include:

(a) The properties of the materials;
(b) The thickness of the material;
(c) The inside radius of the bend.

Table 10.3 lists the approximate positions of the neutral line for some materials.

Table 10.3 Neutral line data for bending sheet metal

MATERIAL (20 to 14 swg)	AVERAGE VALUE OF RATIO $\frac{x}{T}$
Mild steel	0·433
Half-hard aluminium	0·442
Heat-treatable aluminium alloys	0·348
Stainless steel	0·360

In general the position of the neutral line is 0·4 times the thickness of the material in from the inside of the bend.
This means that the radius used for calculating the bend allowance is equal to the sum of the inside bend radius and 0·4 times the thickness of the metal.
The inside bend radius is rarely less than twice the thickness of the material or more than four times.
For the purpose of calculating the required length of blank when forming cylindrical or part-cylindrical work a mean circumference is used — i.e. the neutral line is assumed to be the central axis of the metal thickness.

For general sheet metal work the following values for the radius of the neutral line may be used (where precision is unimportant):

THICKNESS OF MATERIAL		APPROXIMATE VALUE OF NEUTRAL LINE RADIUS
swg	mm	
30 to 19	0·315 to 1·016	One-third metal thickness plus inside bend radius
18 to 11	1·219 to 2·346	Two-fifths metal thickness plus inside bend radius
10 to 1	3·251 to 7·620	One-half metal thickness plus inside bend radius

Definitions

In this section the following British Standard (BS 1649) definitions apply:

Neutral line. The boundary line between the area under COMPRESSION and the area under TENSION in any angle bend.

Radius of bend. The radius of the inside of the bend.

Outside radius of the bend. The inside radius of the bend plus the metal thickness.

Bend allowance. The length of the metal required to produce the radius portion only of the bend.

10.25 Applications of bending allowances

As previously stated the length of the neutral line is represented by an ARC of a CIRCLE.

Arc lengths are dependent upon their SECTOR ANGLES, and can be determined by calculation as follows:
SECTOR ANGLE DIVIDED BY 360° AND MULTIPLIED BY THE CIRCUMFERENCE.

For example, consider an arc of RADIUS 100 mm whose subtended angle is 90°. Then its length will be:

$$\frac{90}{360} \times 2\pi R$$

$$= \frac{1}{4} \times 2 \times 3·142 \times 100 \text{ mm}$$

$$= 50 \times 3·142 = \underline{157·1 \text{ mm}}$$

Alternatively, by inspection the ratio $\frac{2\pi}{360}$ is a constant which may be used for all bend allowance calculations:

$$\frac{2\pi}{360} = \frac{2 \times 3·142}{360}$$

$$= \frac{3·142}{180} = \underline{0·0175}$$

Thus the length of the arc will be:

$$0·017\,5 \times R \times 90$$
$$= 0·017\,5 \times 100 \times 90 \text{ mm}$$
$$= 1·75 \times 90 = \underline{157·5 \text{ mm}}$$

From the above it will be seen that a formula may be derived for calculating bend allowances. It is as follows:

BEND ALLOWANCE
= 0·017 5 multiplied by inside radius to the neutral line multiplied by the subtended angle of the bend.

This can be expressed as follows:

$$A = \theta \times R \times 0.0175$$

where: $\theta = (180 - \text{angle given})$ A = Bend allowance
R = Inside radius to the NEUTRAL LINE

Table 10.4 shows two worked examples for bend allowances using the 'MEAN LINE' method.

Table 10.4(a) Calculation — centre line bend allowance

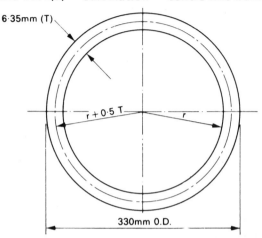

Calculate the length of the blank required to roll the cylinder shown opposite.
The position of the neutral line = $0.5T$, where the thickness of the plate, T = 6.35 mm.

Solution:

The length of the blank required is equal to the MEAN CIRCUMFERENCE. *The MEAN RADIUS 'R' is equal to the INSIDE RADIUS 'r' plus half the thickness 'T' of the plate.*

The outside diameter	=	330 mm
The inside diameter	=	$330 - 2T$
	=	$330 - (2 \times 6.35)$
	=	$330 - 12.7$
	=	317.3 mm

The inside radius

$r = \dfrac{317.3}{2}$ = 158.65 mm

From which $R = 158.6 + (0.5 \times 6.35)$
$= 158.6 + 3.175$
$\therefore R = 161.825$ mm

The mean circumference = $2\pi R$
$= 2 \times 3.142 \times 161.825$
$= 6.284 \times 161.825$

No.	Log
6.284	0.7983
161.825	2.2089
1 016	3.0072

Length of blank required = 1 016 mm

Table 10.4(b) Calculation — centre line bend allowance

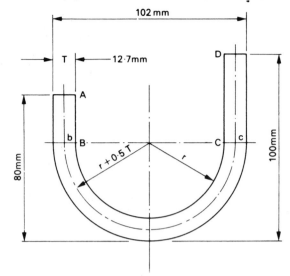

Calculate the length of the blank required to form the 'U-bracket' shown opposite.
The position of the neutral line = $0.5T$, where the thickness of the plate, T = 12.7 mm.

Solution:
The length of the blank required is equal to the sum of the flats, 'A B' and 'C D', plus the length of the mean line, 'bc'.
Thus $L = AB + CD + \text{'}bc\text{'}$.
Now bc represents a semi-circular arc whose mean radius R is equal to the INSIDE RADIUS r plus half the thickness T of the plate.

The outside diameter of the semi-circle = 102 mm
The inside diameter of the semi-circle = $102 - (2T)$
 = $102 - 25.4$
 = 76.6 mm

From which the inside radius $r = \dfrac{76.6}{2}$ = 38.3 mm

The mean radius $R = 38.3 + (0.5 \times 12.7)$
 = $38.3 + 6.35$
$\therefore R = 44.65$ mm

Length of flats
A B = $80 - \dfrac{102}{2}$
 = $80 - 51$ = 29 mm
C D = $100 - \dfrac{102}{2}$
 = $100 - 51$ = 49 mm
Total length of flats = 78 mm

Bend allowance for 'b c'
'Bend Allowance' = πR (semi-circle)
   = 3·142 × 44·65
'Bend Allowance' = 140·3 mm

No.	Log
3·142	0·4971
44·65	1·6498
1403	2·1469

Length of blank
  L = A B + C D + b c
     = 78 + 140·3

Length of blank required = 218·3 mm

Tables 10.5, 10.6 and 10.7 show worked examples using the approximate values for the neutral line from Table 10.3.

Table 10.5 Calculation — neutral line bend allowance

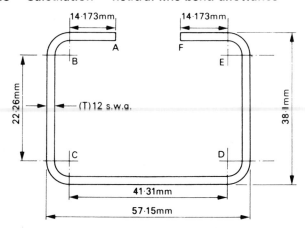

Calculate the length of the blank required to form the four-bend channel section shown in the diagram. Use the approximate value for the neutral line given in Table 10.3. The thickness of the sheet metal, T = 2·64 mm (12 swg) and the inside radius for the bends r = 2T.

Solution:
From the Table 10.3, the approximate value for the neutral line when bending 12 swg

is $\frac{2}{5}T$ $\frac{2}{5}T$ = 0·4T

The inside bend radius, r, may be found as follows:

(1) Subtract inside dimensions from outside dimensions

(a) 57·15 − 41·31
    = 15·84 mm

(b) 38·1 − 22·26
    = 15·84 mm

(2) Subtract two thicknesses of metal from (1) and divide by two (2 bends)

  15·84 − (2 × 2·64)
    = 15·84 − 5·28   = 10·56 mm

Divide by 2
   $\frac{10·56}{2}$ ∴ r = 5·28 mm

It will be seen that $r = 2T$.

Length of flats
  A B = 14·173
  B C = 22·260
  C D = 41·310
  D E = 22·260
  E F = 14·173

Total length of flats = 114·176 mm.

Bend allowances for four quadrant corners
The four quadrant corners, B + C + D + E represent a circle for which the bend allowance radius

  R = r + 0·4T
     = 2T + 0·4T
     = 2·4T
     = 2·4 × 2·64
  ∴ R = 6·336 mm

No.	Log
2·4	0·3802
2·64	0·4216
6336	0·8018

Total bend allowance

'Bend Allowance' = $2\pi R$
   = 2 × 3·142 × 6·336
   = 6·284 × 6·336
∴ Total bend allowance = 39·82 mm

No.	Log
6·284	0·7983
6·336	0·8018
3982	1·6001

Length of blank
   = Total length of flats + total bend allowance
   = 114·176 + 39·82

Length of blank required = 153·996 mm

Table 10.6 Calculation — neutral line bend allowance

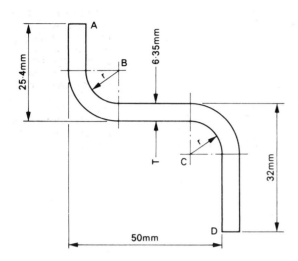

Calculate the length of blank required to form the support bracket shown opposite. Use the approximate value for the neutral line from Table 10.3. The thickness of the steel plate, $T = 6.35$ mm and the inside bend radius $r = 2T$.

Solution:
From the Table 10.3, the approximate value for the neutral line when bending plate of 6.35 mm thickness is given as $0.5T$.

Length of flats

$$\begin{aligned} AB &= 25.4 - (r + T) \\ &= 25.4 - (2T + T) \\ &= 25.4 - 3T \\ &= 25.4 - (3 \times 6.35) \\ &= 25.4 - 19.05 \qquad = 6.35 \text{ mm} \end{aligned}$$

Similarly
$$\begin{aligned} CD &= 32 - 3T \\ &= 32 - 19.05 \qquad = 12.95 \text{ mm} \end{aligned}$$

But
$$\begin{aligned} BC &= 50 - (r + T + r) \\ &= 50 - (2r + T) \\ &= 50 - (4T + T) \\ &= 50 - 5T \\ &= 50 - (5 \times 6.35) \\ &= 50 - 31.75 \qquad = 18.25 \text{ mm} \end{aligned}$$

∴ Total length of flats $= 37.55$ mm

The bend allowance radius

$$\begin{aligned} R &= r + 0.5T \\ &= 2T + 0.5T \\ &= 2.5T \\ &= 2.5 \times 6.35 \end{aligned}$$

∴ $R = 15.88$ mm

No.	Log
2.5	0.3979
6.35	0.8028
1588	1.2007

Bend allowance for bends B and C
Total bend allowance for 2 bends
'Bend Allowance' $= 2 (\theta \times R \times 0.0175)$
(where $\theta = 180° - 90° = 90°$)
$= 2 \times 90 \times 15.88 \times 0.0175$
$= 180 \times 15.88 \times 0.0175$

Total bend allowance $= 50$ mm

Length of blank
$=$ Total length of flats $+$ total bend allowance
$= 37.55 + 50$

Length of blank required $= 87.55$ mm

No.	Log
180	2.2553
15.88	1.2007
0.0175	2.2430
5000	1.6990

Table 10.7 Calculation — neutral line bend allowance

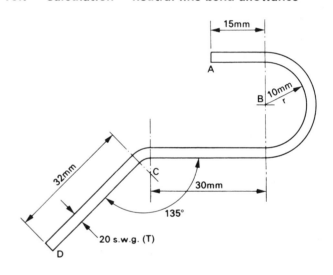

Calculate the length of blank required to form the sheet metal clip shown in the diagram. Use the approximate value for the neutral line from Table 10.3 for bending sheet metal whose thickness $T = 0.914$ (20 swg). The inside bend radius, $r = 2T$ for the bend at C.

Solution:
From the Table 10.3 the value of the neutral line for 20 swg $= \frac{1}{3}T$.

Length of flats

$$\begin{aligned} AB &= 15 \text{ mm} \\ BC &= 30 \text{ mm} \\ CD &= 32 \text{ mm} \end{aligned}$$

Total length of flats $= 77$ mm

The inside bend radius for bend B
$r = 10$ mm
∴ The bend allowance radius for bend B
$$R = r + \frac{1}{3}T$$
$$= 10 + \frac{0.914}{3}$$
$$= 10 + 0.305$$
$$R = 10.305 \text{ mm}$$

Bend allowance for bend B
The bend at B is a semi-circle
Bend allowance = πR
$= 3.142 \times 10.305$
Bend allowance for B $= 32.38$ mm

No.	Log
3·142	0·4971
10·305	1·0132
3238	1·5103

The inside bend radius for bend C
$r = 2T$
Bend allowance radius for bend C
$$R = r + \frac{1}{3}T$$
$$= 2T + \frac{1}{3}T$$
$$= 2\frac{1}{3}T$$
$$= \frac{7}{3} \times 0.914$$
$$= 7 \times 0.305$$
$$\therefore R = 2.135 \text{ mm}$$

Bend allowance for bend C
Bend allowance $= \theta \times R \times 0.0175$
(where $\theta = 180° - 135° = 45°$)
$= 45 \times 2.135 \times 0.0175$
Bend allowance for C $= 1.681$ mm

Length of blank
= Total length of flats + total bend allowance
= 77 + 32·38 + 1·681

Length of blank required = 111·061 mm

No.	Log
45	1·6532
2·135	0·3294
0·0175	2·2430
1681	0·2256

Tables 10.8, 10.9 and 10.10 show worked examples for 'Precision Sheet Metal Work' using the ratio $\dfrac{x}{T} = 0.4$

Table 10.8(a)   Calculation — precision bend allowance

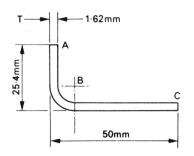

Calculate the length of blank required to make the simple sheet metal bracket shown opposite. The value for the neutral line $= 0.4T$. The thickness of the sheet metal, $T = 1.62$ mm. The inside bend radius $r = 2T$.

Solution:
Length of flats
$$AB = 25.4 - (r + T)$$
$$= 25.4 - (2T + T)$$
$$= 25.4 - 3T$$
$$= 25.4 - (3 \times 1.62)$$
$$= 25.4 - 4.86 \qquad = 20.54 \text{ mm}$$

Similarly $\quad BC = 50 - 3T$
$\qquad\qquad\quad = 50 - 4.86 \qquad = 45.14$ mm
Total length of flats $\qquad = 65.68$ mm

The inside bend radius
$r = 2T$
∴ The bend allowance radius for bend B
$$R = 2T + 0.4T$$
$$= 2.4T$$
$$= 2.4 \times 1.62$$
$$\therefore R = 3.888 \text{ mm}$$

No.	Log
2·4	0·3802
1·62	0·2095
3888	0·5897

Bend allowance for bend at B
'Bend allowance' $= \theta \times R \times 0.0175$
(where $\theta = 180° - 90° = 90°$)
$= 90 \times 3.888 \times 0.0175$
Bend allowance for B $= 6.122$ mm

No.	Log
90	1·9542
3·888	0·5897
0·0175	2·2430
6122	0·7869

Length of blank
= Total length of flats + bend allowance
= 65·68 + 6·122

Length of blank required = 71·802 mm

Table 10.8(b) Calculation — precision bend allowance

Calculate the length of blank required to form the sheet metal detail shown in the diagram.
The value for the neutral line = 0·4T.
The thickness of the metal, T = 2·642 mm
The inside bend radius, r = 2T.

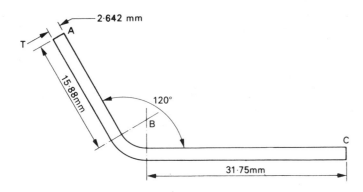

Solution:
Length of flats

AB = 15·88 mm
BC = 31·75 mm
Total length of flats = 47·63 mm

The inside bend radius
$r = 2T$
∴ The bend allowance radius
$R = 2T + 0·4T$
$= 2·4T$
$= 2·4 \times 2·642$
∴ $R = 6·34$ mm

No.	Log
2·4	0·3802
2·642	0·4219
6340	0·8021

Bend allowance for bend at B
'Bend allowance' = $\theta \times R \times 0·0175$
(where $\theta = 180° - 120° = 60°$)
$= 60 \times 6·34 \times 0·0175$
Bend allowance for B = 6·658 mm

No.	Log
60	1·7782
6·34	0·8021
0·0175	2·2430
6658	0·8233

Length of blank
= Total length of flats + bend allowance
= 47·63 + 6·658
Length of blank required = 54·288 mm

Table 10.9 Calculation — precision bend allowance

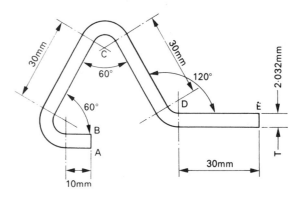

Calculate the length of blank required to form the sheet metal section shown in the diagram.
The value for the neutral line = 0·4T.
The thickness of the metal, T = 2·032 mm
The inside bend radius, r = T

Solution:
Length of flats

AB = 10 mm
BC = 30 mm
CD = 30 mm
DE = 30 mm
Total length of flats = 100 mm

The inside bend radius
$r = T$
∴ The bend allowance radius
$R = T + 0·4T$
$= 1·4T$
$= 1·4 \times 2·032$
∴ $R = 2·844$ mm

No.	Log
1·4	0·1461
2·032	0·3079
2844	0·4540

For bend B $\theta = 180° - 60° = 120°$
Similarly for bend C $\theta = 120°$
For bend D $\theta = 180° - 120° = 60°$
Total sum of bending angles = 300°

Total bending allowance
B.A. = $\theta \times R \times 0·0175$
(where $\theta = 300°$)
= 300 × 2·844 × 0·0175
Total bend allowance = 14·93 mm

No.	Log
300	2·4771
2·844	0·4540
0·0175	2·2430
1493	1·1741

Length of blank
= Total length of flats + total bend allowance
= 100 + 14·93
Length of blank required = 114·93 mm

Table 10.10 Calculation — precision bend allowance

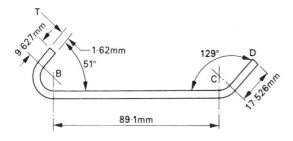

Calculate the length of the blank required to form the sheet metal section shown in the diagram.

The value for the neutral line = 0·4T
The thickness of the metal, T = 1·62 mm
The inside bend radius, r = 2T

Solution:

Length of flats

 A B = 9·627 mm
 B C = 89·100 mm
 C D = 17·526 mm
Total length of flats = 116·253 mm

The inside bend radius
 r = 2T
∴ The bend allowance radius
 R = 2T + 0·4T
  = 2·4T
  = 2·4 x 1·62
∴ R = 3·888 mm

No.	Log
2·4	0·3802
1·62	0·2095
3888	0·5897

For bend B θ = 180°−51° = 129°
For bend C θ = 180°−129° = 51°
Total sum of bending angles = 180°

Total bending allowance
 B.A. = θ x R x 0·0175
  (where θ = 180°)
  = 180 x 3·888 x 0·0175
∴ Total bend allowance = 12·25 mm

No.	Log
180	2·2553
3·888	0·5897
0·0175	2·2430
1225	1·0880

Length of blank
= Total length of flats + total bend allowance
= 116·253 + 12·25
Length of blank required = 128·503 mm

10.26 Deep drawing and pressing

Deep drawing and pressing are among the most common methods of forming sheet metal.

Metals upon which drawing and pressing operations are to be carried out must be ductile.

The simplest of drawing operation is that of 'CUPPING'. In this operation sheet metal blanks are formed into cup shapes by the application of a force.

The basic principle of cupping operations is very simple and the essential tools consist of a die, a blankholder or pressure plate and a punch. These are set up in a suitable power press.

The function of the pressure plate is to prevent 'wrinkling' of the metal.

It is possible, with some materials to produce cups of diameter somewhat less than half that of the original blanks.

Figure 10.44 shows how an article may be formed by drawing.

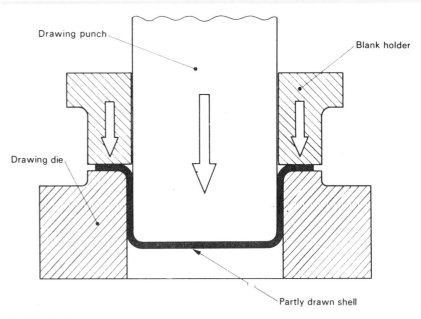

Fig.10.44 Deep drawing

The stresses involved during the drawing operation are:
(a) COMPRESSIVE at the periphery, here the metal tends to thicken;
(b) TENSILE in the walls because of the natural resistance to the action of drawing.

The metal blank retains its original dimensions at the base whilst the rest of the blank is compressed into shape with parallel sides as the metal is drawn from the flange into the walls.

With metals which WORK-HARDEN, the drawing process is generally commenced with the material in the fully annealed (soft) condition.

Some of the most common materials which are suitable for deep drawing are: Mild steel; Austenitic stainless steel; Aluminium and aluminium alloys; Copper and copper alloys (particularly cartridge brass).

10.27 Stiffening of fabricated material (introduction)

The basic principle of stiffening may be illustrated by a popular party trick — that of supporting a tumbler of water on a piece of note paper bridging two other tumblers. This simple trick is explained in Fig. 10.45.

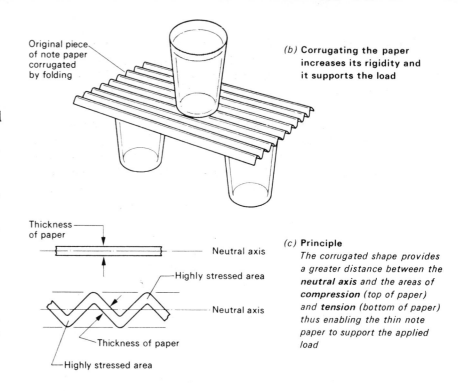

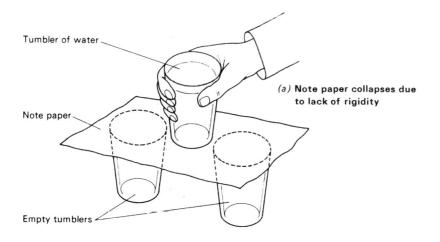

Fig.10.45 The basic principle of stiffening

A sheet-metal panel will not support a very great load due to the thinness of the material. A metal plate of the same surface area will support a fairly substantial load because of its extra thickness. Unfortunately, although the metal plate is much more rigid than the sheet metal panel, this rigidity is obtained at the expense of considerable additional weight. This STRENGTH/WEIGHT ratio is a very important factor in the fabrication industry and, fortunately, it is possible to produce a multiplicity of light fabrications which are very rigid and strong. This may be achieved in a number of ways which involve imparting stiffness to the material itself or by the addition of stiffeners.

10.28 Methods of imparting stiffness to sheet metal

The three main reasons for stiffening sheet metal are:
1. To give strength and rigidity to the material.
2. To produce a safe edge.
3. For decorative purposes.

The simplest method of giving strength to metal is to form angles or flanges along the edges of sheets. A right-angle bend greatly increases the strength of a sheet, as can be demonstrated by forming a right-angle bend in a thin sheet and then trying to bend the sheet across the angle.

Curving the surface of sheet metal by rolling or pressing will increase its strength and rigidity These methods are used to great advantage in the manufacture of car bodies. A curved surface is much stronger than a flat sheet. This is because the metal on the outer surface of the bend is under TENSION whilst the metal on the inner surface of the bend is in COMPRESSION, and these two stressed zones balance one another and will resist any force tending to change the shape or contour of the panel.

A surface with very little curve such as on some panels, bonnets or the top surface of roof panels on car bodies is springy, whilst surfaces with a lot of contour or shape, such as wings or the edges of roof panels, are very resistant to an applied force.

The raw edges of thin sheet metal can be extremely dangerous. Because of their sharpness they can cause nasty cuts when being handled. Components made out of thin sheet metal should have their raw edges made safe. One method of providing a safe edge is to make a return fold or hem. A 'double-returned edge' imparts greater rigidity to the metal than a single hem. Figure 10.46 shows various methods of stiffening metal components.

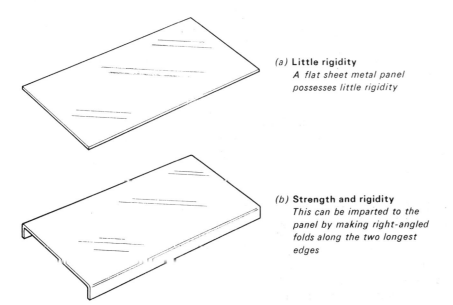

(a) **Little rigidity**
A flat sheet metal panel possesses little rigidity

(b) **Strength and rigidity**
This can be imparted to the panel by making right-angled folds along the two longest edges

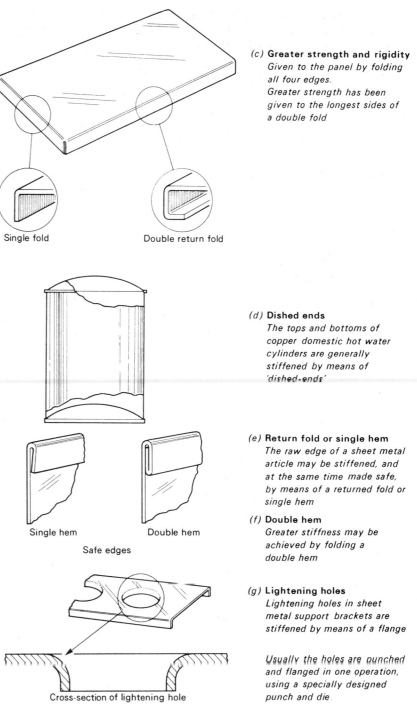

(c) **Greater strength and rigidity**
Given to the panel by folding all four edges.
Greater strength has been given to the longest sides of a double fold

(d) **Dished ends**
The tops and bottoms of copper domestic hot water cylinders are generally stiffened by means of 'dished-ends'

(e) **Return fold or single hem**
The raw edge of a sheet metal article may be stiffened, and at the same time made safe, by means of a returned fold or single hem

(f) **Double hem**
Greater stiffness may be achieved by folding a double hem

(g) **Lightening holes**
Lightening holes in sheet metal support brackets are stiffened by means of a flange

Usually the holes are punched and flanged in one operation, using a specially designed punch and die.

Fig.10.46 Imparting stiffness to sheet metal

10.29 Wiring the edges of sheet metal

Wiring sheet metal not only makes the edge rigid and safe, but also provides a pleasing and decorative appearance.

Wiring or 'beading' is a process of forming a sheet-metal fold around a wire of suitable diameter. Much of the strength of this type of edge is provided by the wire. Additional strength is obtained from the stressed metal which closely follows the exact contour of the wire.

The allowance to be added to the sheet metal edge to be wired is TWO AND A HALF TIMES THE DIAMETER *of the wire used.*

False wired edges

These may be one of two types:
1. *Applied* The APPLIED type is used when the position or metal thickness is unsuitable for normal wiring. These wired edges are attached and fastened in position by a returned flange, riveting, spot-welding or soldering.
2. *Hollow* This type of HOLLOW bead is usually produced by folding the edge around the wire and then withdrawing it. Although hollow beads are rigid due to their form, because they do not contain a wire, they will not withstand an impact blow and can be damaged. Figure 10.47 illustrates types of wired edges, whilst Fig. 10.48 shows how a hollow bead may be produced on a spinning lathe.

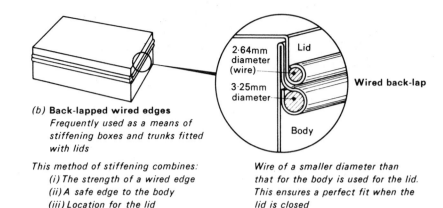

(b) **Back-lapped wired edges**
Frequently used as a means of stiffening boxes and trunks fitted with lids

This method of stiffening combines:
 (i) The strength of a wired edge
 (ii) A safe edge to the body
 (iii) Location for the lid

Wire of a smaller diameter than that for the body is used for the lid. This ensures a perfect fit when the lid is closed

The edge of the metal is hidden on the inside in the case of the 'back-lapped wired edge'

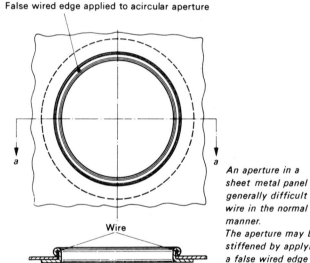

False wired edge applied to a circular aperture

*An aperture in a sheet metal panel is generally difficult to wire in the normal manner.
The aperture may be stiffened by applying a false wired edge*

(c) Cross-section a–a of applied false wired edge

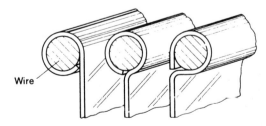

(a) Three common types of wired edges

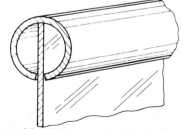

(d) **Split beading**
*This type of false wiring consists basically of split tubing. It may be rolled to required contour and slipped over the raw edge of the sheet metal article and soldered into position.
Split beading is mainly used for decorative purposes*

Fig.10.47 Types of wired edge

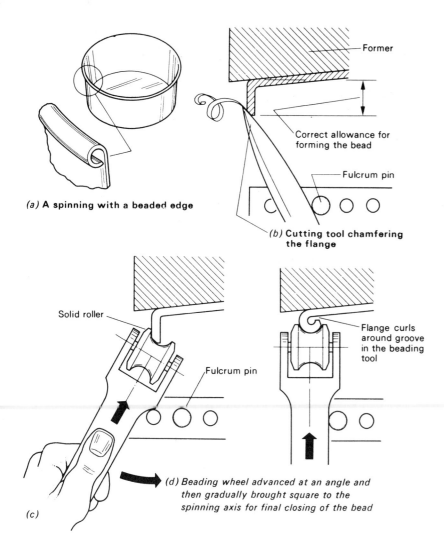

Fig.10.48 Forming a hollow bead on a spinning lathe

10.30 The swaging of sheet metal

Swaging is the operation used to raise up a moulding (SWAGE) on the surface of sheet metal. A 'swage' is produced by means of a pair of special contoured rollers. Swaging rolls are available in a large variety of contours to fit a 'swaging machine' which may be hand or power operated.

Although swaging has many similar functions to that of wired beads, it is not confined to stiffening edges but may be used some distance from the edge of the sheet.

The projecting shape of the swage above the surface imparts considerable strength to sheet metal articles.

The 'OGEE', or return curve swage, is frequently used to strengthen the centre portions of cylindrical containers or drums because of its high resistance to internally or externally applied forces in service.

Examples of the combination of strength with decoration associated with swaging are to be found in the design of circular sheet metal articles such as drums, dustbins, waste paper bins, buckets and water tanks.

The swage is a very important part of the decoration of a motor car body, and a great deal of attention is usually given to it. In motor body production the panels are formed to the correct contour on very large power presses and the swage is formed as an integral part of the panel. Figure 10.49 shows some aspects of swaging.

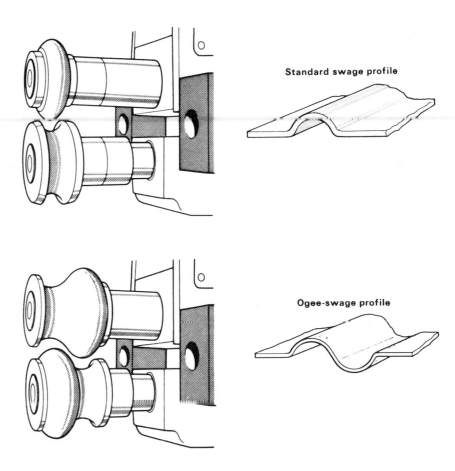

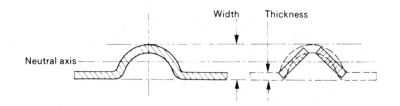

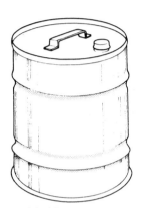

The basic shape of a swaged bead conforms to the principle that a greater force or load is required to bend or deflect a sheet across its width than across its thickness

Fig. 10.49 The swaging of sheet metal

The edges of fabrications constructed in sheet metal which is too thick to wire or hem can be stiffened by the use of flat bar or D-shaped section. It may be attached by spot-welding, brazing, welding or riveting.

One of the most common methods of achieving strength by means of attached stiffeners is the use of 'angle section frames'.

Figure 10.50 shows various methods of stiffening large panels.

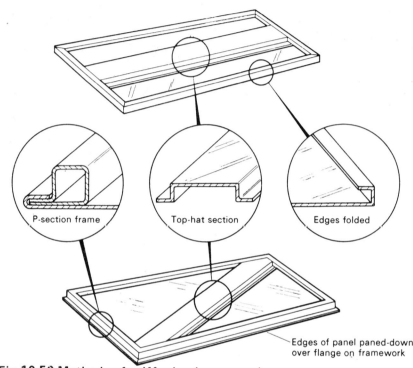

Fig. 10.50 Methods of stiffening large panels

The metal in the swaged bead is very highly stressed. This produces a much greater STRENGTH/THICKNESS ratio than that of the sheet metal with which it is formed. In general the maximum thickness of sheet metal which may be swaged is 1·62 mm.

A container or drum made of thin sheet metal will be of little use if when full of liquid or other substance it distorts and imposes extra load on its seams. Such drums are swaged around the circumference of the body as shown in Fig. 10.49. This adds strength and rigidity to the walls of the drum enabling it to withstand severe rough handling in service.

10.31 The use of stiffeners

Large panels may be reinforced by means of applied stiffeners. Generally, panels are stiffened by virtue of the fact that they are fastened to some sort of framework. These frameworks are usually fabricated from sheet-metal sections which are strong and rigid due to their form. Sheet-metal sections may also be roll-formed for the purpose of internal or external stiffening of large components of cylindrical or circular shape.

A large sheet-metal panel may be stiffened as shown in Fig. 10.50. All four edges are made rigid by folding. 'TOP-HAT SECTION' is used to stiffen the centre section of the panel and is usually secured in position by spot-welding.

Another method of stiffening large sheet-metal panels is to attach them to a rigid framework. The welded frame is fabricated from lengths of 'P-section' which has a very high STRENGTH/WEIGHT ratio for a sheet metal section. All four edges of the panel are folded at 90° to a suitable width. The panel is then placed in position over the frame and the edges 'paned-down' over the flange on the 'P-section, as shown in Fig. 10.50. The centre of the panel is stiffened by means of a diagonal top-hat section. Figure 10.51 shows how circular components may be stiffened.

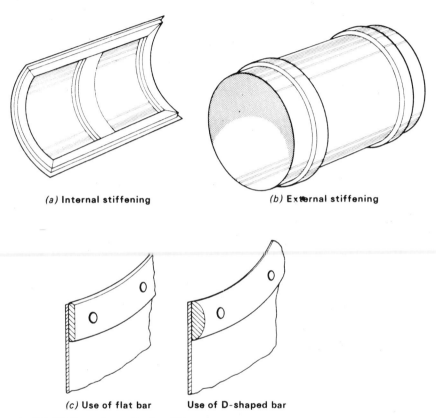

Fig. 10.51 Use of applied stiffeners

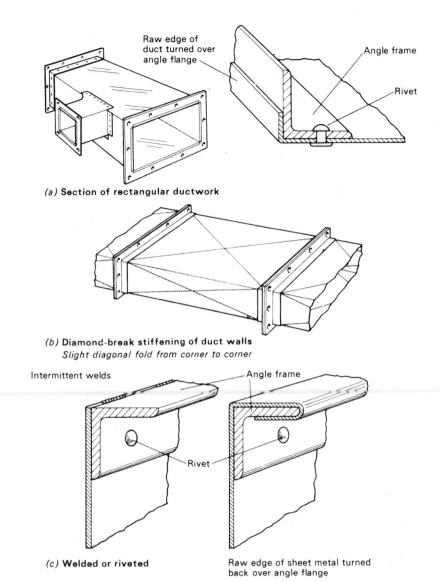

Fig. 10.52 The use of angle stiffeners

Figure 10.51(a) illustrates an application of internal stiffening on a panel of circular shapes. The stiffening section (in this case 'top-hat' sections rolled to correct contour, and attached externally.

When sheet metal is too thick to allow the edge to be wired the raw edge may be stiffened by attaching either flat bar or 'D-shaped' bar as shown in Fig. 10.51(c).

Figure 10.52 shows the use of angle stiffeners.

Welded angle frames are widely used as a means of stiffening and supporting rectangular ducts for high-velocity systems. They also serve as a jointing media when assembling sections together by bolting, as shown in Fig. 10.52(a).

The large sizes of square or rectangular ducting tend to drum as the air pressure passing through them varies. To overcome this drumming it is necessary to provide adequate stiffening to the walls

263

of the duct. This may be achieved by use of swaging, but often a 'diamond-break' is used, as shown in Fig. 10.52(b).

Simple angle frames of welded construction may be used as a means of supporting and stiffening the open ends of tanks or bins fabricated from sheet metal. Two methods of attaching the angle frame are shown in Fig. 10.52(c).

10.32 Fabricated structural members

The material most commonly used in constructional steelwork is Mild steel to B.S. 4360, which is produced by steel rolling mills in a variety of shapes and sizes, some of which are illustrated in Fig. 10.53 and set out in table 10.11.

Table 10.11 Standard sizes hot-rolled sections for structural work

ROLLED STEEL SECTION	RANGE OF SIZES (Dimensions in mm)		FLANGES
	From: Web Flange	To: Web Flange	
Universal beams (U.B.)	609·6 x 304·8 / 203·2 x 133·4	920·5 x 420·5 / 617 x 230·1	Slight taper (2° 52') / Parallel
Universal columns (U.C.)	152·4 x 152·4	474·7 x 424·1	Parallel
Joists (R.S.J.)	76·2 x 50·8	203·2 x 101·6	5° Taper
Channels (R.S.C.)	76·2 x 38·1	431·8 x 101·6	5° Taper
Equal angles	25·4 x 25·4 x 3·175	203·2 x 203·2 x 25·4	
Unequal angles	50·8 x 38·1 x 4·762	228·6 x 101·6 x 22·13	
	Table Stalk	Table Stalk	
Tees (from Universal beams)	133·4 x 101·6	305·5 x 459·2	Parallel or slight taper
Tees (from Universal columns)	152·4 x 76·2	395·0 x 190·5	Parallel
Tees (short stalk)	38·1 x 38·1 x 6·35	152·4 x 152·4 x 15·88	Slight taper (½°)
	Table Stalk	Table Stalk	
Tees (long stalk)	25·4 x 76·2	127 x 254	8° Taper

All Universal beams and Universal columns are specified by their serial size and mass per metre

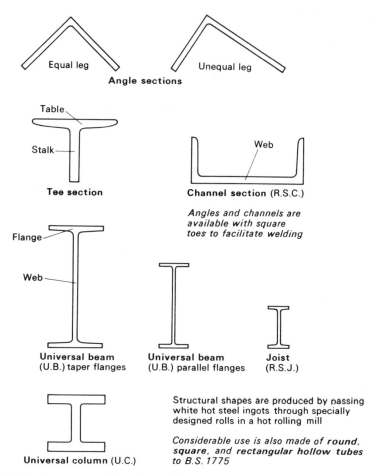

Fig. 10.53 Typical structural steel sections

For efficiency a structural member should use as little material as possible to carry as much load as possible.

The I-section, extensively used in steelwork, is designed to give an efficient STRENGTH/WEIGHT ratio, and the 'universal sections' provide even better efficiency (see *Basic Engineering*, 4.4).

10.33 Typical structural steel connections and assemblies

In constructional steel work the connections of the members are made by one of the following four methods:

1. Use of BLACK BOLTS,
2. Use of HIGH-STRENGTH FRICTION-GRIP BOLTS,
3. Riveting,
4. Welding.

Basically a steel structure is so designed that the sizes of the various components are determined by their ability to withstand the effects of the applied loads in service. The connections between the various members or components by which the forces and any moments are transmitted are designed to comply with a rigid code of safe practice.

In this section brief details will be given of typical connections, all of which combine strength, appearance and safety.

Stanchion bases

There are two types of stanchion base:
1. Bases for stanchions transmitting DIRECT LOAD only to the foundations,
2. Bases required where the stanchion has to transmit to the foundations a considerable bending moment in addition to the vertical loading.

These are illustrated in Fig. 10.54.

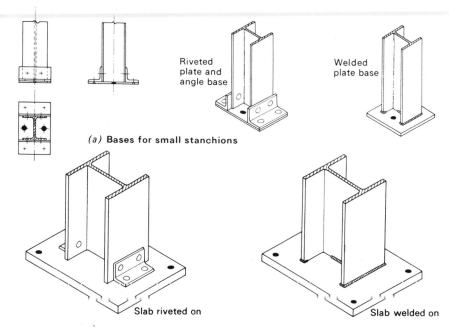

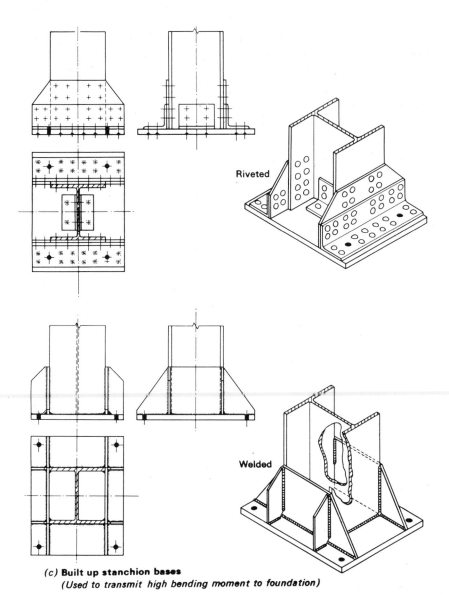

Fig. 10.54 Stanchion bases

Generally the top face only of a slab base is machined to give close contact with the stanchion.

A slab under 50 mm thickness does not normally require machining. The foot of the stanchion must be machined square on an 'end machine'. This ensures that the load is carried direct on to the base plate or slab and into the foundation (footing).

If the foot of the stanchion were not square to the base plate, the total load on the stanchion would be carried through the bolt shanks or rivets, in the case of bolted or riveted assemblies.

Stanchion splices

Splices in stanchions or columns should be arranged at a position above the adjacent floor level so that the joint, including any splice plates are well clear of any beam connection. Splices should never be made on a connection otherwise the bolts making the joint will be subjected to double loading. Figure 10.55 illustrates stanchion splices.

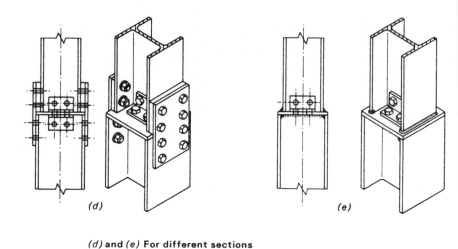

(d) and *(e)* **For different sections**
Web cleats in welded splice are temporary

Fig. 10.55 Stanchion splices

Connections of beams to stanchions

Figure 10.56 illustrates simple beam to stanchion connections. It will be noticed that all the load is transmitted from the beam to the stanchion through the seating cleat and its fixings.

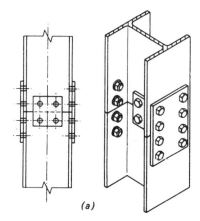

(a) **For equal sections**
Welded splice similar to detail below

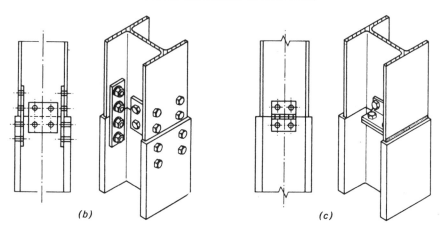

(b) and *(c)* **For different U.C. section in same serial size**
web cleats in welded splice are temporary

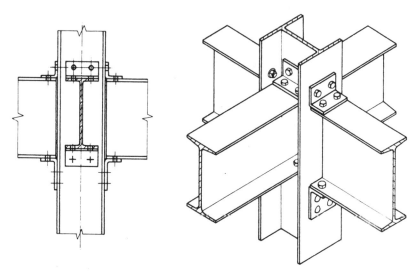

(a) **Simple riveted and bolted detail**

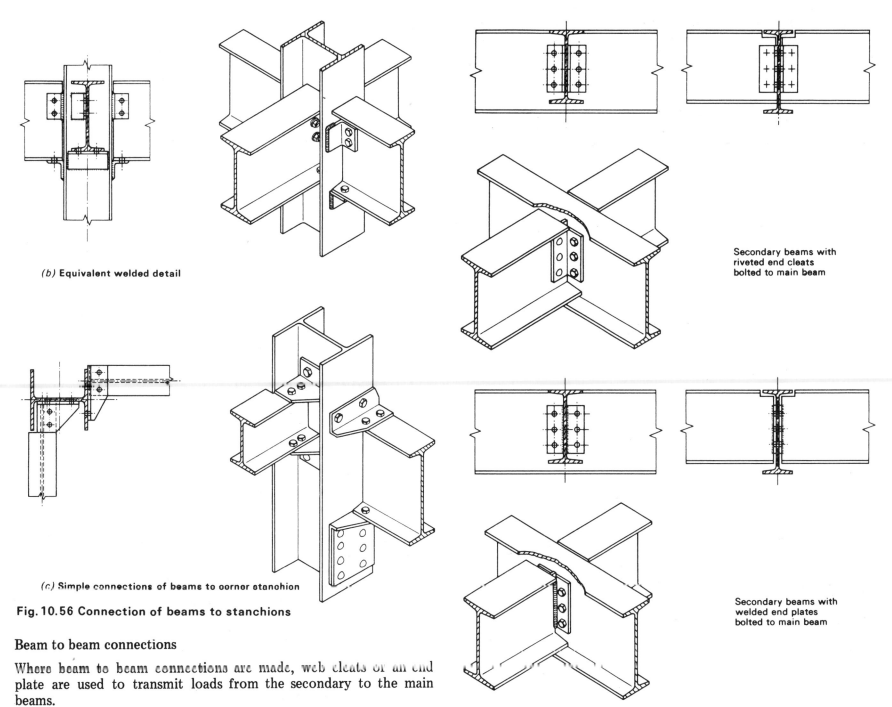

(b) Equivalent welded detail

(c) Simple connections of beams to corner stanchion

Fig. 10.56 Connection of beams to stanchions

Secondary beams with riveted end cleats bolted to main beam

Secondary beams with welded end plates bolted to main beam

Fig. 10.57 Beam to beam connections

Beam to beam connections

Where beam to beam connections are made, web cleats or an end plate are used to transmit loads from the secondary to the main beams.

Figure 10.57 illustrates this type of connection.

267

Trusses and other lattice frames

Roof trusses are plane frames consisting of sloping rafters which meet at the ridge. A main tie connects the feet of the rafters, and the internal bracing members. They are used to support the roof covering in conjunction with purlins. Purlins are secondary members laid longitudinally across the rafters to which the roof covering is attached.

Figure 10.58 gives details of both riveted and welded roof trusses.

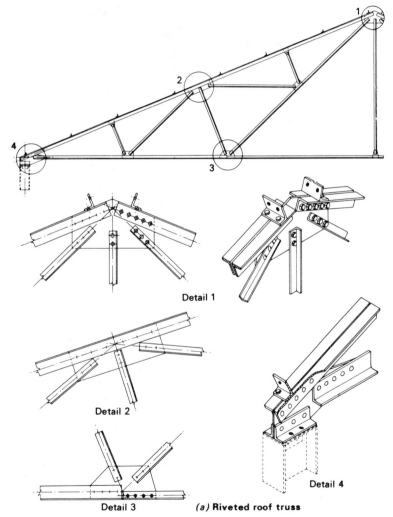

*The **gusset plate** are designed to withstand the forces applied by the members which they connect together*

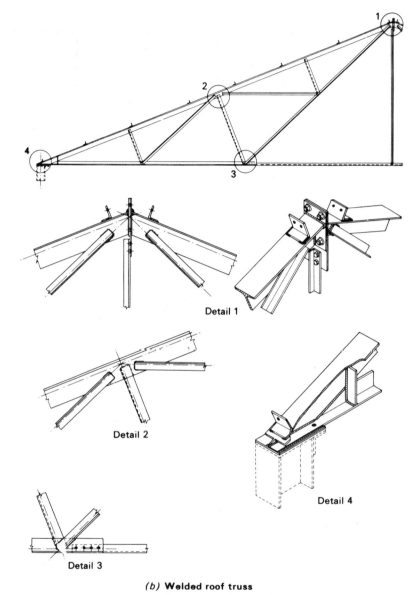

Fig. 10.58 Roof trusses

Lattice girders, also sometimes called 'trusses', are plane frames of open-web construction. They usually have parallel chords or booms which are connected with internal web bracing members. There are two basic types of lattice girder, the 'N' type and the 'Warren' type as shown in Fig. 10.59.

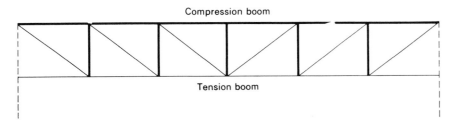

The 'N-type' lattice girder

The diagonal bracing members are arranged so that they act as ties. If reversed they would become struts and the shorter vertical members would be ties

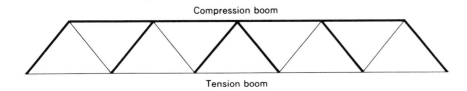

The 'Warren-type' lattice girder

Note: The thick lines in the diagrams represent 'struts'

Fig. 10.59 Lattice girders

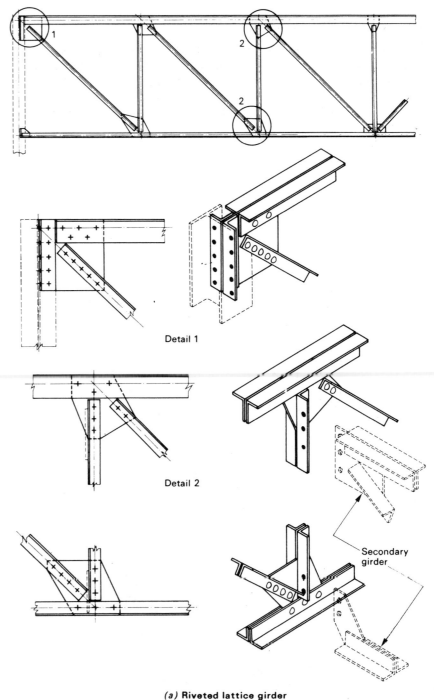

(a) Riveted lattice girder

As with roof trusses, the framing of a lattice girder should be triangulated, taking into account the span and the spacings of the applied loads. The booms are divided into panels of equal length and, as far as possible, the panel points are arranged to coincide with the applied loads. Lattice girders may be used in flat roof construction or in conjunction with trusses or trussed rafters.

Because of the greater forces usually borne by members of lattice girders, shop connections are generally riveted or welded. Site connections are normally made with HIGH-STRENGTH FRICTION-GRIP BOLTS (which have largely replaced rivets for site work) or by welding. Figure 10.60 shows how lattice girders are fabricated from standard sections.

10.34 Web stiffeners

Web stiffeners are required when a beam or plated structure is subjected to a twisting force (TORSION) or a sideways thrust.

The need for web stiffeners, or gussets, increases as the depth of the beam increases, as shown in Fig. 10.61.

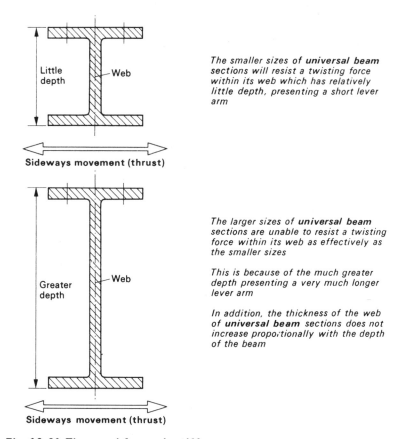

*The smaller sizes of **universal beam** sections will resist a twisting force within its web which has relatively little depth, presenting a short lever arm*

*The larger sizes of **universal beam** sections are unable to resist a twisting force within its web as effectively as the smaller sizes*

This is because of the much greater depth presenting a very much longer lever arm

*In addition, the thickness of the web of **universal beam** sections does not increase proportionally with the depth of the beam*

Fig. 10.61 The need for web stiffeners

Web stiffeners may be welded or riveted. When fabricating stiffeners which are to be welded in position, it is important that the stiffener is an exact fit on the beam. The slope of the tapered flanges should be copied faithfully. With triangular-shaped gussets 'feather edges' must be avoided, the sharp corners should be cut off, for otherwise the strength of the assembly may be reduced rather than increased. Another important reason for cutting off the corners of webs or gussets is to provide ample clearance from fillet welds or bending radii. This allows for ease of assembly and permits welding

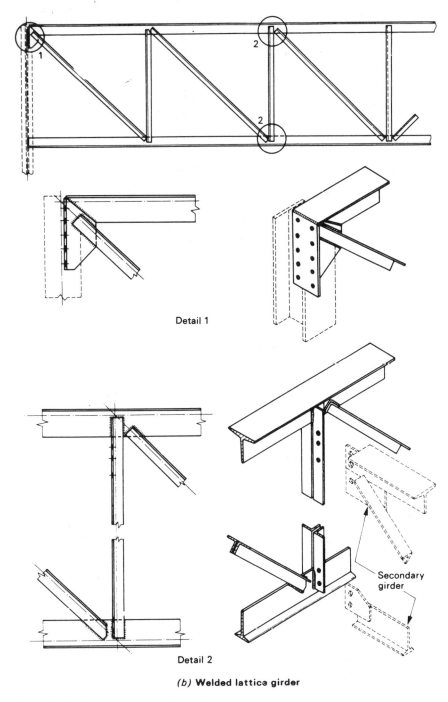

(b) Welded lattice girder

Fig. 10.60 Construction of lattice girders

through the gap with the web, or gusset, in position. It also avoids costly fitting operations at the corners.

Various applications of web stiffeners and gussets are illustrated in Fig. 10.62 and 10.63.

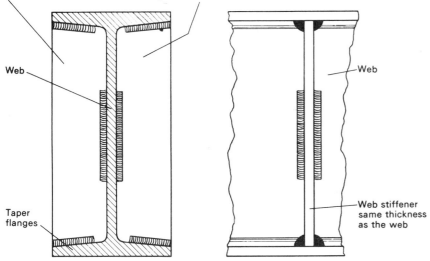

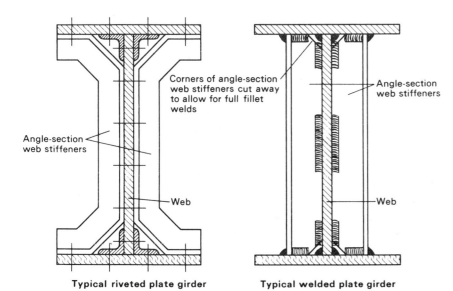

Fig. 10.62 Web stiffeners

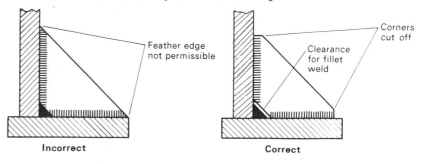

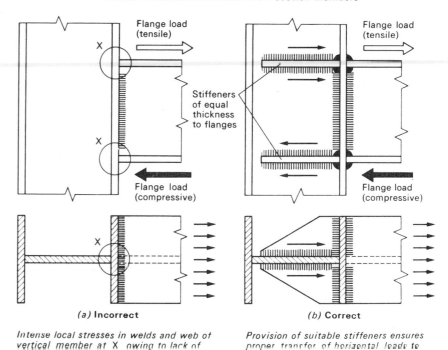

Fig. 10.63 Applications of stiffeners

11 Welding

It is important to set the pressure reducing valve correctly!

PART A. OXY-ACETYLENE WELDING

The principles of *gas welding* have already been introduced in *Basic Engineering*, sections **10.27** and **10.28**. These basic principles will now be examined in greater depth.

11.1 Equipment

In this chapter reference will only be made to the components of 'HIGH PRESSURE' oxy-acetylene welding systems.

The assembled, basic welding equipment is illustrated in Fig. 11.1.

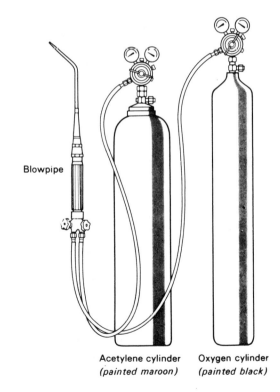

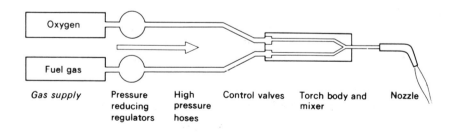

Fig 11.1 Basic gas-welding equipment

11.2 Oxygen and acetylene cylinders

The differences between an oxygen cylinder and an acetylene cylinder are clearly illustrated in Fig. 11.2.

Oxygen

This is supplied to the welding torch from a solid drawn steel cylinder where it is contained in compressed form. Oxygen cylinders are usually supplied in capacities of 3.4 m^3, 5 m^3, and 6.8 m^3. Mild steel cylinders are charged to a pressure of $13\,660 \text{ kN/m}^2$ (136.6 bar) and alloy steel cylinders to $17\,240 \text{ kN/m}^2$ (172 bar).

The oxygen volume in a cylinder is directly proportional to its pressure, and the consumption for a welding job can, therefore, be found by noting the pressure drop during the welding operation. For example, if it were noted that the original pressure of a full oxygen cylinder had dropped 5% during a welding operation, then 1/20 of the cylinder contents would have been consumed.

THE VALVE OUTLET ON AN OXYGEN CYLINDER HAS A RIGHT-HAND SCREW THREAD.

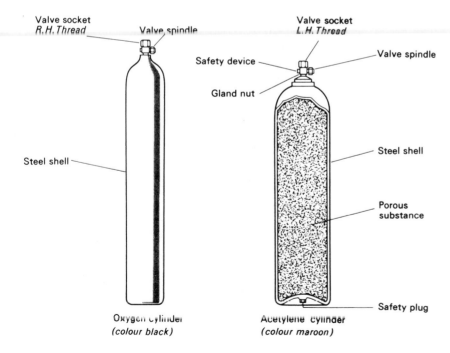

Fig. 11.2 Acetylene and oxygen cylinders

Acetylene

For high pressure welding is supplied in a solid drawn steel cylinders as shown in Fig. 11.2. *High pressure acetylene is not stable and for this reason it is dissolved in* **acetone**, *which has the ability to absorb a large volume of the gas and release it as the pressure falls.* One volume of acetone at atmospheric pressure and at a temperature of 15°C is capable of dissolving about 25 volumes of acetylene. *This dissolving capacity can be increased in proportion to the pressure.* Acetylene cylinders are charged to a pressure of $1\,552 \text{ kN/m}^2$ (15.5 bar), and as one atmosphere is equivalent to one bar it follows that, at the same temperature, one volume of acetone absorbs approximately $25 \times 15 = 375$ times its own volume of acetylene at a pressure of 15 bar.

Because of the danger of explosions to which compressed acetylene is susceptible, the steel cylinder is filled with a porous substance, and the construction of the cylinder, its filling and testing, are all strictly controlled by the manufacturers in the interests of safety. The pores in the filling material divide the space into a large number of very small compartments which are completely filled with 'dissolved acetylene'. *These small compartments prevent the sudden decomposition of the acetylene throughout the mass, should it be started by local heating or other causes.*

THE VALVE OUTLET ON ACETYLENE CYLINDERS IS FITTED WITH A LEFT-HAND SCREW THREAD.

11.3 Discharge rate

Should OXYGEN be withdrawn from a cylinder at too great a rate of consumption, *a rapid drop in pressure will occur, with the result that the cylinder valve may freeze.* When flame cutting heavy cast-iron sections, which involves a high rate of gas consumption, it is advisable to couple together sufficient oxygen cylinders to complete the work — this also applies when repairing large castings which have to be preheated.

The rate of acetylene consumption also has to be kept below a certain limit. Acetylene cylinders should never be discharged at a rate which will empty them in less than 5 hours. This means that *the discharge rate must be limited to less than 20% of the total cylinder content per hour.* This is not because of freezing, as with oxygen, but because should this limit be exceeded acetone will be drawn off and mixed with the acetylene. Acetylene which contains small quantities of acetone vapour will lower the flame temperature. Unlike oxygen, the volume and pressure of acetylene are not in linear proportion which means that the gas consumption cannot be reliably found from the loss in pressure in the cylinder during a welding operation.

The sizes of acetylene cylinders in general use are 2·8 m³ and 5·7 m³.

In workshops, where welding gases are needed in several places, or at high rates of consumption, it is of considerable advantage to use a manifold system. Instead of having cylinders at each place of work, they are assembled in one centralised position in specially designed racks and connected by a manifold, as shown in Fig. 11.3. These gases are then distributed by means of a pipeline to the different work-places.

ADVANTAGES OF MANIFOLDS:

More space available at work-place.
No replacement of cylinders inside the workshop.
Less transportation.
More effective use of the gas.
The cylinders are easily reached in case of fire.

11.4 Automatic pressure regulators

These are fitted to the oxygen and acetylene cylinders to reduce the pressure and control the flow of the welding gases. Examples are shown in Fig. 11.4. They are fitted with two pressure gauges. One indicating the gas pressure in the cylinder, and the other indicating the reduced outlet pressure. The operation of the pressure gauge is explained in Section 11.5.

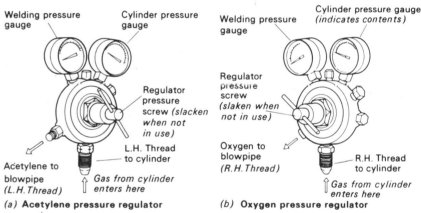

(a) Acetylene pressure regulator (b) Oxygen pressure regulator

Fig.11.3 Pressure regulators

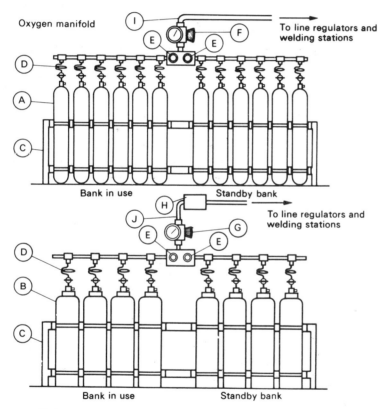

A Oxygen cylinders
B Acetylene cylinders
C Storage rack
D High-pressure coupling pipe ('pig-tail')
E Separate valve for each bank
F Oxygen output regulator and pressure gauge (Line pressure 4.15 bar)
G Acetylene output regulator and (Line pressure 620 millibar)
H Anti-Flashback device (acetylene)
I Oxygen supply line (copper pipe)
J Acetylene supply line (steel pipe)

Fig.11.3 The manifold system for gas welding

The four principal elements which constitute a pressure-reducing regulator are:
1. A *valve* consisting of a nozzle and a mating seat member.
2. An *adjustable screw* which controls the thrust of the cover spring.
3. A *cover spring* which transmits to a diaphragm the thrust created by the adjusting screw.
4. A *diaphragm* connected with the mating seat member.

Figure 11.5 illustrates two basic types of 'single-stage' regulator.

In the 'needle type' regulator the inlet pressure tends to close the seat member against the nozzle. The outlet pressure on this type of regulator has a tendency to increase somewhat as the inlet pressure decreases. *This increase is caused by a decrease of the force produced by the gas pressure against the seating area as the inlet pressure decreases.* This type of single-stage regulator is sometimes referred to as 'inverse' or 'negative' type.

In the *'nozzle type' regulator* the inlet pressure tends to move the seat member away from the nozzle, therefore the outlet pressure decreases somewhat as the inlet pressure decreases. *This is because the force tending to move the seat member away from the nozzle is reduced as the inlet pressure decreases.* This type of single-stage regulator is referred to as 'direct acting' or 'positive' type.

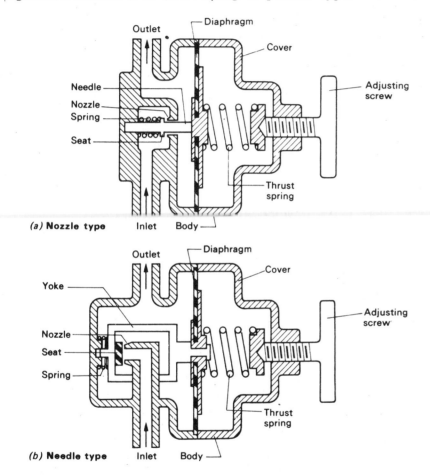

Fig. 11.5 Single-stage regulators

THE GAS OUTLET PRESSURE FOR ANY PARTICULAR SETTING OF THE ADJUSTING SCREW IS REGULATED BY A BALANCE OF FORCES.

This balance is between the cover spring thrust and the opposing forces created by a combination of the outlet pressure against the underside of the diaphragm and the inlet pressure against the seating area.

When the inlet pressure decreases, its force against the seat member decreases, allowing the cover spring force to move the seat member away from the nozzle. Thus more gas pressure is allowed to build up to re-establish the balanced condition.

A smaller outlet pressure on the underside of the diaphragm is all that is necessary to close the seat member against the nozzle. The opening between the seat members and the nozzle is reduced, which results in less gas flow.

11.5 The pressure gauge

Inside a pressure gauge there is a BOURDON TUBE. This is a copper-alloy tube of oval section, bent in a circular arc. One end of the tube is sealed shut and attached by light linkage to a mechanism which operates a pointer. The other end is fixed, and is open for the application of the pressure which is to be measured. The internal pressure tends to change the section of the tube from oval to circular and this causes it to straighten out slightly. The resultant movement of the tube causes the pointer to move over a suitably calibrated scale. An example of a bourdon tube pressure gauge is shown in Fig. 11.6.

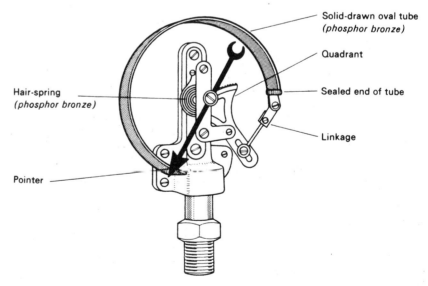

Fig. 11.4 Bourdon tube pressure gauge

11.6 Welding torches

The gases having been reduced in pressure by the gas regulators are fed through suitable hoses to a welding torch.

The WELDING TORCH *is a specially designed piece of equipment used for mixing and controlling the flow of gases to the* WELDING NOZZLE *or* TIP. The torch provides a means of holding and directing the welding nozzle.

The basic elements of a high-pressure welding torch are shown in the simplified drawing, Fig. 11.7.

The fuel gas hose fitting on all welding torches has a left-hand thread, making it possible only to screw on the left-hand grooved nuts used on fuel gas hose. The other fitting, used for oxygen, has a right-hand thread.

There are two CONTROL VALVES usually positioned at the rear end of the torch.

After passing the valves, the gases flow through metal tubes inside the handle and are brought together by the GAS MIXER at the front end.

The NOZZLE is shown as a simple tube tapered down at the outlet end to produce a suitable welding cone.

This type of torch is provided with sealing rings in the torch head, or in the mixer seats to facilitate a hand-tight assembly. These rings are normally made of natural rubber or synthetic materials.

An essential feature of the high pressure welding and cutting torch is the mixing chamber. This ensures that the oxygen and acetylene are thoroughly mixed before they enter the nozzle. Figure 11.8 shows a typical gas mixing chamber.

The two gases, controlled with respect to volumetric rate by needle valves are fed in at points marked A (FUEL GAS) and B (OXYGEN).

The mixing of the gases commences at point C, and continues throughout the chamber (as indicated by the small arrows) and forward to the welding nozzle where it is ignited.

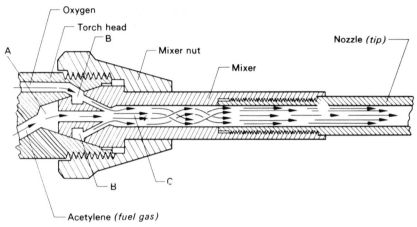

Fig.11.6 Mixing chamber for a welding torch

A well designed gas mixer will perform the following essential functions:
1. Mix gases thoroughly for proper combustion.
2. Arrest 'flashbacks' which may occur as a result of improper operation or welding procedure.
3. Stop any flame from travelling back farther than the mixer.
4. Permit a wide range of nozzle sizes to be used with one particular size of mixer.

11.7 Welding nozzles

The welding nozzle or tip is that portion of the torch through which the gases pass prior to their ignition and combustion. A welding nozzle enables the operator to guide the flame and direct it with the maximum ease and efficiency.

Nozzles are made from a NON-FERROUS metal such as COPPER or a COPPER ALLOY. *These materials possess a* HIGH THERMAL

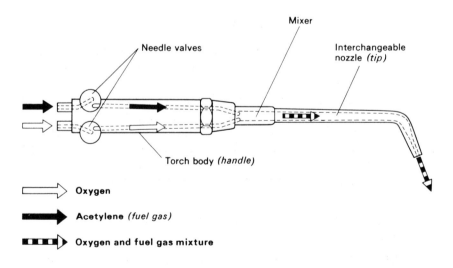

Fig.11.5 High-pressure welding torch

CONDUCTIVITY *and their use greatly reduces the danger of burning the nozzle at high temperatures.*

Welding nozzles are available in a variety of sizes, shapes, and construction. However, there are two main classes of *nozzle and mixer combinations* in general use:
1. A SEPARATE NOZZLE AND MIXER UNIT — each size of nozzle has the proper mixer which provides efficient combustion of the gases and optimum heating efficiency.
2. ONE OR MORE MIXERS USED FOR THE ENTIRE RANGE OF NOZZLE SIZES — The nozzle is screwed into the appropriate mixer, and each size of mixer has a particular thread size to ensure the correct combination of the proper nozzle and mixer.

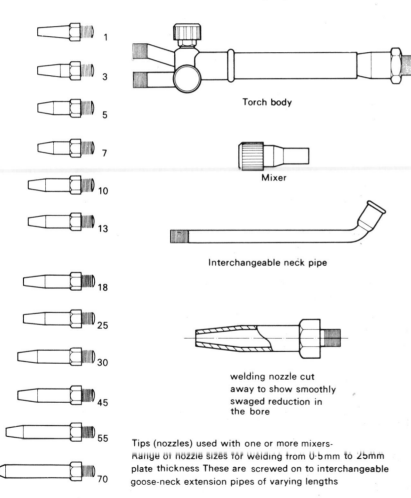

Fig. 11.9 Welding torch and nozzle/mixer combinations

Figure 11.9 shows a welding torch combination in this category. In this case use is made of interchangeable 'goose-neck' extensions which are screwed into the mixer portion of the torch body. A wide range of nozzles or tips may be fitted to these extensions.

HEAT RADIATION *of the flame produced with the larger sizes of nozzles is great, therefore it is advisable to have the torch body at a comfortable distance from the flame.* In this respect, nozzle mixer units or interchangeable neck pipes will vary in length according to the thickness of plate being welded.

Care and maintenance of nozzles

Since nozzles are made from materials which are relatively soft, care must be taken to guard them against damage. The following simple precautions are recommended:
1. Make sure that the nozzle seat and threads are absolutely free from foreign matter in order to prevent any scoring when tightening on assembly.
2. Nozzles should only be cleaned with tip cleaners, which are specially designed for this purpose. A special *nozzle cleaning compound* is available for dirty nozzles. The correct strength of cleaning solution is obtained by using approximately 50 grammes of compound to 1 litre of water. Dirty nozzles should be immersed in this solution for a period of *at least* two hours.
3. Nozzles should never be used for moving or holding the work.

11.8 Gas welding hose

For welding and cutting, hose should be used which is specially manufactured for the purpose. *Other types of hose may cause considerable gas losses and accidents.*

Welding hose has a seamless lining which is manufactured from rubber (or a rubber compound) which is reinforced with canvas or wrapped cotton plies. It is resistant to the action of gases normally used in welding and cutting. The outer casing is made of tough abrasion-resistant rubber. The hose is very robust and capable of withstanding high pressure; it is available in two colours, black and red.

BLACK HOSE is used for OXYGEN and other 'non-combustible' gases.

RED HOSE is used for ACETYLENE and other 'combustible' or fuel gases.

Table 11.1 lists the sizes of welding hose in general use:

Table 11.1 Sizes of welding hose in general use

INTERNAL DIAMETER		APPROXIMATE OUTSIDE DIAMETER		USED FOR OXYGEN AND ACETYLENE SUPPLIES WITH:
mm	(in)	mm	(in)	
4·8	(3/16)	13·5	(17/32)	Light-duty high-pressure torches
6·3	(1/4)	15·1	(19/32)	
8·0	(5/16)	16·7	(21/32)	Heavy-duty high-pressure torches
9·5	(3/8)	18·3	(23/32)	Large cutting torches and where *low pressure generated acetylene* is used

Care must be taken not to damage welding hose by dropping heavy weights where it crosses the workshop floor, playing the torch flame on to it, or allowing it to come in contact with hot or molten metal.

Hose clips For all welding and cutting operations the hose 'union nipples and nuts' must be secured with the aid of reliable hose clips. Standard lengths of hose are available on which the coupling nipples are permanently attached by means of special clips, as shown in Fig. 11.10.

Hose couplers These are used when joining welding hoses of equal or unequal sizes (Fig. 11.10).

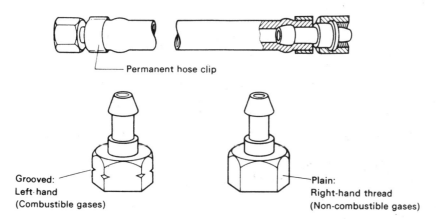

Union nuts and nipples

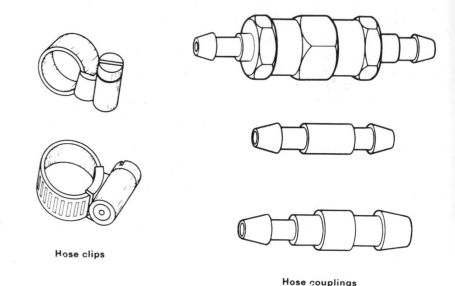

Fig.11.10 Welding hose fittings

11.9 Miscellaneous accessories (equipment)

Safety goggles and protective clothing have been fully described in Chapter 1.

In addition to the equipment previously described there are numerous pieces of auxiliary equipment which may be used in gas welding and cutting processes. For example, cylinder trucks, jigs, and fixtures, devices for introducing liquid flux into the acetylene stream for brazing operations, tip cleaners, friction lighters, and gas economisers for automatically shutting off the torch flame. Of these only the last two items will be described in this chapter.

1. *Gas lighters* A spark lighter is a very useful accessory which provides a convenient and inexpensive means of lighting the torch. *From a safety point of view, it is advisable to use such friction lighters rather than matches to ignite the gas.*

2. *Gas economiser* This unit provides an efficient means of conserving gas to the user who frequently needs to interrupt the welding or cutting operation. The unit may be mounted on a light metal stand, or fastened to the welding table, in some convenient position.

The basic elements and function of a gas economiser is shown in Fig. 11.11.

Using a gas economiser
The operator does not have to re-adjust the flame each time he lights the torch

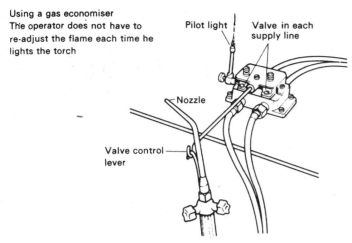

Fig. 11.11 Gas economiser unit

The two lengths of hose from the regulators are attached to the inlet connections. The outlets are connected to the torch with additional lengths of hose. The torch is ignited, and the flame set in the normal manner, but at any interruption of the welding operation, or after completing a weld, it is simply hooked to the lever arm control of the economiser.

The weight of the torch depresses the lever, causing the valve to operate and cut off the supply of gas to the torch and the flame is extinguished. When the torch is required again, it is lifted from the lever arm and can be ignited by a small pilot flame on the economiser.

11.10 Structure of the oxy-acetylene flame

The welding flame is produced by burning approximately equal volumes of the oxygen and acetylene which are supplied to the torch. Figure 11.12 illustrates the basic structure of the oxy-acetylene welding flame.

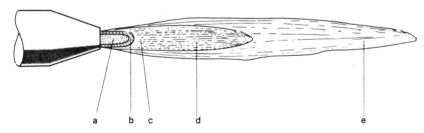

Fig. 11.12 The structure of the oxy-acetylene welding flame

The oxy-acetylene flame, like most oxy-gas flames used for welding, is characterised by the following zones:

(a) The innermost cone of MIXED UNBURNT GASES leaving the nozzle. It appears intensely white and clearly defined with the correct NEUTRAL setting. *Any small excess of ACETYLENE is indicated by a white 'feather' around the edge of this cone. With excess OXYGEN the cone becomes shorter, more pointed, and tinged with blue.*

(b) A very narrow stationary zone wherein the chemical reaction of the FIRST STAGE OF COMBUSTION takes place *producing a sudden rise in temperature.*

(c) The REDUCING ZONE, appearing dark to light blue, in which the PRIMARY COMBUSTION PRODUCTS are concentrated. *The nature of these products determines the* CHEMICAL NATURE *of the flame, i.e., whether 'NEUTRAL', 'OXIDISING' or 'CARBURISING'.*

(d) Within the region of PRIMARY COMBUSTION and approximately 3 mm from the tip of the nicely defined cone of unburnt gases there is a REGION OF MAXIMUM TEMPERATURE, this is the zone used for welding.

(e) The yellow to pinkish outer zone or 'plume' around and beyond the previous zones represents the *chemical reaction* of the SECOND STAGE OF COMBUSTION. *Here the two* COMBUSTIBLE *gases* CARBON MONOXIDE *and* HYDROGEN, *which are the products of primary combustion, combine with* OXYGEN FROM THE SURROUNDING ATMOSPHERE. This part of the flame is always OXIDISING and contains large amounts of NITROGEN *The presence of a high percentage of nitrogen in the* OXIDISING OUTER ZONE *is not surprising because the* ATMOSPHERE *contains approximately 80% NITROGEN and 20% OXYGEN by volume.*

For the oxy-acetylene welding flame the CHEMICAL REACTIONS can be summarised as follows:

Primary combustion Produces the REDUCING ZONE consisting of approximately 60 per cent CARBON MONOXIDE and 40 per cent HYDROGEN:

ACETYLENE + OXYGEN = CARBON MONOXIDE + HYDROGEN

Secondary combustion The products of the primary reaction *react with* OXYGEN *in the surrounding air* producing the outer envelope or plume of the flame:

1. CARBON MONOXIDE + AIR (oxygen & nitrogen) = CARBON

DIOXIDE + NITROGEN

2. HYDROGEN + AIR (oxygen & nitrogen) = WATER VAPOUR + NITROGEN

Thus the products of complete combustion can be said to be CARBON DIOXIDE *and* WATER VAPOUR.

With the oxy-acetylene welding it is essential to maintain a NEUTRAL FLAME for, except in special cases, an OXIDISING FLAME or a CARBURISING FLAME will result in welds with unsatisfactory mechanical properties. *The presence of even the smallest quantity of an* OXIDISING AGENT, *such as* CARBON DIOXIDE, WATER VAPOUR, *or* OXYGEN, *will rapidly destroy the* REDUCING PROPERTIES *of the flame.*

11.11 The three oxy-acetylene flame conditions

The correct type of flame is essential for the production of satisfactory welds, and the characteristics of the three flames are shown in Fig. 11.13.

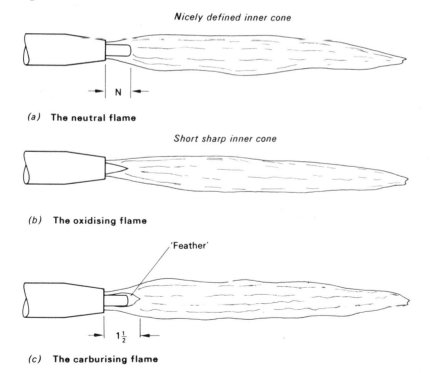

(a) The neutral flame

(b) The oxidising flame

(c) The carburising flame

Fig. 11.13 Oxy-acetylene welding flame conditions

1. *The neutral flame* For most applications the *neutral flame* condition is used, as shown in Fig. 11.13(*a*). This is produced when approximately equal volumes of oxygen and acetylene are mixed in the welding torch. It is termed NEUTRAL because it effects no chemical change on the molten metal and therefore will not oxidise or carburise the metal.

   It is easily recognised by its characteristic clearly defined white inner cone at the tip of the torch nozzle. Correct adjustment is indicated by a slight white flicker (feather) on the end of this cone resulting from a slight excess of acetylene. It is less desirable, when setting this condition, to have a slightly oxidising flame than one which is reducing.

   The neutral flame setting is most commonly used for the welding of mild steel, stainless steels, cast iron, and copper.

2. *Oxidising flame* When an *oxidising flame* is required, as shown in Fig. 11.13(*b*), the flame is first set to the neutral condition and the acetylene supply is slightly *reduced* by the control valve on the welding torch.

   This flame can be easily recognised by the inner cone of the flame which is shorter and more pointed than that of the neutral flame.

   The oxidising flame is *undesirable* in most cases as it oxidises the molten metal, although this can be an advantage with some brasses and bronzes. The oxidising flame gives the highest possible temperatures providing the oxygen : acetylene ratio does not exceed 1·5:1.

3. *Reducing flame* When a *reducing flame* is required, as shown in Fig. 11.13(*c*), the flame is first set to the neutral condition and the acetylene supply is slightly *increased* by the control valve on the welding torch.

   This flame is recognised by the 'feather' of incandescent carbon particles between the inner cone and outer envelope.

   The reducing flame is used for carburising (surface hardening) and for the 'flame brazing' of aluminium.

11.12 Nozzle size

For a given welding torch, the NOZZLE OUTLET SIZE has a much greater influence on governing the flame size than changing the gas pressures or adjusting the control valves.

The manufacturers of gas welding equipment have adopted various methods of indicating nozzle sizes, such as:

1. By the approximate consumption of each gas per hour.

2. By the nozzle outlet bore size (orifice diameter).
3. By a reference number corresponding to a metal thickness range which may be welded with a specific nozzle.

Whatever the method employed for indicating nozzle sizes *there is a definite relationship between the sizes of welding nozzles and the metal thicknesses.* Figure 11.14 indicates this relationship where the thickness of steel plate is plotted against the orifice diameter of the welding nozzle.

Manufacturer's recommendations should always be followed with regard to nozzle sizes and gas pressures for a particular application.

Table 11.2 shows the nozzle sizes, gas consumption, and working pressures required for welding various thicknesses of mild steel using a neutral flame. The values given are for butt welds made in the downhand position.

Table 11.2 Downhand butt welds in steel

	THICKNESS OF METAL	DIAMETER OF FILLER ROD mm (in)	JOINT EDGE PREPARATION	PLATE THICKNESS	NOZZLE SIZE	APPROXIMATE CONSUMPTION OF EACH GAS:	
						Litres per hour	Cubic Feet per hour
LEFTWARD WELDING TECHNIQUE	Less than 20 swg	1·2 (3/64) to 1·6 (1/16) for square each preparation	No filler rod	0·8	1	28	1
				1·2	2	57	2
				1·6	3	86	3
	20 swg to 3·2 mm (1/8 in)	1·6 (1/16) to 3·2 (1/8)	0·8–3·2 mm Gap	2·4	5	140	5
				3·2	7	200	7
	3·2 mm to 5·0 mm (1/8 in to 3/16 in)	3·2 (1/8) to 4·0 (5/32)	80° 1·6–3·2 mm Gap	4·0	10	280	10
				5·0	13	370	13
RIGHTWARD WELDING TECHNIQUE	5·0 mm to 8·2 mm (3/16 in to 5/16 in)	3·2 (1/8) to 4·0 (5/32)	3·2–4·0 mm Gap	6·5	18	520	18
				8·2	25	710	25
	8·2 mm to 16·2 mm (5/16 in to 5/8 in)	4·0 (5/32) to 6·5 (1/4)	60° 3·2–4·0 mm Gap	10·0	35	1000	35
				13·0	45	1300	45
				16·2	55	1600	55
	16·2 mm and over	6·5 (1/4)	60° 80° 3·2–4·0 mm Gap	19·0	60	1700	60
				25·0	70	2000	70
				Over 25·0	90	2500	90

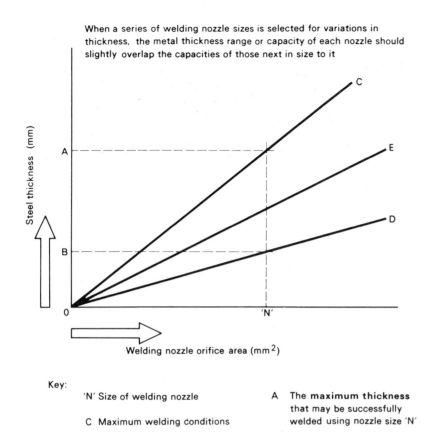

When a series of welding nozzle sizes is selected for variations in thickness, the metal thickness range or capacity of each nozzle should slightly overlap the capacities of those next in size to it

Key:

'N' Size of welding nozzle

C Maximum welding conditions

D Minimum welding conditions

E Optimum welding conditions

A The **maximum thickness** that may be successfully welded using nozzle size 'N'

B The **minimum thickness** that may be successfully welded using nozzle size 'N'

Fig. 11.14 Relationships between nozzle size and metal thickness

11.13 Gas velocities

Flame cones are produced by the velocity gradient which exists across the circular orifice of the nozzle when the high pressure gases are flowing. Since the gas velocity is greatest in the centre of the stream it follows that the flame length at the centre is similarly the greatest. Likewise, since the gas velocity is lowest at the periphery of the nozzle orifice the flame around the edge is shortest. This is because *friction* to the gas stream is greatest near the walls of the bore. The velocity of the gas increases towards the centre of the stream producing a nicely defined flame cone, as illustrated in Fig. 11.15(*a*). The effect of nozzle size on the cone profile is shown in Fig. 11.15(*b*).

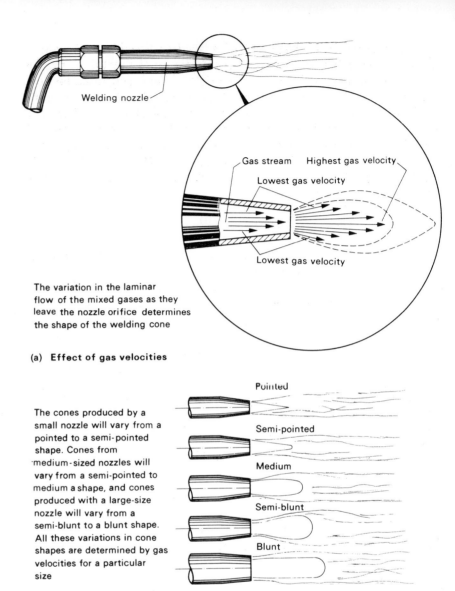

The variation in the laminar flow of the mixed gases as they leave the nozzle orifice determines the shape of the welding cone

(a) **Effect of gas velocities**

The cones produced by a small nozzle will vary from a pointed to a semi-pointed shape. Cones from medium-sized nozzles will vary from a semi-pointed to medium a shape, and cones produced with a large-size nozzle will vary from a semi-blunt to a blunt shape. All these variations in cone shapes are determined by gas velocities for a particular size

(b) **Welding flame-cone shapes**

Fig. 11.15 Factors affecting the welding flame-cone profile

In practice, the efficiency of a welding nozzle can be determined by observing the action of the flame on the metal. If the gas velocity is too high a violent flame results (easily recognised because it burns with a harsh hissing sound) which has the effect of tending to blow the metal out of the weld pool. A harsh flame condition can readily be corrected by simply reducing the volumetric rates of oxygen and acetylene to a point where a satisfactory weld can be performed. THIS POINT WOULD INDICATE THE MAXIMUM VOLUMETRIC RATE WHICH WOULD GIVE OPTIMUM EFFICIENCY FOR A SPECIFIC SIZE OF NOZZLE.

11.14 Leftward welding

In this method of gas welding the flame is directed away from the finished weld, i.e. towards the unwelded part of the joint. Filler rod, when used, is directed towards the welded part of the joint.

Although this technique is termed 'leftward welding', it is not confined to right-handed operators. Normally, with right-handed persons, the welding torch is held in the right hand and welding proceeds from right to left, i.e., LEFTWARDS. With a left-handed operator the torch is held in the left hand and welding proceeds from left to right.

Leftward welding is used on normal low carbon steels for the following:
(a) Flanged edge welds for thicknesses less than 20 s.w.g. These welds are made without the addition of filler rod.
(b) Square-butt welds on unbevelled steel plates up to 3·2 mm thicknesses.
(c) Vee-butt welds on bevelled steel plates over 3·2 mm and up to 5 mm thickness.

The leftward method of welding is not considered to be economical for thicknesses above 5 mm thickness for reasons which will be explained later.

The angles of the torch and filler rod are clearly illustrated in Fig. 11.16.

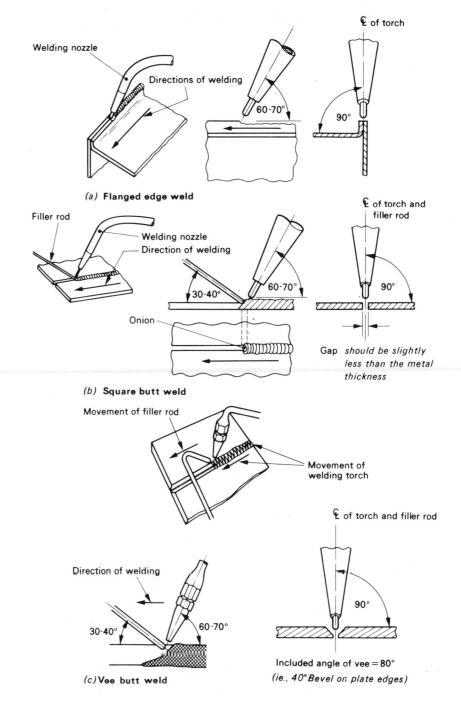

(a) Flanged edge weld

(b) Square butt weld

Gap *should be slightly less than the metal thickness*

(c) Vee butt weld

Included angle of vee = 80°
(*ie., 40° Bevel on plate edges*)

Flanged edge weld (Figure 11.16(a)) No filler rod required. The flame should be manipulated with steady semi-circular sideways movement and progressively forwards only as fast as the edges of the sheet metal are melted. *The tip of the flame cone must be kept about 3 mm from the weld pool.*

Square butt weld (Figure 11.16(b)) As the filler rod is melted it should be fed forward to build up the molten pool of weld metal and then retracted slightly. The flame is guided progressively forwards with small sideways movements as fast as the plate edges are melted.

A pear-shaped melted area (often referred to as an 'onion') should be maintained ahead of the weld pool. Care must be exercised to ensure uniform melting of both plate edges.

Vee butt weld (Figure 11.16(c)) The tip of the flame cone should never touch the weld metal or the filler rod.

As the filler rod is melted it should be fed forward into the molten weld pool in order to build it up. The rod is then retracted slightly to enable the heat to fuse the bottom edges of the 'Vee'. *This ensures full penetration of the weld to the bottom of the plate edges.*

The side-to-side movement of the welding torch should only be sufficient to melt the sides of the 'Vee'.

This weld deposit is built up as the torch and filler rod move progressively forwards filling the 'Vee' to a level slightly higher than the edges of the plate. Table 11.3 gives the data for leftward welding.

Table 11.3 Welding speeds and data for leftward welding

THICKNESS OF METAL		EDGE PREPARATION	GAP		DIAMETER OF FILLER ROD		POWER OF TORCH		RATE OF WELDING		FILLER ROD USED	
							(Gas consumption)				per metre	per foot
mm	(in)		mm	(in)	mm	(in)	Litres/hr	Cubic ft/hr	Metres/hr	Feet/hr	(m)	(ft)
0·8	1/32	Square	—	—	1·6	1/16	28 — 57	1 — 2	6·0—7·6	20—25	0·3	1
1·6	1/16	Square	1·6	1/16	1·6	1/16	57 — 86	2 — 3	7·6—9·0	25—30	0·53	1·75
2·4	3/32	Square	2·4	3/32	1·6	1/16	86 140	3 — 5	6·0—7·6	20—25	0·84	2·75
3·2	1/8	Square	3·2	1/8	2·4	3/32	140—200	5 — 7	5·4—6·0	18—20	0·50	1·65
4·0	5/32	80° Vee	3·2	1/8	3·2	1/8	200—280	7 —10	4·6—5·4	15—18	0·64	2·1
5·0	3/16	80° Vee	3·2	1/8	4·0	5/32	280—370	10—13	3·6—4·6	12—15	1·40	4·8

11.15 Rightward welding

With this technique the weld is commenced at the lefthand end of the joint and the welding torch moves towards the right. The direction of welding is opposite to that when employing the leftward technique. The torch flame in this case is directed towards the metal being deposited, unlike the leftward method in which the flame is directed away from the deposited metal.

Rightward welding should be used for steel plates which exceed 5 mm thickness as follows:
(a) Square butt welds on unbevelled steel plates between 5 mm and 8·2 mm thickness (inclusive).
(b) Vee-butt welds on bevelled steel plates over 8·2 mm thickness. The angles of the torch and filler rod are shown in Fig. 11.17.

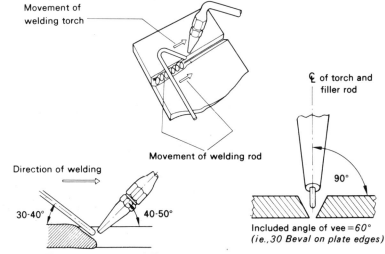

Rightward welding is sometimes termed 'backward' or 'back-hand' welding

Fig.11.17 Rightward welding

It is important that the flame cone is always kept just clear of the filler rod and the deposited weld metal.

The welding torch is moved steadily to the right along the joint.

By comparison with leftward welding [Fig. 11.16(c)] it will be noticed that the cone of the flame is deeper in the 'Vee'.

The filler rod is given an elliptical looping movement as it travels progressively to the right.

The filler rod and torch must be maintained in the same vertical plane as the weld, otherwise unequal fusion of the two sides of the weld will result. When a weld is completed, examination of the back should show an UNDER BEAD which should be perfectly straight and uniform. THE QUALITY OF THE WELD CAN BE JUDGED BY THE APPEARANCE OF THIS UNDERBEAD. Table 11.4 gives the data for rightward welding.

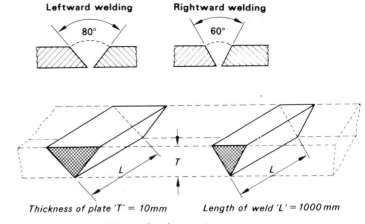

Fig.11.18 Amounts of deposited metal

Table 11.4 Welding speeds and data for rightward welding

THICKNESS OF METAL		EDGE PREPARATION	GAP		DIAMETER OF FILLER ROD		POWER OF TORCH (Gas consumption)		RATE OF WELDING		FILLER ROD USED	
mm	(in)		mm	(in)	mm	(in)	Litres/hr	Cubic ft/hr	Metres/hr	Feet/hr	per metre (m)	per foot (ft)
5·0	3/16	Square	2·4	3/32	2·4	3/32	370 — 520	13 — 18	3·6 — 4·6	12 — 15	1·03	3·4
6·5	1/4	Square	3·2	1/8	3·2	1/8	520 — 570	18 — 20	3·0 — 3·6	10 — 12	1·03	3·4
8·2	5/16	Square	4·0	5/32	4·0	5/32	710 — 860	25 — 30	2·1 — 2·4	7 — 8	1·03	3·4
10·0	3/8	60° Vee	3·2	1/8	5·0	3/16	1000 — 1300	35 — 45	1·8 — 2·1	6 — 7	1·22	4·6
13·0	1/2	60° Vee	3·2	1/8	6·5	1/4	1300 — 1400	45 — 50	1·3 — 1·5	4·5 — 5	1·69	4·75
16·2	5/8	60° Vee	3·2	1/8	6·5	1/4	1600 — 1700	55 — 60	1·1 — 1·3	3·75 — 4·25	2·05	6·75
19·0	3/4	60° Vee (Top) 80° Vee (Bottom)	4·0	5/32	6·5	1/4	1700 — 2000	60 — 70	0·9 — 1·0	3 — 3·25	2·96	9·75
25·0	1	60° Vee (Top) 80° Vee (Bottom)	4·0	5/32	6·5	1/4	2000 — 2500	70 — 90	0·6 — 0·7	2 — 2·25	5·08	16·75

11.16 The advantages of rightward welding

1. No bevel is necessary for plates up to 8·2 mm thickness. This saves the cost of preparation and reduces the filler rod consumption.

2. When bevelling of the plate edges becomes necessary the included angle of the Vee need only be 60°, which needs less filler rod than would be required to fill the 80° Vee preparation for leftward welding, as illustrated in Fig. 11.18.

For simplicity consider a 1 metre Vee-butt weld, made without a gap on 10 mm thick steel plate:

For LEFTWARD WELDING the included angle of the Vee will be 80° (although in practice 90° is often used), whilst for RIGHTWARD WELDING the included angle will be 60°. *In each case the volume of metal deposited can be calculated by multiplying the cross-sectional area by the length.*

THE VOLUME OF WELD METAL DEPOSITED WILL BE:

Leftward
$L \times T \times T \tan 40°$
$1\,000 \times 10 \times 10 \times 0.8391 = \underline{83\,910\text{ mm}^3}$

Rightward
$L \times T \times T \tan 60°$
$1\,000 \times 10 \times 10 \times 0.5774 = \underline{57\,740\text{ mm}^3}$

The above results indicate that in the case of rightward welding the amount of filler rod is less than 69 per cent of that required for leftward welding.

3. Larger welding nozzles must be used which results in higher welding speeds. With the use of more powerful nozzles the force of the flame 'holds back' the molten metal in the weld pool and allows more metal to be deposited so that welds in plates up to 10 mm thickness can be completed in one pass.

4. The operator's view of the weld pool and the sides and bottom of the Vee is unobstructed, enabling him to control the molten metal. This ensures that full fusion of the plate edges, particularly of the bottom edges, is always maintained, and results in an adequate and continuous bead of penetration.

5. The quality and appearance of the weld is better than that obtained with the leftward technique. This is due, to a large extent, to the fact that the deposited metal is protected by the envelope of the flame which retards the rate of cooling.

6. Compared with leftward welding, the smaller total volume of deposited metal with rightward welding reduces shrinkage and distortion. The graph in Fig. 11.19 shows clearly how, by adopting the rightward technique, the heat of the flame is localised in the joint and is not allowed to spread across the plate. The blowpipe movement associated with leftward welding causes a greater spread of heat. The amount of distortion depends on the amount of heat put in, therefore it is important that this heat be confined as far as possible to the weld-seam itself.

The graph shows the maximum temperatures attained at various distances from welds made in 10 mm thick steel plate by the rightward and leftward techniques in the downhand position. A study of the graph will clearly indicate that in the case of RIGHTWARD WELDING:

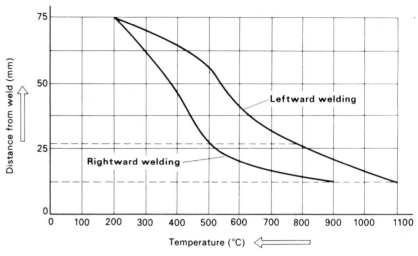

Fig. 11.19 Comparison of heat spread

(a) The temperature of the plate is considerably lower, even 12.5 mm from the weld seam, than the corresponding area on the plate which was leftward welded.

(b) The temperature falls away much more rapidly, so that at approximately 28 mm from the weld the plate is almost 300°C cooler than the corresponding area for leftward welding.

7. The cost of welding is lower than the leftward technique, despite the use of more powerful welding nozzles. The greatest economy is in the consumption of filler rods owing to the smaller amount of metal deposited with the rightward technique.

11.17 Comparison of welding techniques

A basic comparison of the leftward and rightward welding techniques has been made by illustrating butt welds in the flat or downhand position (Figs. 11.16 and 11.17). In order to make a much broader comparison, consideration will now be given to other types of weld made in the flat position.

Figure 11.20 illustrates the basic comparison of each welding

technique when applied to the following joints:
- (*a*) Lap joint
- (*b*) Tee fillet joint
- (*c*) Open corner joint
- (*d*) Closed corner joint.

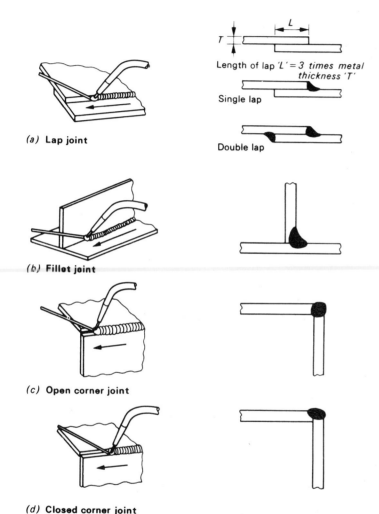

Fig.11.20 Types of joint

PART B. METAL-ARC WELDING

The principles of 'metal-arc' welding have already been introduced in *Basic Engineering*, section **10.28**. These basic principles will now be examined in greater depth. Examples of typical arc-welding plant have been illustrated in Chapter 1.

11.18 The metal-arc process

The arc is produced by a low-voltage, high-amperage electric current jumping an air gap between the electrode and the joint to be welded. The heat of the electric arc is concentrated on the edges of two pieces of metal to be joined. This causes the metal edges to melt. While these edges are still molten additional molten metal, transferred across the arc from a suitable electrode, is added. This molten mass of metal cools and solidifies into one solid piece.

As soon as the arc is struck, the tip of the electrode begins to melt, thus increasing the gap between electrode and work. Therefore it is necessary to cultivate a continuous downward movement with the electrode holder in order to maintain a constant arc length of 3 mm during the welding operation. The electrode is moved at a uniform rate along the joint to be welded, melting the metal as it moves.

11.19 The electrode

The greatest bulk of electrodes used with manual arc welding are *coated* electrodes. A coated electrode consists mainly of a core wire of closely controlled composition having a concentric covering of flux and/or other material, which will melt uniformly, with the core wire forming a partly vaporised and partly molten screen around the arc stream. This shield protects the arc from contamination by atmospheric gases.

The liquid slag produced performs three important functions:
1. Protects the solidifying weld metal from further contamination from the atmosphere.
2. Prevents rapid cooling of the weld metal.
3. Controls the contour of the completed weld.

11.20 The 'arc stream'

The function of an electrode is more than simply to carry the current to the arc. The core wire melts in the arc and tiny globules of molten metal shoot across the arc into the molten pool (arc crater in parent metal) during welding. These tiny globules are explosively forced through the arc stream. They are not transferred across the arc by the force of gravity, otherwise it would not be possible to use the manual arc welding process for overhead welding.

The chemical coating surrounding the core wire melts or burns in the arc. It melts at a slightly higher temperature than the metal core and therefore extends a little beyond the core and directs the arc. This extension also prevents sideways arcing when welding in deep grooves.

The arc stream and other basic features of manual *Gas-shielded metal-arc welding* are illustrated in Fig. 11.21.

11.21 Functions of the electrode coating

Uncoated electrodes

Arc welding has not always been as easy as one finds it today using the shielded electrode. Before coated electrodes were produced commercially, arc welders used bare or lightly coated wire. Bare (uncoated) wires are still used on some applications where maximum strength is not essential, or where complete slag removal is difficult. Because they have no flux and cannot produce a shielded arc, the arc is less powerful and will 'short out' easily as droplets of molten electrode bridge the arc gap. This makes them more difficult to use than a shielded electrode. Because of the unstable arc they are unsuitable for use with a.c., but they can be satisfactorily used on d.c. equipment using electrode negative polarity.

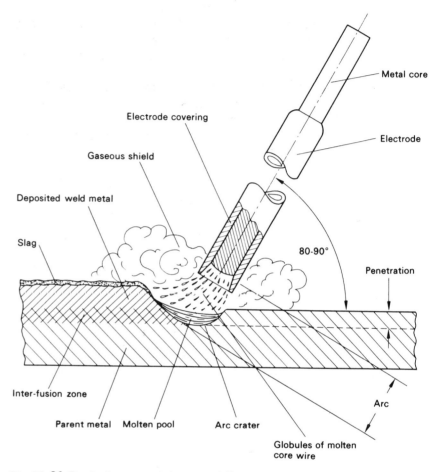

Fig.11.21 Basic features of arc-welding

Coated electrodes

The coating on electrodes has several functions, some of these are:
(a) It facilitates striking the arc and enables it to burn stably.
(b) It serves as an insulator for the core wire.
(c) It provides a flux for the molten pool, which picks up impurities and forms a protective slag which is easily removed.
(d) It stabilises and directs the arc and the globules of molten core metal as shown in Fig. 11.21.
(e) It provides a protective non-oxidising or reducing gas shield (smoke-like gas) around the arc to keep oxygen and nitrogen in the air away from the molten metal.
(f) It increases the rate of melting (i.e., metal deposition) and so speeds up the welding operation.
(g) It enables the use of alternating current.
(h) Additions to the coating can be made (during manufacture) which will replace any alloying constituents of the core wire or the parent metal which are likely to be lost during the welding process.
(i) It gives good penetration.
(j) It increases or decreases the fluidity of the slag for special purposes. It can, for example, reduce the slag fluidity of electrodes used for overhead welding.

11.22 Welding with direct current

The basic equipment used for welding with d.c. operates on an 'OPEN CIRCUIT' VOLTAGE which is much lower than that of an a.c. plant.

With DIRECT CURRENT *the electron current always flows in the same direction.* The POSITIVE pole therefore becomes somewhat more heated than the NEGATIVE. From a practical point of view this is very important to the welder because *the* HEAT *generated by d.c. is not distributed equally between the two poles (i.e., the electrode and the work).* TWO-THIRDS OF THE HEAT IS GENERATED AT THE POSITIVE POLE AND ONLY ONE-THIRD AT THE NEGATIVE POLE. This has considerable influence on welding procedure. By connecting the electrode to the appropriate pole the heat input may be increased or reduced. This variation may be used to advantage as shown in Fig. 11.22.

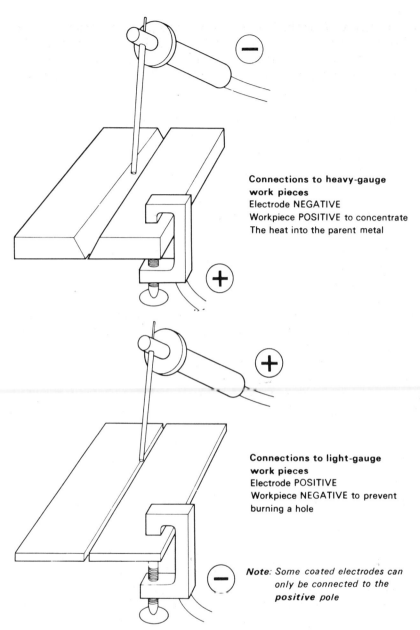

Connections to heavy-gauge work pieces
Electrode NEGATIVE
Workpiece POSITIVE to concentrate
The heat into the parent metal

Connections to light-gauge work pieces
Electrode POSITIVE
Workpiece NEGATIVE to prevent burning a hole

Note: Some coated electrodes can only be connected to the *positive pole*

Fig. 11.22 Connections for direct current arc-welding

The polarity of the electrode is most important when welding with d.c. and the manufacturer's instructions should be rigidly adhered to. In order to prevent any misuse, electrodes to be connected to the positive pole are termed 'ELECTRODE POSITIVE'.

11.23 Welding with alternating current

With ALTERNATING CURRENT *the electron current is changing its magnitude and direction 50 times per second*, and therefore there are no positive and negative poles in the ordinary sense. THE SAME AMOUNT OF HEAT IS PRODUCED AT THE ELECTRODE AND THE WORK.

Thus, while manufacturers specify that a particular electrode should be connected either to the negative or the positive pole of a d.c. supply, no polarity is quoted for electrodes to be used with a.c.

The arc is extinguished each time the current changes direction, but re-establishes itself because the arc atmosphere is heavily ionised. Figure 11.23 shows the comparison between the two types of electric current used for arc welding.

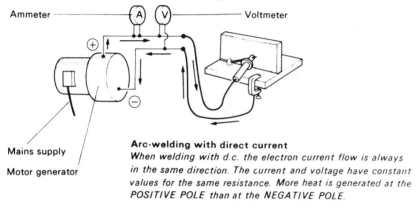

Arc-welding with direct current
When welding with d.c. the electron current flow is always in the same direction. The current and voltage have constant values for the same resistance. More heat is generated at the POSITIVE POLE than at the NEGATIVE POLE.

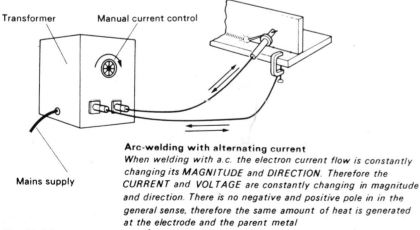

Arc-welding with alternating current
When welding with a.c. the electron current flow is constantly changing its MAGNITUDE and DIRECTION. Therefore the CURRENT and VOLTAGE are constantly changing in magnitude and direction. There is no negative and positive pole in in the general sense, therefore the same amount of heat is generated at the electrode and the parent metal

Fig. 11.23 Comparison of alternating current and direct current arc-welding

11.24 Welding current values

All electrodes are designed for use with a specific welding current, and for this reason the values quoted by the manufacturers must always be closely adhered to. Any alteration will affect the behaviour of the electrode, the appearance of the weld, and possibly the properties of the deposited metal. Variations in current values and their resultant effects will be discussed later in this chapter.

The approximate currents to be used when welding mild steel plates are shown in Table 11.5

Table 11.5 Welding currents

MINIMUM THICKNESS (MILD STEEL PLATE)		WELDING CURRENT VALUE	DIAMETER OF ELECTRODE	
mm	(s.w.g.)	Amps.	mm	(in)
1·62	16	40—60	1·6	1/16
2·03	14	60—80	2·4	3/32
2·64	12	100	3·2	1/8
3·18	1/8 (in)	125	3·2	1/8
3·25	10	125	3·2	1/8
4·06	8	160	4·8	3/16
4·76	3/16 (in)	190	4·8	3/16
4·88	6	190	4·8	3/16
5·89	4	230	6·4	1/4
6·35	1/4 (in)	250	6·4	1/4
7·01	2	275—300	7·9	5/16
8·23	0	300—400	7·9	5/16
8·84	00	400—600	9·5	3/8
9·53	3/8 (in)	400—600	9·5	3/8

Note: The diameter of the electrode is the size of the core wire.

11.25 Open circuit and welding voltages

In order to produce an arc which is suitable for welding it is necessary to have an ELECTRO-MOTIVE FORCE (VOLTAGE) to drive the CURRENT through the circuit. Depending on the type of welding set used (i.e., a.c. or d.c.) a voltage between 60 and 100 V is necessary to start the arc. Once the arc has been started, the arc voltage drops and only 20 to 45 V is necessary to maintain it.

The OPEN CIRCUIT VOLTAGE is often referred to as the 'STRIKING VOLTAGE' and is the e.m.f. required when striking the arc. Most a.c. welding transformers operate at an open circuit voltage of 90 V. Welding sets providing d.c. operate with an open circuit voltage of between 60 and 70 V.

The ARC VOLTAGE is the e.m.f. required to maintain the

arc during welding and usually varies between 20 and 45 V.

Figure 11.24 illustrates the changes which occur in producing the arc.

11.26 Welding current stability

Direct current from a generator

As the arc voltage depends upon the size and type of electrode being used, the generator is designed so that the arc voltage can adjust itself to these conditions. This is done by use of a d.c. generator having a 'drooping voltage' characteristic, as shown in Fig. 11.25.

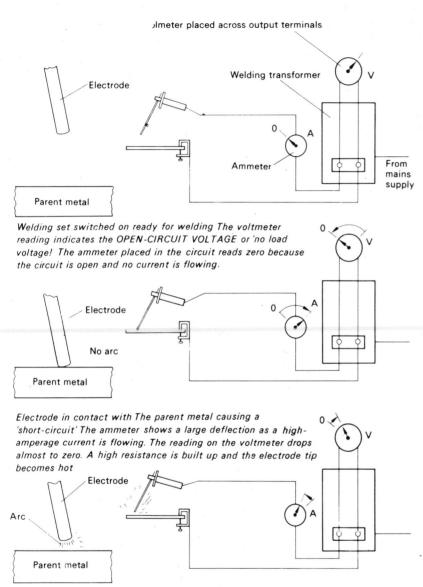

Fig. 11.24 Effect on current and voltage when striking the arc

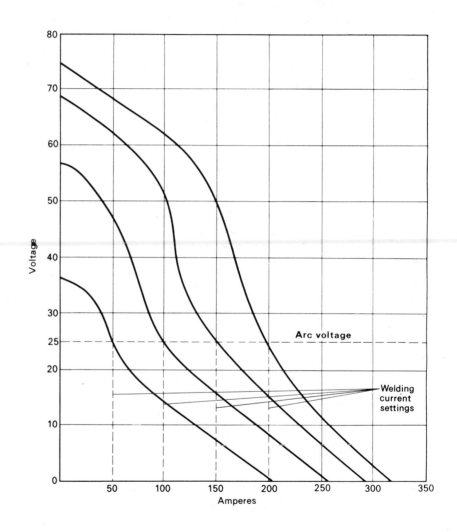

Fig. 11.25 Static characteristic curves of a direct current welding generator

291

The diagram shows that RELATIVELY LARGE CHANGES IN ARC VOLTAGE WILL ONLY RESULT IN SMALL CHANGES IN WELDING CURRENT. Changes in the arc voltage occur with any variation in the length of the arc. With manual arc-welding this is due to the inability of the welder to maintain a constant arc length. *These curves (Fig. 11.25) show that the welding arc will remain stable and easy to hold, and that the welding current will remain practically constant despite slight variations in the length of the arc.*

Alternating current from a transformer

As in the case of d.c. equipment arc voltages control the welding current. Whether using a.c. or d.c. plant, unless the current output remains constant the operator will have difficulty in controlling the slag and will therefore produce unsound welds.

11.27 Welding electrodes (specific)

During manufacture, coated electrodes for welding are subjected to a very high standard of quality control at all stages of the manufacturing process. Most electrodes are made by solid extrusion, whereby the flux covering is applied direct to the core wire under high pressure. The covering size is controlled to very precise limits of concentricity. If the covering were not concentric to the core wire an unstable arc would result.

Covering constituents The covering of the electrodes for the welding of mild and high tensile steels consists of various mixtures of the substances shown in table 11.6.

Table 11.6 Electrode coating materials

CONSTITUENT	REMARKS
Titanium dioxide	Available in the form of natural sands as RUTILE containing 96% titanium oxide. Forms a highly fluid and quick freezing slag. Is a good ionizing agent
Cellulose	Provides a reducing gas shield for the arc. Increases the arc voltage
Iron oxide and *manganese oxide*	Used to adjust the fluidity and the properties of the slag
Potassium aluminium silicate	Is a good ionizing agent, also gives strength to the coating
Mineral silicates and *asbestos*	Provides slag and adds strength to the coating
Clays and *gums*	Used to produce the necessary plasticity for extrusion of the coating paste
Iron powder	Increases the amount of metal deposited for a given size of core wire
Calcium fluoride	Used to adjust the basicity of the slag
Metal carbonates	Provides a reducing atmosphere at the arc. Adjusts the basicity of the slag
Ferro-manganese and *ferro-silicate*	Used to deoxidize and supplement the manganese content of the weld metal

Type of flux covering By using different properties and combinations of the constituents listed in Table 11.6 it is possible to produce an infinite variety of electrode types. In practice, practically all the electrodes at present in use conform to one of six main well defined classes in accordance with BS 1719. These six classes are listed in Table 11.7

Table 11.7 Coding of coverings and data for six classes of electrodes

BS 1719 CODING	TYPE OF COVERING	RESULTANT SLAG	REMARKS
1— —	Cellulose	Thin	Produces a voluminous gas shield. Has a deeply penetrating arc and rapid burn-off rate. The weld appearance is slightly coarse — the ripples are pronounced and unevenly spaced. Easy to use in any welding position. Suitable for all types of mild steel welding and deep penetration welding. Mainly used with d.c. with the electrode connected to the positive pole.
2— —	Rutile	Fairly viscous	The slag is dense but is easily detached, except from the first run in a deep Vee. Suitable for butt and fillet welds in all positions. Particularly easy to use for fillet welds in the horizontal-vertical position. Fillet welds have a convex profile with medium root penetration. Suitable for a.c., and with d.c. the electrode may be connected to either pole.

3 — —	*Rutile*	Fluid	The slag is easy to detach, even from the first run in a deep Vee. Particularly suitable for welding in the vertical and over-head positions. Useful electrode for site work. Suitable for a.c., and for d.c. with the electrode connected to either pole.
4 — —	*Iron oxide*	Inflated	The covering consists basically of oxides and carbonates of iron and manganese, together with silicates. The electrode has a thick covering and is used for deep groove welding in the flat position only. Produces a fluid voluminous slag which freezes with a characteristic internal honeycomb of holes — hence the term inflated slag. The weld profile is very smooth and is concave. The electrode is suitable for use with d.c. with the electrode connected to the positive pole, but may also be used with a.c.
5 — —	*Iron oxide*	Solid	The covering is thick and consists basically of iron oxides with or without oxides of manganese. A heavy solid slag is produced which is sometimes self-detaching. Used for single-run fillet welds where smooth contour is more important than high mechanical properties in the weld metal. Covering melts with a distinctive 'cupped' effect, enabling the electrode to be used touching the work — touch welding. The weld has a smooth concave profile. These electrodes are often referred to as dead soft electrodes because the weld metal has a low carbon and manganese content. Suitable for a.c. and on d.c. the electrode may be connected to either pole.
6 — —	*Lime flurospar*	Basic	This class of electrode is often termed low hydrogen or basic type. Suitable for welding in all positions. The slag is fairly fluid. The weld deposit is usually convex to flat in profile. Is less sensitive to variation in plate quality than any other class of electrode. Used for welding of heavy sections and highly restrained joints. Can be used with a.c. but with some types d.c. is preferred with electrode positive.

Storage of electrodes Because electrode coatings tend to absorb moisture from the atmosphere it is essential that they be kept dry. *If the electrode coating becomes damp, steam will be generated during the welding process causing excessive spatter, and porosity in the weld.* Low hydrogen electrodes should be dried in an oven prior to their use. Specially designed electric ovens are available which can be thermostatically controlled over a temperature range of 38 to 275°C.

Table 11.8 Definitions of welding positions

POSITION	BS 1719 SYMBOL	LIMITS OF SLOPE	LIMITS OF ROTATION
Flat	F	0° to 5°	0° to 10° / 0° to 10°
Horizontal-Vertical	H	0° to 5°	30° to 90° / 30° to 90°
Vertical-Up	V	80° to 90°	0° to 180° / 0° to 180°
Vertical-Down	D		
Overhead	O	0° to 10°	115° to 180° / 115° to 180°

Any intermediate position not specified above is undefined but the general term 'inclined' is used.

Table 11.8 cont

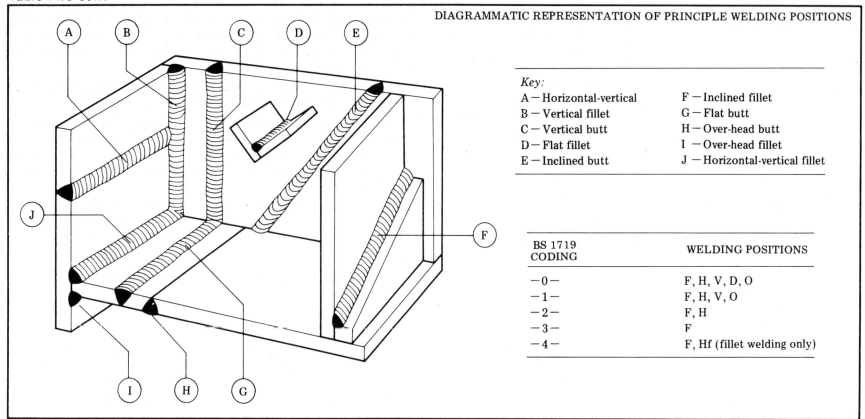

11.28 Effect of variable factors in metal-arc welding

The variable factors in metal-arc welding are:

1. *The welding current* If the welding current is set *too low* the electrode core wire is not melted sufficiently. This results in poor penetration due to the lack of heat necessary to complete fusion. The molten weld metal tends to pile up instead of flowing smoothly, producing a weld bead which is irregular in width and contour, as shown in Fig. 11.26(a). There is a tendency for slag to become entrapped within the weld bead itself.

The weld crater is not very well defined. The arc is unsteady and burns with an irregular spluttering sound.

An *excessive current* can cause the weld metal to become too fluid and difficult to control. *There is considerable spatter and unnecessary wastage of electrode*, as shown in Fig. 11.26(b). When the welding current is set too high the electrode tends to become red hot and the arc burns with a regular explosive sound.

Too high a current can often result in blow holes being formed in the parent metal accompanied by excessive penetration and some undercutting along the edge of the weld.

The slag produced is usually difficult to remove. Table 11.9 gives the British Standard Coding for welding currents.

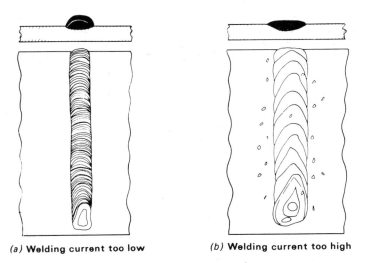

Fig.11.26 Effect of variation in welding current

(a) Welding current too low (b) Welding current too high

Table 11.9 Coding for welding current conditions

The welding current conditions as defined in BS 1719 are as follows:

WELDING CURRENT CONDITIONS	SYMBOL
D.C. with electrode positive	D +
D.C. with electrode negative	D −
D.C. with electrode positive or negative	D ±
A.C. with an open-circuit voltage not less than 90 volts	A_{90}
A.C. with an open-circuit voltage not less than 70 volts	A_{70}
A.C. with an open-circuit voltage not less than 50 volts	A_{50}

BS 1719 CODING	WELDING CURRENT	BS 1719 CODING	WELDING CURRENT
− − 0	D +	− − 4	$D + A_{70}$
− − 1	$D + A_{90}$	− − 5	$D \pm A_{90}$
− − 2	$D - A_{70}$	− − 6	$D \pm A_{70}$
− − 3	$D - A_{50}$	− − 7	$D \pm A_{50}$

2. *The arc length*   *Too short* an arc length causes irregular build up of the molten metal. The ripples on the weld bead are not uniform with regard to both width and height, as shown in Fig. 11.27(a). With too short an arc length it is difficult to control and maintain the arc, and there is a tendency for the electrode to 'freeze' or stick to the weld pool.

Note   A normal arc length should be slightly less than the diameter of the electrode, and is generally considered to be between 1·6 and 3·2 mm.

Too long an arc causes an appreciable increase in spatter, and penetration is poor. The weld bead is wide and of poor appearance because the core wire metal is deposited in large globules instead of a steady stream of fine particles. The arc crater is flat and blistered as shown in Fig. 11.27(b).

Note   To ensure complete penetration the arc length should be kept as short as possible to enable the heat from the arc to melt the parent metal and the electrode core-wire simultaneously. If the arc length is too great a considerable amount of this heat is lost to the air.

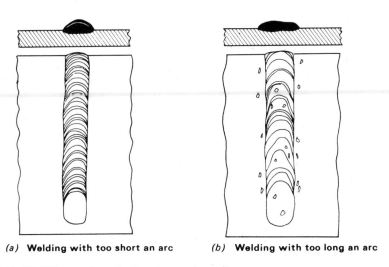

(a) Welding with too short an arc   (b) Welding with too long an arc

Fig.11.27 Effect of variation in arc length

3. *Speed of travel*   When the speed of travel is too fast a narrow and thin weld deposit, longer than normal, is produced. Penetration is poor, the weld crater being small and rather well defined, as shown in Fig. 11.28(a). The surface of the weld bead has elongated ripples and there is 'undercutting'. *The reduction in bead size and amount of undercutting depends on the ratio of speed and current.*

Note Most manufacturers specify the length of run which may be obtained with electrodes of different types, sizes, and lengths, used under their recommended current conditions. Thus, if the length of deposit per electrode is smaller or greater than is specified, the speed of travel is either slower or faster than that intended by the manufacturer.

As the rate of travel decreases the width and thickness of the deposit increases, as shown in Fig. 11.28(*b*). The weld deposit is much shorter than the normal length. The weld crater is flat. The surface of the weld bead has coarse evenly spaced ripples. With a slower rate of travel the molten weld metal will tend to pile up and cause excessive overlap of the weld bead.

Slower speeds cause the parent metal and the weld bead to become hotter than when the movement is faster. This factor must be taken into consideration when welding certain metals, or when trying to minimise distortion. Slow rates of travel also result in a reduction in penetration, and there is a tendency for the slag to flood the molten weld pool making it difficult to control the weld deposit.

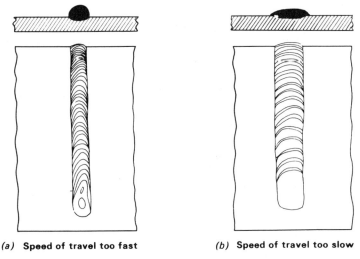

(a) Speed of travel too fast (b) Speed of travel too slow

Fig. 11.28 Effect of variation in rate of travel

4. *Angle of electrode* The more upright the electrode is held relative to the parent metal the greater the depth of 'PENETRATION'. This is because the full force of the arc is directed and concentrated on to the parent metal.

In the vertical position the force of the arc stream tends to drive the molten metal from the edge of the weld pool outwards, and in doing so produces undesirable 'UNDERCUTTING' of the weld profile.

When an electrode is held in an upright position, 'SLAG' will build up and surround the weld metal deposit. This will cause a problem as soon as the electrode is moved in the directions shown by the arrows.

These effects are shown in Fig. 11.29(*a*), and the diagram clearly indicates that *some slag must become entrapped in the resultant weld.*

The more upright the electrode is held, the more difficult it is for the operator to be able to clearly observe the weld pool and control it.

When the electrode is inclined at an angle from the vertical as shown in Fig. 11.29(*b*), the direction of the arc stream causes the slag to form away from the molten weld pool and ahead of the arc. *No slag is built up in advance of the direction of welding, thus eliminating the problem of slag entrapment.*

The adoption of a suitable electrode angle enables the operator to easily observe the weld pool and control the slag. In practice, the angle of inclination of the electrode to the parent metal varies between 60° and 90° in a vertical plane.

One factor which influences the choice of electrode angle is the class of electrode being used. *Some classes of electrodes have coatings which produce a very fluid slag, while others produce a viscous slag.* This factor is important because with a fluid slag it is much more difficult to control, and there is a greater danger of slag entrapment as the angle of the electrode approaches the upright position.

Between the flat and the upright positions there is one which provides the optimum welding conditions, with the following result:
(*a*) Adequate penetration.
(*b*) Correct weld profile.
(*c*) Correct width of weld bead.
(*d*) Minimum spatter.
(*e*) Minimum difficulty in controlling the slag.

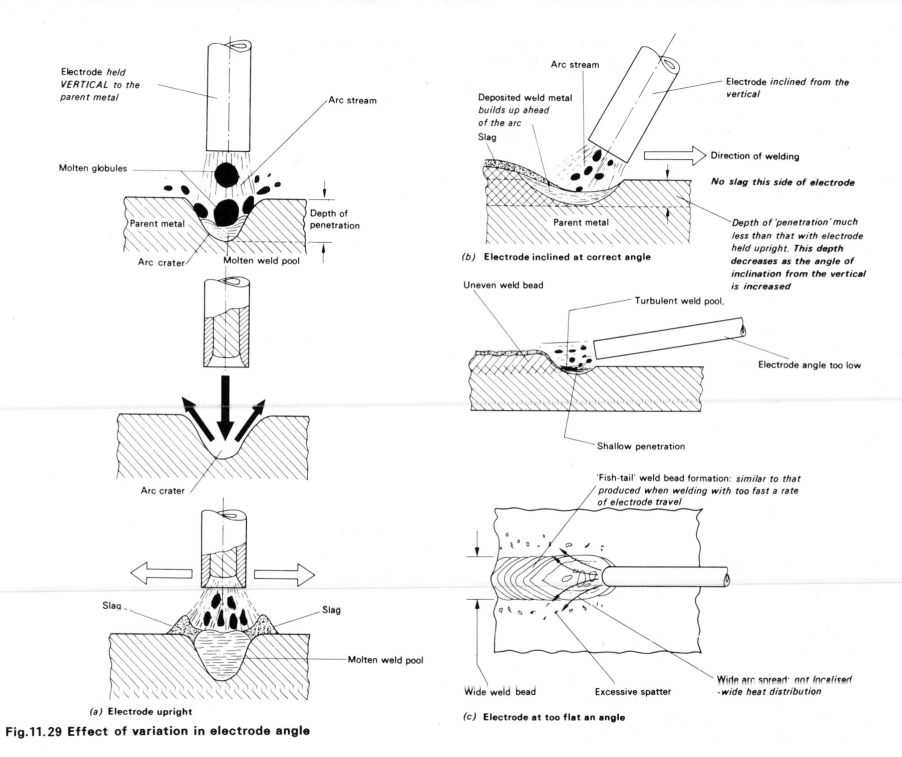

Fig.11.29 Effect of variation in electrode angle

PART C. WELDING DEFECTS

11.29 British Standard 1295 (1959)

British Standard 1295 (1959) *Tests for use in the training of welders* is subtitled 'manual, metal-arc, and oxy-acetylene welding of mild steel'. Referring to this standard, some of the main factors to be considered when making an assessment of weld quality are:

(*a*)   Shape of profile.
(*b*)   Uniformity of surface.
(*c*)   Degree of undercut.
(*d*)   Smoothness of join where weld is recommended.
(*e*)   Freedom from surface defects.
(*f*)   Penetration bead.
(*g*)   Degree of fusion.
(*h*)   Degree of root penetration.
(*i*)   Non-metallic inclusions and gas cavities.

Figure 11.30 illustrates desirable and undesirable weld profiles.

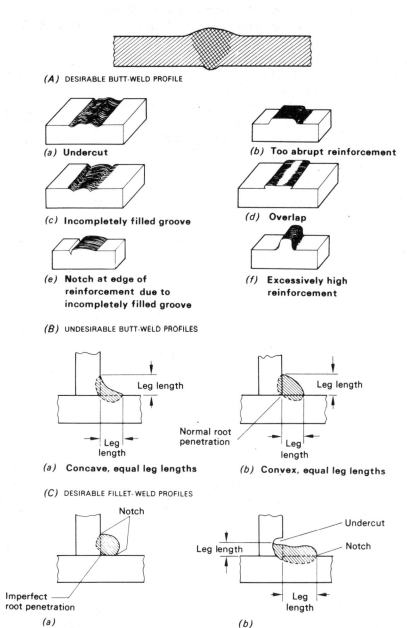

Fig. 11.30 Weld profiles

The weld metal and the parent metal affected by the welding process can be divided into THREE DISTINCT ZONES which can be clearly seen in a polished and etched section through the weld. These are as follows:
1. ACTUAL WELD METAL ZONE
2. WELD PENETRATION ZONE — consisting of parent metal which has been fused during the welding process.
3. HEAT-AFFECTED ZONE — where the parent metal, although not fused has been affected by welding heat (see Chapter 6).

Figure 11.31 shows these zones diagrammatically.

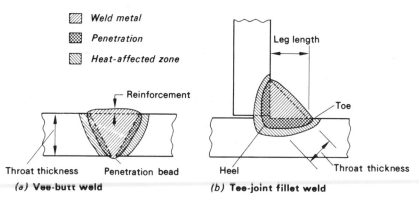

Fig. 11.31 Weld zones

11.30 Welding defects

Undercutting

This is a term used to denote either the burning away of the side walls of the joint recess, or the reduction in parent metal thickness at the line where the weld bead is joined to the surface, as shown in Fig. 11.32.

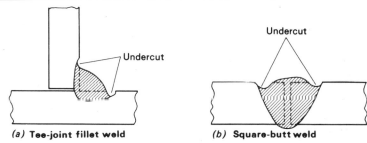

Fig. 11.32 Undercutting

CAUSES OF UNDERCUTTING

1. *Manual metal-arc welding process* Use of excessive welding currents will melt a relatively large amount of parent metal and cause it to sag under its own weight which results in the crater not being filled. Too fast a travel with the electrode will also result in the crater not being filled.

   Excessive weaving of the electrode, or the wrong type of electrode, can produce undercutting. If the electrode is not held at the correct angle in relationship to joint, one-sided undercutting may result through the heat of the arc being concentrated on one side of the joint.

   The formation of undercuts is particularly undesirable because it tends to weaken the structure. *With regard to multi-pass welds, there is a strong possibility of slag entrapment on subsequent runs and any undercutting will make slag removal very difficult.*

2. *Gas welding process* Too much heat results in excessive melting of the parent metal.

   Inadequate filler rod feed will result in the crater not being filled. Poor manipulation of the welding torch and filler rod, incorrect angle of torch relative to the joint will result in one-sided undercutting. Heavy mill-scale on the parent metal is also a possible cause of undercutting.

Smoothness of join where weld is recommenced

Whenever a welding run has to be interrupted and then restarted it is very important that where the weld is recommenced the join should be as smooth as possible. The join should show no pronounced hump or crater in the weld surface. The joins at the ends of the weld runs are liable to have poor strength. This is caused by crater cracks producing stress concentrations. The welded joint may also be weakened by overlap and lack of fusion.

Surface defects

Surface defects in welds are generally due to the use of unsuitable materials and/or incorrect techniques. The surface should be free from porosity, cavities, and either burnt-on scale (in the case of gas welding) or trapped slag (in the case of metal-arc welding). Surface cavities may be caused by lack of fusion, gas bubbles, or trapped slag.

Penetration

One of the more common causes of faulty welding is the *lack of penetration*, often termed 'incomplete penetration' or 'lack of fusion'. When viewed from the underside, a sound weld should have a slight penetration bead. The size of this penetration bead will be influenced by the welding process used, the type of joint preparation, and the skill of the welder. Figure 11.33 shows lack of penetration.

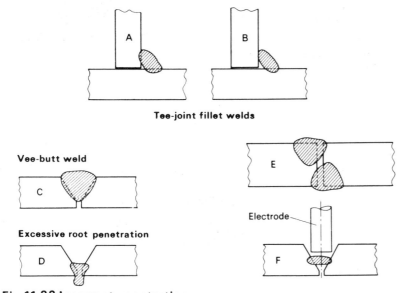

Fig. 11.33 Incorrect penetration

With manual *metal-arc welding* lack of penetration can be attributed to the welding current being set too low for the particular type of electrode. With *gas welding* the use of too large a diameter filler rod or incorrect angles of the welding torch. Use of the wrong polarity when arc *welding with* DIRECT CURRENT will result in lack of penetration.

Excessive penetration in the root of the joint will result with too high a welding current or concentration of heat in the case of gas welding, and with the use of an unsuitable edge preparation.

Unsatisfactory penetration may result if the welds are not located directly opposite each other, as shown in the double-side square-butt weld 'D'.

Lack of penetration can also be caused by using an electrode for the first run which is not small enough in diameter to reach down to the ROOT of the weld, as shown at 'D'.

Incorrect joint design, including the type of preparation, can be the cause of incomplete penetration. The joint must allow entry of the electrode and permit unrestricted manipulation. Too small a diameter of electrode in a root run will result in inadequate fusion or penetration.

Other causes of inadequate penetration with the manual metal-arc welding process which can usually be attributed to the welder and must be avoided are:
(a) *Bad incorporation of tack welds* (This also applies to gas welding).
(b) Inadequate removal of the slag.
(c) *The use of too low an arc* — the arc must always be kept as short as possible.

Lack of fusion

Lack of fusion can be defined as the failure to fuse together adjacent layers of the weld metal or adjacent weld metal and parent metal as shown in Fig. 11.34. This condition may be caused by failure to raise the temperature of the parent metal to its melting point. Another cause is the failure to remove oxides or other foreign matter.

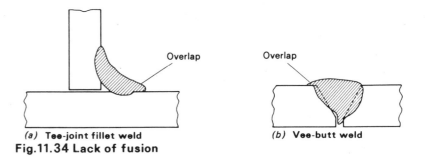

Fig. 11.34 Lack of fusion

Non-metallic inclusions and gas cavities

1. *Slag inclusions* The term 'slag inclusions', or merely 'inclusion', refers to slag or other non-metallic foreign matter entrapped in the weld metal. The usual source of inclusions is the slag formed by the electrode covering, although other substances may be present due to welding on dirty surfaces. Figure 11.35 illustrates the problem of slag inclusion.

Inclusions may also be caused by contamination from the atmosphere by oxidising conditions, faulty manipulation of the welding torch or filler rod, or fluxes used in welding operations With gas welding they are usually caused by dirty parent meta surfaces or millscale.

Slag inclusions in manual metal-arc welding, are generally caused by using either too high or too low a welding current. Other factors include incorporation of bad tack welds, too-long an arc or too-high a speed of travel, lack of penetration, or the use of too large a diameter electrode which is also a relevant factor.

Carelessness in slag removal prior to putting down another run of weld metal when producing multi-run weld may be the cause, or the inaccessibility of the weld not allowing the removal of slag resulting in 'slag traps'.

IN GENERAL, THE WELDING CHARACTERISTICS OF AN ELECTRODE ARE SUCH THAT THE MOLTEN SLAG FLOATS FREELY TO THE SURFACE OF THE MOLTEN WELD METAL AND IS EASILY REMOVED WITH THE AID OF A CHIPPING HAMMER.

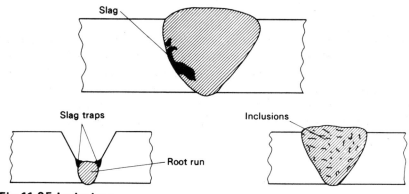

Fig. 11.35 Inclusions

2. *Porosity* Porosity consists of a group of small cavities caused by gas entrapped in the weld metal. When these cavities come out to the surface of the weld metal they are commonly termed 'blow-holes'. Figure 11.36 illustrates the meaning of porosity. *One of the main factors contributing to this fault is excessive moisture in the electrode or in the joint.* ALWAYS USE DRY ELECTRODES AND DRY MATERIALS.

Porosity may be scattered uniformly throughout the entire weld, isolated in small groups, or concentrated at the root. Other causes of porosity are:

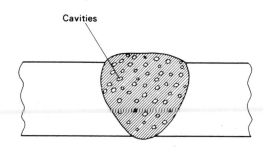

Fig. 11.36 Porosity

Cause:	Remedy:
1. High rate of weld freezing.	Increase the heat input.
2. Oil, paint, or rust on the surface of the parent metal.	Clean the joint surfaces.
3. Improper arc-length, current, or manipulation.	Use proper arc-length, (within recommended voltage range), control welding technique.
4. Heavy galvanised coatings.	Remove sufficient zinc on both sides of the joint.

11.31 Workshop testing of welds

The correct procedure for preparing test welds is fully explained in BS 1295, and for the purpose of bend tests two specimens are cut from the test weld. *One specimen is tested with the face of the weld in tension and the other with the root in tension.* The method of bend testing is simply explained by Fig. 11.37.

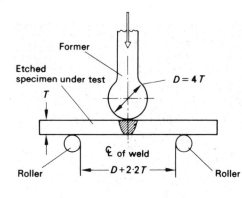

*The specimen is bent over a former, as shown, through an angle of 180° i.e., until the two ends are parallel to each other. A simple **hydraulic press** is generally used to supply the necessary bending force.*

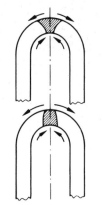

One specimen is tested with the face of the weld in TENSION as shown opposite

One specimen is tested with the root of the weld in TENSION, as shown opposite

Each specimen tested should show no crack or defect due to poor workmanship
Slight cracking at the ends of the weld metal occuring during bending should be disregarded.

Fig. 11.37 Bend testing of welds

The upper and lower surfaces of the weld may be filed, ground, or machined level with the surface. The direction of machining should be along the specimen and across the weld. The sharp corners of the test piece should be rounded to a radius not exceeding one tenth of the thickness. The centre line of the weld must be accurately aligned with the centre line of the former, as shown. The specimen is bent over a former through an angle of 180° — i.e., until the two ends are parallel to each other. A simple HYDRAULIC PRESS is generally used to supply the necessary bending force.

11.32 Inspection of etched sections and bend specimens

Etched sections and weld fractures will give information of the following defects if present:
1. Incorrect profile.
2. Undercutting.
3. Slag inclusions.
4. Porosity and cracks.
5. Poor root penetration and lack of fusion.

It is recommended that the examination of etched specimens should be made with the aid of, at least, a hand magnifying glass.

11.33 Method of preparing etched specimens

1. *Preparation of surface for etching* The surface should first be filed using a coarse file until all the deep marks are removed. A smooth file is then used to file the surface at right angles to the initial coarse file marks. This smooth filed surface should then be polished with successively finer grades of emery paper, the direction of polishing being at right angles to the marks made by the previous grade in each case. *It is important to polish with each grade of emery paper until the scratches made by the previous paper have been removed before proceeding to the next finer grade.* This will ensure a first-class finish to the surface to be etched.

2. ETCHING FOR MACRO-EXAMINATION   A suitable etching solution is as follows:

10—15 ml NITRIC ACID (specific gravity 1·42)

90 ml ALCOHOL (industrial spirit)

The specimen may be immersed in the etching solution until a well defined macro-structure is obtained, or the surface of the specimen may be swabbed with cotton wool saturated with the solution. In either case, the specimen should be thoroughly washed in hot water, followed by rinsing with acetone or industrial spirit and drying in a current of air. It is possible to preserve etched specimens by a coating of clear lacquer.

Examples of macro-structures of weld specimens are shown in Appendix 1.

Appendix 1

Weld defects

(b)   The defect magnified

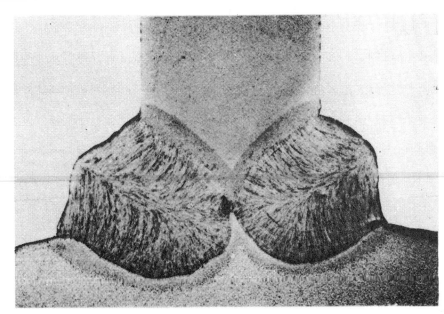

(a)   Fillet welds in low alloy steel, showing slag inclusion at the root of two runs

A manual metal arc single 'V' butt weld, showing lack of penetration and misalignment of the plate edges

A manual metal arc single 'V' butt weld, showing undercut and misalignment of the plate edges

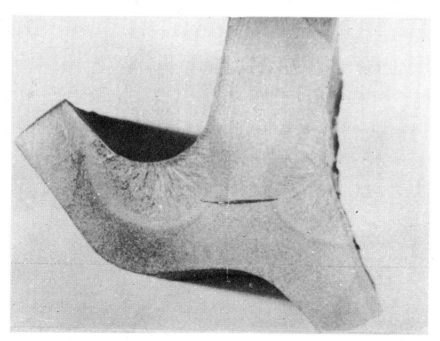

Submerged arc welds, showing lack of penetration

Index

A

Abrasives; *see* material removal
Angle section work; *see* bending, fabrication
Arc welding; *see* metal arc welding

B

Bending
  air, 226
  allowances for sheet metal, 250—7 inc.
  angle section, rolls for, 237, 238
  basic methods, 226, 227
  machines, 220
  mechanics of, 223, 224
  pressure tools, 230
  press brake
    operations, 232—4 inc.
    principles of, 231
    types of, 231, 232
  pyramid type roll bending machine, 236
  roll bending
    machines, 235 and 237
    operations, 234 *et seq.*
    plate, 235
  spring back, 225, 226
Blanking, principles of; *see* press work

C

Calculations
  bar components, 65, 66
  density, 61—4 inc.
  mass of contents of a container, 66, 67
  mass and weight, 60
  sheet metal components, 64, 65
  trigonometry, 67—9 inc.
Chemical changes in metal, 73—6 inc.
Chemical reactions during welding, 70
Chisels
  blacksmiths', 125
  cold, 124
Colour coding, compressed gas cylinders, 38
Connections; *see* fabrication
Constructions; *see* geometry

D

Development; *see* geometry
Dimensioning, 33—5 inc.
  selection of, 36, 37
Down cut (climb) milling; *see* milling techniques
Drilling process, 166

F

Fabrication
  angle section work, 211—16 inc.
  beam to beam connection, 267
  cone rolling, 237
  lattice frame, 269, 270
  slip rolls, 236
  stanchion
    bases, 265
    splices, 266
    to beam connection, 266, 267
  structural connections and assemblies, 264 *et seq.*
  structural members, 264
  trusses, 268
  web stiffeners, 270, 271
Flame cutting; *see* material removal
Flanges
  locating hole centres, 217, 218
  forming; *see* sheet metal work
Fly press; *see* presswork
Folding machine operations, 227-9 inc.
Folding machines, 229
Forging
  aluminium alloys, 115
  anvil, 218
  bending, 221
  copper alloys, 115
  drawing down, 220
  forming tools, 219
  hammers, 219
  hearth, 218
  hot and cold, 113
  material properties, effect on, 113—5 inc.
  punching and drifting, 220
  swaging, 221
  tongs, 219
  up-setting, 220
  welding (pressure), 222
Forming; *see* sheet metal work

G

Geometry
  angles, setting out
    use of compasses, 42, 43
    use of set squares, 42
  construction
    miscellaneous, 50—2 inc.
    ovals and ellipses, 46—9 inc.
    plane figures, 43—6 inc.
  development
    of surfaces, 53
    parallel line, 53—5 inc.
    radial line, 55—7 inc.
    triangulation, 57—9 inc.
Grinding and grinding machines; *see* material removal

H

Heat treatment
  carbon content, effect of, 111
  overheating, effect of, 111, 112
  quench hardening, 110
  rate of cooling, 111
  tempering, 111
Heat treatment examples
  annealing, 112, 113
  hardening, 112
  normalising, 113
  tempering, 112
Heat straightening; *see* press work
Hollowing; *see* sheet metal work

L

Lathe, spinning; *see* spinning sheet metal
Linear expansion, 76
  at high temperatures, 77
  in welding processes, 77, 78
Location and restraint; *see* tool holding
Lubrication, spinning; *see* spinning sheet metal

M

Marking out
  angle sections, holes in, 105
  bevel, use of the, 95
  centre punch, use of the, 96
  channel sections, holes in, 106
  columns and beams, holes in, 107
  flanges, holes in, 93—5 inc.
  large plates, 91—3 inc.
  patterns, arrangement of, 98
  pipe square, use of, 95
  plumb lines, 97
  rule, use of, 88
  scratch gauge, 96
  scribing, 89
  spirit level, 97
  steel tape, 92
  'tee' sections, holes in, 106, 107
  template
    box type, 104
    large, 89—91 inc.
    making, 99—101 inc.
    plain and bushed, 104, 105
  templates, use of, 101—4 inc.
  templates for inspection and checking, 101
Materials
  cold reduced sheets, 186
  hot rolled plates, 187
  hot rolled sheets, 186
  market forms of supply, 185
Material removal
  abrasives, 143—5 inc.
  abrasive wheel cutting off machines, 146
  flame cutting
    applications of, 155—8 inc.
    distortion caused by, 85
    effect on metal properties, 122, 123
    equipment, 151, 152
    factors influencing quality of cut, 154, 155
    machines, 159—65 inc.
    principles, 150—65 inc.
  grinding machines, portable, 147—50 inc.
  milling cutter
    action of, 138
    tooth form of, 138
    mounting, *see* tool holding
    types of, 139
  milling machine
    horizontal, 138
    vertical, 137
  milling techniques, 140
Material removal
  rate of, 140—3 inc.
  weld grinding, 145, 146
Mathematical tables, appendix 2
Measurement
  direct eye, 88
  tensioned wire, 97
Metal arc welding
  alternating current, use of, 290
  angle of electrode, effect of, 296, 297
  arc length, effect of, 295
  arc stream, 288
  direct current, use of, 289
  electrode coating, function of, 288, 289
  electrodes, 287, 288; *see* 292, 293
  excess current, effect of, 294, 295
  fluxes, use of, 73
  open circuit voltage, 290, 291
  positions, definitions of, 293, 294
  process, principles of, 287
  shielding gases, 72
  speed of travel, effect of, 295, 296

variable factors, effect of, 294 *et seq.*
welding current values, 290
welding voltage, 290—2 inc.
Metal inert gas (MIG) welding; *see* welding
Milling and milling machines; *see* material removal
Moments of forces, 127 and 129

N
Nibbling machines, 133, 134

O
Oblique drawing, *see* pictorial drawing
Oxy-fuel gas welding
   acetylene cylinders, 273
   discharge rate, cylinders, 273, 274
   economiser, gas, 278, 279
   equipment, welding, 272
   flame condition, 280
   flame, structure of, 279, 280
   hoses, 277, 278
   leftward welding, 283, 284
   lighter, gas, 278
   manifold, 274
   mixing chamber, 276
   nozzle size, 280, 281
   nozzle, welding torch, 276
   oxygen cylinders, 273
   pressure gauge, 275
   pressure regulator (automatic), 274, 275
   rightward welding, 284, 285
      advantages of, 285, 286
   torch, welding, 276
   velocity, gas, 282
   welding techniques, comparison of, 286, 287
Oxidation of welds, 71
Orthographic drawing (3rd angle), 24, 25
Orthographic drawing, selection of views, 26, 27

P
Pictorial drawing
   isometric
      construction, 28—30 inc.
      paper, use of, 32, 33
      scale, 30
   oblique construction, 27, 28
   squared paper, use of, 30—2 inc.
Piercing; *see* press work
Power press; *see* press work
Press work
   blanking, principles of, 134—6 inc.
   corner notching tool, 206
   cropping tool, 206
   deep drawing, 257, 258
   fly press, 135 and 202
   master template, use of, 203, 204
   piercing, principles of, 134—6 inc.
   power press, 135
   sheet metal operations, 202—7 inc.
   'vee' bend tool, 204—5 inc.

R
Raising; *see* sheet metal work
Restraint and location; *see* tool holding, work holding
Riveting; *see* sheet metal work

S
Safety (Cranes), precautions when using, 21—3 inc.
Safety (Forging), 19—20 inc.
Safety (Metal arc welding)
   body protection, 13, 14
   cable connectors, 11
   circuit diagram, 8, 9
   electrode holders, 11, 12
   external welding circuit, 10, 11
   eye and head protection, 12, 13
   fire hazards, 14
   mains-operated equipment, 8
      hazards, 9, 10
   mobile welding plant, 12
   return current clamps, 10
   screens, 14
   ventilation, 15
   welding cables, 10, 11
Safety (oxy-fuel gas welding and cutting)
   explosion risks, 17
   fire hazards, 16
   gas cylinders, use of, 18
   goggles, 15
   personal protection, 15 *et seq.*
   testing for leaks, 18
   welding in confined spaces, 19
Safety (pressure vessels)
   testing, 20
   welded repairs, 21
Safety (site)
   ladders, care and use of, 6, 7
   protective clothing, 1, 2
   protective foot wear, 4
   protective gloves, 3
   protective helmets, 3, 4
   safety blocks
      self contained, 5, 6
      static, 5
   working aloft, 5 *et seq.*
Setting out — roof trusses, 108, 109
Shears, bench, 127
   circle cutting, 210, 211
   guillotine
      operations on the, 208—10 inc.
      setting, 207, 208
   hand (snips), 126, 127
   portable (hand), 132, 133
   rotary, 131, 132
   summary of requirements, 131
   universal steel machine, 213
Shearing, principles of, 128—30 inc.
Sheet metal work
   applied stiffeners, 263
   bench tools, use of, 241—3 inc.
   bend allowance; *see* bending
   breaking the grain, 199
   cold-forming operations, comparison of, 223
   flanging methods, 238, 239
   fly press operations; *see* press work
   forming operations, 238 *et seq.*
   hand lever punch, 192
   hatchet stake, use of, 193
   hollow bead (false wiring), 260, 261
   hollowing, 243—5 inc.
   jennying machine use of, 200—2 inc.
   notched corners, 189—90 inc.
   patterns, use of, 189
   pittsburgh lock, 187, 188
   planning, need for, 187
   spinning, *see* spinning sheet metal
   spring back; *see* bending
   raising, 243—5 inc.
   riveting, methods of, 193
   rivets, removal of, 125, 126
   stiffening
      introduction to, 258
      methods of, 258—63 inc.
   swaging, 261, 262
   wired edge faults, 195, 196
   wiring edges (cylinders and cones), 196—8 inc.
   wiring straight edges, 260
Sheet metal work examples
   funnel, tin plate domestic, 198—200 inc.
   simple box, 191—5 inc.
Skip welding, 84
Spinning sheet metal
   back stick, use of, 249
   hollow bead, forming a, 261
   introduction to, 246
   lathes for, 246, 247
   lubrication for, 250
   metal removal, 136, 137
   process of, 247—50 inc.
   spindle speeds for, 250
Stiffening; *see* sheet metal work

T
Tempering; *see* heat treatment
Template; *see* marking out
Tool holding
   location and restraint of end mills, 171
   location and restraint of drills, 166, 167
   location and restraint of shell end mills, 172
   mounting milling cutters, 139
Trammels, 90
Trigonometry; *see* calculations
Tungsten Insert Gas (TIG) welding; *see* welding

U
Up-cut (conventional) Milling; *see* milling techniques

W
Welding
   aluminium, 121
   copper, 120
   distortion
      causes of, 78, 79
      minimising, 81—4 inc.
      types of, 80—1 inc.
   effect on parent metal, 117, 188
   forge (pressure); *see* forging
   grinding; *see* material removal
   manufacturing stresses, 79
   metal inert gas (MIG), 72
   mild steel, 119, 120
   processes; *see* oxy-fuel gas welding, metal arc welding, forging
   relationship with material properties, 115—17 inc.
   sequence of operations, 84
   symbols, 38—42 inc.
   tungsten inert gas (TIG), 72
Welding defects
   cracking, 122
   etched specimens
      inspection of, 302
      preparation of, 302
   fusion, lack of, 300
   illustrations of (appendix 1), 303, 304
   inclusions (slag), 300
   penetration, lack of, 300
   porosity, 301
   profiles of, 298, 299
   surface, 299
   undercutting, 299
   work shop testing for, 301, 302
Work holding
   bending bars, sheet metal, 173
   bridge pieces, 182
   chain and bar, 183
   clamps, sheet metal, 173, 174
   cleats, 181
   dogs and blocks, 181
   drilled components
      machine table, 169
      vee blocks, 169
      vice, 167, 168
   drilling sheet metal, 170
   glands, 181
   jack clamps, 183
   machine clamps, 176
   magnetic clamps, 184
   milling machine
      dividing head, 170
      table, 170, 171
      vice, 170
   plate dogs and pins, 178, 179
   riveting, whilst, 177
   skin pegs, 177, 178
   soft iron wire, use of, 178
   spiders, 183, 184
   strong backs, 180
   welding clamps, 175, 176, *see also* 179, 180

Natural sines

Degrees	0' 0.0	6' 0.1	12' 0.2	18' 0.3	24' 0.4	30' 0.5	36' 0.6	42' 0.7	48' 0.8	54' 0.9	Mean Differences				
											1	2	3	4	5
0	·0000	0017	0035	0052	0070	0087	0105	0122	0140	0157	3	6	9	12	15
1	·0175	0192	0209	0227	0244	0262	0279	0297	0314	0332	3	6	9	12	15
2	·0349	0366	0384	0401	0419	0436	0454	0471	0488	0506	3	6	9	12	15
3	·0523	0541	0558	0576	0593	0610	0628	0645	0663	0680	3	6	9	12	15
4	·0698	0715	0732	0750	0767	0785	0802	0819	0837	0854	3	6	9	12	15
5	·0872	0889	0906	0924	0941	0958	0976	0993	1011	1028	3	6	9	12	14
6	·1045	1063	1080	1097	1115	1132	1149	1167	1184	1201	3	6	9	12	14
7	·1219	1236	1253	1271	1288	1305	1323	1340	1357	1374	3	6	9	12	14
8	·1392	1409	1426	1444	1461	1478	1495	1513	1530	1547	3	6	9	12	14
9	·1564	1582	1599	1616	1633	1650	1668	1685	1702	1719	3	6	9	12	14
10	·1736	1754	1771	1788	1805	1822	1840	1857	1874	1891	3	6	9	12	14
11	·1908	1925	1942	1959	1977	1994	2011	2028	2045	2062	3	6	9	11	14
12	·2079	2096	2113	2130	2147	2164	2181	2198	2215	2232	3	6	9	11	14
13	·2250	2267	2284	2300	2317	2334	2351	2368	2385	2402	3	6	8	11	14
14	·2419	2436	2453	2470	2487	2504	2521	2538	2554	2571	3	6	8	11	14
15	·2588	2605	2622	2639	2656	2672	2689	2706	2723	2740	3	6	8	11	14
16	·2756	2773	2790	2807	2823	2840	2857	2874	2890	2907	3	6	8	11	14
17	·2924	2940	2957	2974	2990	3007	3024	3040	3057	3074	3	6	8	11	14
18	·3090	3107	3123	3140	3156	3173	3190	3206	3223	3239	3	6	8	11	14
19	·3256	3272	3289	3305	3322	3338	3355	3371	3387	3404	3	5	8	11	14
20	·3420	3437	3453	3469	3486	3502	3518	3535	3551	3567	3	5	8	11	14
21	·3584	3600	3616	3633	3649	3665	3681	3697	3714	3730	3	5	8	11	14
22	·3746	3762	3778	3795	3811	3827	3843	3859	3875	3891	3	5	8	11	14
23	·3907	3923	3939	3955	3971	3987	4003	4019	4035	4051	3	5	8	11	14
24	·4067	4083	4099	4115	4131	4147	4163	4179	4195	4210	3	5	8	11	13
25	·4226	4242	4258	4274	4289	4305	4321	4337	4352	4368	3	5	8	11	13
26	·4384	4399	4415	4431	4446	4462	4478	4493	4509	4524	3	5	8	10	13
27	·4540	4555	4571	4586	4602	4617	4633	4648	4664	4679	3	5	8	10	13
28	·4695	4710	4726	4741	4756	4772	4787	4802	4818	4833	3	5	8	10	13
29	·4848	4863	4879	4894	4909	4924	4939	4955	4970	4985	3	5	8	10	13
30	·5000	5015	5030	5045	5060	5075	5090	5105	5120	5135	3	5	8	10	13
31	·5150	5165	5180	5195	5210	5225	5240	5255	5270	5284	2	5	7	10	12
32	·5299	5314	5329	5344	5358	5373	5388	5402	5417	5432	2	5	7	10	12
33	·5446	5461	5476	5490	5505	5519	5534	5548	5563	5577	2	5	7	10	12
34	·5592	5606	5621	5635	5650	5664	5678	5693	5707	5721	2	5	7	10	12
35	·5736	5750	5764	5779	5793	5807	5821	5835	5850	5864	2	5	7	10	12
36	·5878	5892	5906	5920	5934	5948	5962	5976	5990	6004	2	5	7	9	12
37	·6018	6032	6046	6060	6074	6088	6101	6115	6129	6143	2	5	7	9	12
38	·6157	6170	6184	6198	6211	6225	6239	6252	6266	6280	2	5	7	9	11
39	·6293	6307	6320	6334	6347	6361	6374	6388	6401	6414	2	4	7	9	11
40	·6428	6441	6455	6468	6481	6494	6508	6521	6534	6547	2	4	7	9	11
41	·6561	6574	6587	6600	6613	6626	6639	6652	6665	6678	2	4	7	9	11
42	·6691	6704	6717	6730	6743	6756	6769	6782	6794	6807	2	4	6	9	11
43	·6820	6833	6845	6858	6871	6884	6896	6909	6921	6934	2	4	6	8	11
44	·6947	6959	6972	6984	6997	7009	7022	7034	7046	7059	2	4	6	8	10
45	·7071	7083	7096	7108	7120	7133	7145	7157	7169	7181	2	4	6	8	10
46	·7193	7206	7218	7230	7242	7254	7266	7278	7290	7302	2	4	6	8	10
47	·7314	7325	7337	7349	7361	7373	7385	7396	7408	7420	2	4	6	8	10
48	·7431	7443	7455	7466	7478	7490	7501	7513	7524	7536	2	4	6	8	10
49	·7547	7558	7570	7581	7593	7604	7615	7627	7638	7649	2	4	6	8	9
50	·7660	7672	7683	7694	7705	7716	7727	7738	7749	7760	2	4	6	7	9
51	·7771	7782	7793	7804	7815	7826	7837	7848	7859	7869	2	4	5	7	9
52	·7880	7891	7902	7912	7923	7934	7944	7955	7965	7976	2	4	5	7	9
53	·7986	7997	8007	8018	8028	8039	8049	8059	8070	8080	2	3	5	7	9
54	·8090	8100	8111	8121	8131	8141	8151	8161	8171	8181	2	3	5	7	8
55	·8192	8202	8211	8221	8231	8241	8251	8261	8271	8281	2	3	5	7	8
56	·8290	8300	8310	8320	8329	8339	8348	8358	8368	8377	2	3	5	6	8
57	·8387	8396	8406	8415	8425	8434	8443	8453	8462	8471	2	3	5	6	8
58	·8480	8490	8499	8508	8517	8526	8536	8545	8554	8563	2	3	5	6	8
59	·8572	8581	8590	8599	8607	8616	8625	8634	8643	8652	1	3	4	6	7
60	·8660	8669	8678	8686	8695	8704	8712	8721	8729	8738	1	3	4	6	7
61	·8746	8755	8763	8771	8780	8788	8796	8805	8813	8821	1	3	4	6	7
62	·8829	8838	8846	8854	8862	8870	8878	8886	8894	8902	1	3	4	5	7
63	·8910	8918	8926	8934	8942	8949	8957	8965	8973	8980	1	3	4	5	6
64	·8988	8996	9003	9011	9018	9026	9033	9041	9048	9056	1	3	4	5	6
65	·9063	9070	9078	9085	9092	9100	9107	9114	9121	9128	1	2	4	5	6
66	·9135	9143	9150	9157	9164	9171	9178	9184	9191	9198	1	2	3	5	6
67	·9205	9212	9219	9225	9232	9239	9245	9252	9259	9265	1	2	3	4	6
68	·9272	9278	9285	9291	9298	9304	9311	9317	9323	9330	1	2	3	4	5
69	·9336	9342	9348	9354	9361	9367	9373	9379	9385	9391	1	2	3	4	5
70	·9397	9403	9409	9415	9421	9426	9432	9438	9444	9449	1	2	3	4	5
71	·9455	9461	9466	9472	9478	9483	9489	9494	9500	9505	1	2	3	4	5
72	·9511	9516	9521	9527	9532	9537	9542	9548	9553	9558	1	2	3	3	4
73	·9563	9568	9573	9578	9583	9588	9593	9598	9603	9608	1	2	2	3	4
74	·9613	9617	9622	9627	9632	9636	9641	9646	9650	9655	1	2	2	3	4
75	·9659	9664	9668	9673	9677	9681	9686	9690	9694	9699	1	1	2	3	4
76	·9703	9707	9711	9715	9720	9724	9728	9732	9736	9740	1	1	2	3	3
77	·9744	9748	9751	9755	9759	9763	9767	9770	9774	9778	1	1	2	3	3
78	·9781	9785	9789	9792	9796	9799	9803	9806	9810	9813	1	1	2	2	3
79	·9816	9820	9823	9826	9829	9833	9836	9839	9842	9845	1	1	2	2	3
80	·9848	9851	9854	9857	9860	9863	9866	9869	9871	9874	0	1	1	2	2
81	·9877	9880	9882	9885	9888	9890	9893	9895	9898	9900	0	1	1	2	2
82	·9903	9905	9907	9910	9912	9914	9917	9919	9921	9923	0	1	1	2	2
83	·9925	9928	9930	9932	9934	9936	9938	9940	9942	9943	0	1	1	1	2
84	·9945	9947	9949	9951	9954	9954	9956	9957	9959	9960	0	1	1	1	2
85	·9962	9963	9965	9966	9968	9969	9971	9972	9973	9974	0	0	1	1	1
86	·9976	9977	9978	9979	9980	9981	9982	9983	9984	9985	0	0	1	1	1
87	·9986	9987	9988	9989	9990	9990	991	9992	9993	9993	0	0	1	1	1
88	·9994	9995	9995	9996	9996	9997	9997	9997	9998	9998	0	0	0	1	1
89	·9998	9999	9999	9999	9999	1·000	1·000	1·000	1·000	1·000	0	0	0	0	0
90	1·00														

Natural cosines

Numbers in difference columns to be subtracted, not added

Degrees	0′ 0°·0	6′ 0°·1	12′ 0°·2	18′ 0°·3	24′ 0°·4	30′ 0°·5	36′ 0°·6	42′ 0°·7	48′ 0°·8	54′ 0°·9	Mean Differences 1	2	3	4	5
0	1·000	1·000	1·000	1·000	1·000	1·000	·9999	9999	9999	9999	0	0	0	0	0
1	·9998	9998	9998	9997	9997	9997	9996	9996	9995	9995	0	0	0	0	0
2	·9994	9993	9993	9992	9991	9990	9990	9989	9988	9987	0	0	0	1	1
3	·9986	9985	9984	9983	9982	9981	9980	9979	9978	9977	0	0	1	1	1
4	·9976	9974	9973	9972	9971	9969	9968	9966	9965	9963	0	0	1	1	1
5	·9962	9960	9959	9957	9956	9954	9952	9951	9949	9947	0	1	1	1	2
6	·9945	9943	9942	9940	9938	9936	9934	9932	9930	9928	0	1	1	1	2
7	·9925	9923	9921	9919	9917	9914	9912	9910	9907	9905	0	1	1	2	2
8	·9903	9900	9898	9895	9893	9890	9888	9885	9882	9880	0	1	1	2	2
9	·9877	9874	9871	9869	9866	9863	9860	9857	9854	9851	0	1	1	2	2
10	·9848	9845	9842	9839	9836	9833	9829	9826	9823	9820	1	1	2	2	3
11	·9816	9813	9810	9806	9803	9799	9796	9792	9789	9785	1	1	2	2	3
12	·9781	9778	9774	9770	9767	9763	9759	9755	9751	9748	1	1	2	3	3
13	·9744	9740	9736	9732	9728	9724	9720	9715	9711	9707	1	1	2	3	3
14	·9703	9699	9694	9690	9686	9681	9677	9673	9668	9664	1	1	2	3	4
15	·9659	9655	9650	9646	9641	9636	9632	9627	9622	9617	1	2	2	3	4
16	·9613	9608	9603	9598	9593	9588	9583	9578	9573	9568	1	2	2	3	4
17	·9563	9558	9553	9548	9542	9537	9532	9527	9521	9516	1	2	3	3	4
18	·9511	9505	9500	9494	9489	9483	9478	9472	9466	9461	1	2	3	4	5
19	·9455	9449	9444	9438	9432	9426	9421	9415	9409	9403	1	2	3	4	5
20	·9397	9391	9385	9379	9373	9367	9361	9354	9348	9342	1	2	3	4	5
21	·9336	9330	9323	9317	9311	9304	9298	9291	9285	9278	1	2	3	4	5
22	·9272	9265	9259	9252	9245	9239	9232	9225	9219	9212	1	2	3	4	6
23	·9205	9198	9191	9184	9178	9171	9164	9157	9150	9143	1	2	3	5	6
24	·9135	9128	9121	9114	9107	9100	9092	9085	9078	9070	1	2	4	5	6
25	·9063	9056	9048	9041	9033	9026	9018	9011	9003	8996	1	3	4	5	6
26	·8988	8980	8973	8965	8957	8949	8942	8934	8926	8918	1	3	4	5	6
27	·8910	8902	8894	8886	8878	8870	8862	8854	8846	8838	1	3	4	5	7
28	·8829	8821	8813	8805	8796	8788	8780	8771	8763	8755	1	3	4	6	7
29	·8746	8738	8729	8721	8712	8704	8695	8686	8678	8669	1	3	4	6	7
30	·8660	8652	8643	8634	8625	8616	8607	8599	8590	8581	1	3	4	6	7
31	·8572	8563	8554	8545	8536	8526	8517	8508	8499	8490	2	3	5	6	8
32	·8480	8471	8462	8453	8443	8434	8425	8415	8406	8396	2	3	5	6	8
33	·8387	8377	8368	8358	8348	8339	8329	8320	8310	8300	2	3	5	6	8
34	·8290	8281	8271	8261	8251	8241	8231	8221	8211	8202	2	3	5	7	8
35	·8192	8181	8171	8161	8151	8141	8131	8121	8111	8100	2	3	5	7	8
36	·8090	8080	8070	8059	8049	8039	8028	8018	8007	7997	2	3	5	7	9
37	·7986	7976	7965	7955	7944	7934	7923	7912	7902	7891	2	4	5	7	9
38	·7880	7869	7859	7848	7837	7826	7815	7804	7793	7782	2	4	5	7	9
39	·7771	7760	7749	7738	7727	7716	7705	7694	7683	7672	2	4	6	7	9
40	·7660	7649	7638	7627	7615	7604	7593	7581	7570	7559	2	4	6	8	9
41	·7547	7536	7524	7513	7501	7490	7478	7466	7455	7443	2	4	6	8	10
42	·7431	7420	7408	7396	7385	7373	7361	7349	7337	7325	2	4	6	8	10
43	·7314	7302	7290	7278	7266	7254	7242	7230	7218	7206	2	4	6	8	10
44	·7193	7181	7169	7157	7145	7133	7120	7108	7096	7083	2	4	6	8	10
45	·7071	7059	7046	7034	7022	7009	6997	6984	6972	6959	2	4	6	8	10
46	·6947	6934	6921	6909	6896	6884	6871	6858	6845	6833	2	4	6	8	11
47	·6820	6807	6794	6782	6769	6756	6743	6730	6717	6704	2	4	6	9	11
48	·6691	6678	6665	6652	6639	6626	6613	6600	6587	6574	2	4	7	9	11
49	·6561	6547	6534	6521	6508	6494	6481	6468	6455	6441	2	4	7	9	11
50	·6428	6414	6401	6388	6374	6361	6347	6334	6320	6307	2	4	7	9	11
51	·6293	6280	6266	6252	6239	6225	6211	6198	6184	6170	2	5	7	9	11
52	·6157	6143	6129	6115	6101	6088	6074	6060	6046	6032	2	5	7	9	12
53	·6018	6004	5990	5976	5962	5948	5934	5920	5906	5892	2	5	7	9	12
54	·5878	5864	5850	5835	5821	5807	5793	5779	5764	5750	2	5	7	9	12
55	·5736	5721	5707	5693	5678	5664	5650	5635	5621	5606	2	5	7	10	12
56	·5592	5577	5563	5548	5534	5519	5505	5490	5476	5461	2	5	7	10	12
57	·5446	5432	5417	5402	5388	5373	5358	5344	5329	5314	2	5	7	10	12
58	·5299	5284	5270	5255	5240	5225	5210	5195	5180	5165	2	5	7	10	12
59	·5150	5135	5120	5105	5090	5075	5060	5045	5030	5015	3	5	8	10	13
60	·5000	4985	4970	4955	4939	4924	4909	4894	4879	4863	3	5	8	10	13
61	·4848	4833	4818	4802	4787	4772	4756	4741	4726	4710	3	5	8	10	13
62	·4695	4679	4664	4648	4633	4617	4602	4586	4571	4555	3	5	8	10	13
63	·4540	4524	4509	4493	4478	4462	4446	4431	4415	4399	3	5	8	11	13
64	·4384	4368	4352	4337	4321	4305	4289	4274	4258	4242	3	5	8	11	13
65	·4226	4210	4195	4179	4163	4147	4131	4115	4099	4083	3	5	8	11	13
66	·4067	4051	4035	4019	4003	3987	3971	3955	3939	3923	3	5	8	11	14
67	·3907	3891	3875	3859	3843	3827	3811	3795	3778	3762	3	5	8	11	14
68	·3746	3730	3714	3697	3681	3665	3649	3633	3616	3600	3	5	8	11	14
69	·3584	3567	3551	3535	3518	3502	3486	3469	3453	3437	3	5	8	11	14
70	·3420	3404	3387	3371	3355	3338	3322	3305	3289	3272	3	5	8	11	14
71	·3256	3239	3223	3206	3190	3173	3156	3140	3123	3107	3	6	8	11	14
72	·3090	3074	3057	3040	3024	3007	2990	2974	2957	2940	3	6	8	11	14
73	·2924	2907	2890	2874	2857	2840	2823	2807	2790	2773	3	6	8	11	14
74	·2756	2740	2723	2706	2689	2672	2656	2639	2622	2605	3	6	8	11	14
75	·2588	2571	2554	2538	2521	2504	2487	2470	2453	2436	3	6	8	11	14
76	·2419	2402	2385	2368	2351	2334	2317	2300	2284	2267	3	6	9	11	14
77	·2250	2233	2215	2198	2181	2164	2147	2130	2113	2096	3	6	9	11	14
78	·2079	2062	2045	2028	2011	1994	1977	1959	1942	1925	3	6	9	11	14
79	·1908	1891	1874	1857	1840	1822	1805	1788	1771	1754	3	6	9	11	14
80	·1736	1719	1702	1685	1668	1650	1633	1616	1599	1582	3	6	9	12	14
81	·1564	1547	1530	1513	1495	1478	1461	1444	1426	1409	3	6	9	12	14
82	·1392	1374	1357	1340	1323	1305	1288	1271	1253	1236	3	6	9	12	14
83	·1219	1201	1184	1167	1149	1132	1115	1097	1080	1063	3	6	9	12	14
84	·1045	1028	1011	0993	0976	0958	0941	0924	0906	0889	3	6	9	12	14
85	·0872	0854	0837	0819	0802	0785	0767	0750	0732	0715	3	6	9	12	15
86	·0698	0680	0663	0645	0628	0610	0593	0576	0558	0541	3	6	9	12	15
87	·0523	0506	0488	0471	0454	0436	0419	0401	0384	0366	3	6	9	12	15
88	·0349	0332	0314	0297	0279	0262	0244	0227	0209	0192	3	6	9	12	15
89	·0175	0157	0140	0122	0105	0087	0070	0052	0035	0017	3	6	9	12	15
90	·0000														

Natural tangents

Degrees	0' 0°.0	6' 0°.1	12' 0°.2	18' 0°.3	24' 0°.4	30' 0°.5	36' 0°.6	42' 0°.7	48' 0°.8	54' 0°.9	Mean Differences				
											1	2	3	4	5
0	·0000	0017	0035	0052	0070	0087	0105	0122	0140	0157	3	6	9	12	15
1	·0175	0192	0209	0227	0244	0262	0279	0297	0314	0332	3	6	9	12	15
2	·0349	0367	0384	0402	0419	0437	0454	0472	0489	0507	3	6	9	12	15
3	·0524	0542	0559	0577	0594	0612	0629	0647	0664	0682	3	6	9	12	15
4	·0699	0717	0734	0752	0769	0787	0805	0822	0840	0857	3	6	9	12	15
5	·0875	0892	0910	0928	0945	0963	0981	0998	1016	1033	3	6	9	12	15
6	·1051	1069	1086	1104	1122	1139	1157	1175	1192	1210	3	6	9	12	15
7	·1228	1246	1263	1281	1299	1317	1334	1352	1370	1388	3	6	9	12	15
8	·1405	1423	1441	1459	1477	1495	1512	1530	1548	1566	3	6	9	12	15
9	·1584	1602	1620	1638	1655	1673	1691	1709	1727	1745	3	6	9	12	15
10	·1763	1781	1799	1817	1835	1853	1871	1890	1908	1926	3	6	9	12	15
11	·1944	1962	1980	1998	2016	2035	2053	2071	2089	2107	3	6	9	12	15
12	·2126	2144	2162	2180	2199	2217	2235	2254	2272	2290	3	6	9	12	15
13	·2309	2327	2345	2364	2382	2401	2419	2438	2456	2475	3	6	9	12	15
14	·2493	2512	2530	2549	2568	2586	2605	2623	2642	2661	3	6	9	12	16
15	·2679	2698	2717	2736	2754	2773	2792	2811	2830	2849	3	6	9	13	16
16	·2867	2886	2905	2924	2943	2962	2981	3000	3019	3038	3	6	9	13	16
17	·3057	3076	3096	3115	3134	3153	3172	3191	3211	3230	3	6	10	13	16
18	·3249	3269	3288	3307	3327	3346	3365	3385	3404	3424	3	6	10	13	16
19	·3443	3463	3482	3502	3522	3541	3561	3581	3600	3620	3	7	10	13	16
20	·3640	3659	3679	3699	3719	3739	3759	3779	3799	3819	3	7	10	13	17
21	·3839	3859	3879	3899	3919	3939	3959	3979	4000	4020	3	7	10	13	17
22	·4040	4061	4081	4101	4122	4142	4163	4183	4204	4224	3	7	10	14	17
23	·4245	4265	4286	4307	4327	4348	4369	4390	4411	4431	3	7	10	14	17
24	·4452	4473	4494	4515	4536	4557	4578	4599	4621	4642	4	7	11	14	18
25	·4663	4684	4706	4727	4748	4770	4791	4813	4834	4856	4	7	11	14	18
26	·4877	4899	4921	4942	4964	4986	5008	5029	5051	5073	4	7	11	15	18
27	·5095	5117	5139	5161	5184	5206	5228	5250	5272	5295	4	7	11	15	18
28	·5317	5340	5362	5384	5407	5430	5452	5475	5498	5520	4	8	11	15	19
29	·5543	5566	5589	5612	5635	5658	5681	5704	5727	5750	4	8	12	15	19
30	·5774	5797	5820	5844	5867	5890	5914	5938	5961	5985	4	8	12	16	20
31	·6009	6032	6056	6080	6104	6128	6152	6176	6200	6224	4	8	12	16	20
32	·6249	6273	6297	6322	6346	6371	6395	6420	6445	6469	4	8	12	16	20
33	·6494	6519	6544	6569	6594	6619	6644	6669	6694	6720	4	8	13	17	21
34	·6745	6771	6796	6822	6847	6873	6899	6924	6950	6976	4	9	13	17	21
35	·7002	7028	7054	7080	7107	7133	7159	7186	7212	7239	4	9	13	18	22
36	·7265	7292	7319	7346	7373	7400	7427	7454	7481	7508	5	9	14	18	23
37	·7536	7563	7590	7618	7646	7673	7701	7729	7757	7785	5	9	14	18	23
38	·7813	7841	7869	7898	7926	7954	7983	8012	8040	8069	5	9	14	19	24
39	·8098	8127	8156	8185	8214	8243	8273	8302	8332	8361	5	10	15	20	24
40	·8391	8421	8451	8481	8511	8541	8571	8601	8632	8662	5	10	15	20	25
41	·8693	8724	8754	8785	8816	8847	8878	8910	8941	8972	5	10	16	21	26
42	·9004	9036	9067	9099	9131	9163	9195	9228	9260	9293	5	11	16	21	27
43	·9325	9358	9391	9424	9457	9490	9523	9556	9590	9623	6	11	17	22	28
44	·9657	9691	9725	9759	9793	9827	9861	9896	9930	9965	6	11	17	23	29

Natural tangents

Degrees	0' 0°.0	6' 0°.1	12' 0°.2	18' 0°.3	24' 0°.4	30' 0°.5	36' 0°.6	42' 0°.7	48' 0°.8	54' 0°.9	Mean Differences				
											1	2	3	4	5
45	1·0000	0035	0070	0105	0141	0176	0212	0247	0283	0319	6	12	18	24	30
46	1·0355	0392	0428	0464	0501	0538	0575	0612	0649	0686	6	12	18	25	31
47	1·0724	0761	0799	0837	0875	0913	0951	0990	1028	1067	6	13	19	25	32
48	1·1106	1145	1184	1224	1263	1303	1343	1383	1423	1463	7	13	20	27	33
49	1·1504	1544	1585	1626	1667	1708	1750	1792	1833	1875	7	14	21	28	34
50	1·1918	1960	2002	2045	2088	2131	2174	2218	2261	2305	7	14	22	29	36
51	1·2349	2393	2437	2482	2527	2572	2617	2662	2708	2753	8	15	23	30	38
52	1·2799	2846	2892	2938	2985	3032	3079	3127	3175	3222	8	16	24	31	39
53	1·3270	3319	3367	3416	3465	3514	3564	3613	3663	3713	8	16	25	33	41
54	1·3764	3814	3865	3916	3968	4019	4071	4124	4176	4229	9	17	26	34	43
55	1·4281	4335	4388	4442	4496	4550	4605	4659	4715	4770	9	18	27	36	45
56	1·4826	4882	4938	4994	5051	5108	5166	5224	5282	5340	10	19	29	38	48
57	1·5399	5458	5517	5577	5637	5697	5757	5818	5880	5941	10	20	30	40	50
58	1·6003	6066	6128	6191	6255	6319	6383	6447	6512	6577	11	21	32	43	53
59	1·6643	6709	6775	6842	6909	6977	7045	7113	7182	7251	11	23	34	45	56
60	1·7321	7391	7461	7532	7603	7675	7747	7820	7893	7966	12	24	36	48	60
61	1·8040	8115	8190	8265	8341	8418	8495	8572	8650	8728	13	26	38	51	64
62	1·8807	8887	8967	9047	9128	9210	9292	9375	9458	9542	14	27	41	55	68
63	1·9626	9711	9797	9883	9970	2·0057	2·0145	2·0233	2·0323	2·0413	15	29	44	58	73
64	2·0503	0594	0686	0778	0872	0965	1060	1155	1251	1348	16	31	47	63	78
65	2·1445	1543	1642	1742	1842	1943	2045	2148	2251	2355	17	34	51	68	85
66	2·2460	2566	2673	2781	2889	2998	3109	3220	3332	3445	18	37	55	73	92
67	2·3559	3673	3789	3906	4023	4142	4262	4383	4504	4627	20	40	60	79	99
68	2·4751	4876	5002	5129	5257	5386	5517	5649	5782	5916	22	43	65	87	108
69	2·6051	6187	6325	6464	6605	6746	6889	7034	7179	7326	24	47	71	95	119
70	2·7475	7625	7776	7929	8083	8239	8397	8556	8716	8878	26	52	78	104	131
71	2·9042	9208	9375	9544	9714	9887	3·0061	3·0237	3·0415	3·0595	29	58	87	116	145
72	3·0777	9961	1146	1334	1524	1716	1910	2106	2305	2506	32	64	96	129	161
73	3·2709	2914	3122	3332	3544	3759	3977	4197	4420	4646	36	72	108	144	180
74	3·4874	5105	5339	5576	5816	6059	6305	6554	6806	7062	41	81	122	163	204
75	3·7321	7583	7848	8118	8391	8667	8947	9232	9520	9812	46	93	139	186	232
76	4·0108	0408	0713	1022	1335	1653	1976	2303	2635	2972	53	107	160	213	267
77	4·3315	3662	4015	4374	4737	5107	5483	5864	6252	6646					
78	4·7046	7453	7867	8288	8716	9152	9594	5·0045	5·0504	5·0970	Mean differences cease to be sufficiently accurate				
79	5·1446	1929	2422	2924	3435	3955	4486	5026	5578	6140					
80	5·6713	7297	7894	8502	9124	9758	6·0405	6·1066	6·1742	6·2432					
81	6·3138	3859	4596	5350	6122	6912	7720	8548	9395	7·0264					
82	7·1154	2066	3002	3962	4947	5958	6996	8062	9158	8·0285					
83	8·1443	2636	3863	5126	6427	7769	9152	9·0579	9·2052	9·3572					
84	9·5144	9·677	9·845	10·02	10·20	10·39	10·58	10·78	10·99	11·20					
85	11·43	11·66	11·91	12·16	12·43	12·71	13·00	13·30	13·62	13·95					
86	14·30	14·67	15·06	15·46	15·89	16·35	16·83	17·34	17·89	18·46					
87	19·08	19·74	20·45	21·20	22·02	22·90	23·86	24·90	26·03	27·27					
88	28·64	30·14	31·82	33·69	35·80	38·19	40·92	44·07	47·74	52·08					
89	57·29	63·66	71·62	81·85	95·49	114·6	143·2	191·0	286·5	573·0					
90	∞														